Autoaffection

Autoaffection

Unconscious Thought
in the Age of Teletechnology

Patricia Ticineto Clough

University of Minnesota Press
Minneapolis
London

Published by the University of Minnesota Press
111 Third Avenue South, Suite 290
Minneapolis, MN 55401-2520
http://www.upress.umn.edu

Library of Congress Cataloging-in-Publication Data

Clough, Patricia Ticineto, 1945–
    Autoaffection : unconscious thought in the age of teletechnology /
Patricia Ticineto Clough.
        p.   cm.
    Includes bibliographical references and index.
    ISBN 0-8166-2888-2 (alk. paper) — ISBN 0-8166-2889-0 (pbk. : alk.
paper)
    1. Technology — Social aspects.   2. Telecommunication — Social aspects.
3. Thought and thinking.   4. Subconsciousness.   5. Cognition and
culture.   6. Psychoanalysis and culture.   7. Poststructuralism.
8. Postmodernism.   I. Title.
HM846.C56   2000
3.3.48'33 — dc21   99-050993

Printed in the United States of America on acid-free paper

The University of Minnesota is an equal-opportunity educator and employer.

11 10 09 08 07 06 05 04 03 02 01 00   10 9 8 7 6 5 4 3 2 1

For my parents,
Felix Ticineto and Josephine Ticineto

Contents

Acknowledgments

This book was written when much was changing in and outside the academy (even the meaning of *inside* and *outside* changed once again). Writing became rewriting in the strong sense of the term. A book first planned to be a summary review of poststructuralist thought became instead a search for the future of thought beyond poststructuralism and the cultural criticisms influenced by poststructuralism. What started as a relatively easy project became a seriously challenging one. Through it all, there were those who were an inspiration and a support, friends and colleagues as well as thinkers whose imagination and intellectual daring matter so much to me and have shaped much of what follows. I want especially to thank Stanley Aronowitz, Barbara Bowen, Judith Butler, Lynn Chancer, Adele Clarke, Jonathan Cutler, Norman K. Denzin, Richard Dienst, Hester Eisentein, Nicole Fermon, Ann Galligan, Martha Gever, Barry Glassner, Liz Grosz, Zali Gurevitch, Donna Haraway, Elizabeth Harriss, Barbara Heyl, Anne Hoffman, Anahid Kassabian, David Kazanjian, Steven Kruger, Charles Lemert, Kathy Lord, Randy Martin, Humberto Maturana, Michal McCall, Chet Meeks, Nancy K. Miller, Mary Jo Neitz, Linda Nicholson, Tony O'Brien, Virginia Olesen, Jackie Orr, Paul Pangaro, Stephen Pfohl, Francesca Poletta, Amit Rai, Joseph Schneider, Steven Seidman, Catherine Silver, Charlie Smith, Gayatri Chakravorty Spivak, Jyotsna Uppal, Heinz von Foerster, and Judith Wittner. And a most special thanks to Margaret Cerullo and Ron Lembo for the sheer pleasure of our times together, which made writing this book possible at all.

I want to thank the students at The Graduate Center, City University of New York. They have been my best first readers. They have given me much pleasure with their fresh insight and gentle corrections. I want to thank especially Dong-Ho Cho, Chunmei Chuang, Barry Davidson, Melissa Ditmore, Randall Doane, Ariel Ducey, Karen Gilbert, Manolo Guzman, Jean Halley, Mark Halling, Kristen Lawler, Nadine Lemmon, Jack Levinson, Freddie Marrero, Carmella Marrone, Ananya Mukherjea, Gina Neff, Michael Roberts, Jennifer Smith, David Staples, Aleksandra Wagner, and Betsy Wissinger.

I also want to thank my family, my mother and father, to whom this book is dedicated with much love. I am grateful to my sister, Virginia Steppe, and to Daniel Steppe and Alissa and Kevin Maples. And special thanks to my son, Christopher, who makes everything I do more joyful. Thanks, Christo, especially for your wit and also for your poems.

I would like to thank all those who have been so helpful at the University of Minnesota Press: Douglas Armato, Pieter Martin, Marilyn Martin, and Craig Davidson. Also, thanks to William Murphy.

And, last but not least, my thanks to Jill C. Herbert.

INTRODUCTION

Thought's Reach to the Future

In an interview published nearly two decades ago, Michel Foucault announced that much of what he had written — what was part of a body of criticism already known as poststructuralism — would best be understood as putting an end to a certain tradition of thought rather than providing a way for thought to begin anew. It was an arresting comment. True, Foucault had treated an established tradition of thought, what he referred to as the modern western discourse of Man. But his writings, as well as those of the other so-called poststructuralists, had opened to consideration a number of assumptions that up to then had gone without question. In doing so, the poststructuralists had forced invention. Surely a certain intellectual stamina would be required just to remain open to the various cultural criticisms, which, over the last three decades of the twentieth century, became engaged with poststructuralism and invited scholars into disciplines other than their own, even inviting them to explore the policed spaces of silence in and in between the disciplines.

But the excited and exciting debates that poststructuralism provoked, and which for some time have characterized academic and intellectual discourses, seem finally to have calmed. If cultural criticism has been drawn back from invention, Foucault's comment still haunts, insistently raising the question: have the various cultural criticisms elaborated over the last three decades of the twentieth century made way for thought to begin anew; have they given thought a future? Propelled by this question, the chapters that follow look back over the past three decades in order to trace the future of thought, which, I want to argue, has drawn

1

poststructuralism to it from the start and which has been further elaborated in the cultural criticisms engaged with poststructuralism, such as feminist theory, postcolonial theory, queer theory, critical race theory, marxist cultural studies, cultural studies of science, and the criticism of ethnographic writing. It is this future that I am calling the age of teletechnology.

Although the cultural criticisms engaged with poststructuralism usually have been treated as elaborations of the linguistic turn, focusing on the literary or the literarization of philosophical thought, I want to treat them in relationship to the becoming of the teletechnological. I want to propose that the development of teletechnology in the late twentieth century not only has drawn cultural criticism to the reconfiguration of the social, the political, and the economic conditions of human agency. The development of teletechnology also has drawn cultural criticism to the deconstruction of the opposition of nature and technology, the human and the machine, the virtual and the real, the living and the inert, thereby giving thought over to the ontologization of agencies other than human agency. The chapters that follow, therefore, are less about the influence of teletechnology on societies of the late twentieth century than they are about the cultural criticisms of the late twentieth century and the way their engagement with poststructuralism has drawn them to teletechnology and the future of thought, albeit often without full awareness of it.

But to suggest, as I am, that cultural critics have been drawn by teletechnology to give thought over to a future that they themselves have not always fully grasped is to propose that thought is not given by individual thinkers so much as it is given to them as they are drawn to the future by it. One thinks of "thought" in this way, as Jacques Derrida once put it, when "one cannot say philosophy, theory, logic, structure, scene or anything else; when one can no longer use any word of this sort...."[1] It is in this sense that thought is unconscious and not simply a rational process. It was, of course, against the normative idealization of thought as rational that poststructuralism aimed its critique. The noncoincidence of the subject with consciousness, realized with the deconstruction of the subject, not only gives thought over to an individual's unconscious, the mark in the individual of the noncoincidence of its subjectivity with a conscious self; it also puts thought outside subjectivity, even outside human intersubjectivity, giving thought over to its own movement, intensities, and affects. A more general unconscious than that of the subject or of intersubjectivity is implied; it is the unconscious of thought. But this way of thinking about thought, what Rosi Braidotti characterizes as "postpersonal"[2] in her treatment of Gilles Deleuze's philosophical efforts to think of thought as unconscious or as a desiring machine, is a way of thinking that is itself already drawn to the future, to the age of teletechnology.

By teletechnology I mean to refer to the realization of technoscience, technoculture, and technonature — that is, to the full interface of computer technology and television, promising globalized networks of information and communication whereby layers of electronic images, texts, and sounds flow in real time, so that the speeds of the territorialization, deterritorialization, and reterritorialization of social spaces, as well as the adjustment to the vulnerabilities of exposure to media event-ness, are beyond any user's mere decision to turn "it" on or off. Teletechnology, therefore, refers to all matter of "knowledge objects," — technoscientific productions, from computer devices to intelligent machines to genomes — such that teletechnology is both a register and an actualization of postpersonal thought.

In this sense teletechnology refers not only to an environment or a set of objects, but also to agencies other than human agency, so that the teletechnological joins, if not displaces, what sociologists of western modernity have referred to as the social structural. This displacement demands a rethinking of the determination of human agency that the idea of social structure has implied — that is, the derivation of human agency out of that certain structural configuration of family and national ideologies, the state and civil society, and the public and private spheres presumed in subject-centered, nation-centric discourses, such as the modern western discourse of Man.

In the age of teletechnology this configuration of social spaces is being "smoothed out" or "ungrounded," to use Gilles Deleuze's terms, or "unbundled," to use Saskia Sassen's term.[3] Even as the transnational or the global become visible, proposing themselves as far-flung extensions of social structure, they are ungrounded by that upon which they depend: the speed of the exchange of information, capital, bodies, and abstract knowledge and the vulnerability of exposure to media event-ness.[4] This transformation not only involves postmodern western or northern societies, where the arrangement of social spaces presumed in subject-centered, nation-centric discourses has been characteristic, at least until recently challenged by the teletechnological. It also involves societies in neocolonialism, where this arrangement of social spaces is not necessarily presumed, and, if imposed, not necessarily accepted, but where, nonetheless, the teletechnological speeds of territorialization, deterritorialization, and reterritorialization, as well as the vulnerabilities of exposure to media event-ness, are having their effect in the "glocalization" of cultures and the production of technoculture and technonature.

But I do not mean to suggest that the teletechnological refers simply to the denationalization of the state or to the disappearance of any distinction between the public and private spheres, the family and the nation, and surely not to the deterritorialization of all social spaces, because what is to be expected instead is various reterritorializations in the reconfiguration of social spaces brought along with the transnationalization of capital and the globalization of teletechnology, such that the transnational and the global become nodes in various networks, alongside the local, the singular, the immanent. I do mean to suggest, however, that no matter how social spaces are being reconfigured in the age of teletechnology, there is an increased possibility of the release of the subject's agency from nonreflexive relationships to tradition, community, and large social structures. There is an increased probability of the reconfiguration of the modern "sociological imagination," which has been thought to link the individual subject to a national collectivity through the translation of "personal troubles" into "social problems," as C. Wright Mills famously put it.[5]

Furthermore, the agency of the subject is not to be rethought only in terms of the possibility of an increased reflexivity or complexity in relationship to tradition, community, and social structure. It is to be rethought also in terms of an increased reflexivity and complexity in relationship to the social situation of knowledge objects. But here agency refers not only to the subject, but to an interobjectivity, the limits of which reach to the widening recognition of the agencies immanent to matter, what Pheng Cheah has referred to as "the dynamism of matter" or "mattering."[6] Here "matter-energy flows," as Manuel DeLanda argues, displace the "structural" as the ungrounding ground of agency.[7] Agencies rather inhere in the singular, subindividual, finite forces of mattering.

In referring to the agencies immanent to matter, as I will be doing, I do not mean to suggest a mere return of thought to the forces of nature as opposed to the conditions of culture. Although nature is not to be conceived merely as a cultural construction, nature also is not separable from culture or technology. That is to say, the agencies of the singular, subindividual, finite forces of mattering refer to an interpenetration of nature and technoculture all the way down; after all, the forces of mattering are realized as agencies through a technoscientific production. As DeLanda argues, what "has allowed us to 'see' matter as self-organizing is the advance in technology that materially supports the (non-linear) mathematics, and with it mathematical technology."[8] Similarly, Donna Haraway proposes that agencies such as those belonging to the fetus, the chip, the genome, or the database are realizable only as and through "material-semiotic objects"; that is, they are "forged by heterogeneous practices ... of technocscience."[9] Although insisting on the political, economic, psychic, and cultural complexity of these practices, Haraway argues that material-semiotic objects in no meaningful way can be simply or only referred to human agency.

I want to suggest that it is the realization of the interpenetration of nature and technoculture, as well as the teletechnological transformation of the social spaces in terms of which human agency has been conceived in modern western discourse, that poststructuralism has both registered and referred to the domain of ontology. That is to say, against the usual treatment of poststructuralism as provoking an epistemological shift, I want to suggest that poststructuralism's reach to the future of thought is in its ontological implications. Derrida's treatment of *différance*, Foucault's treatment of the force relations of power, and Deleuze and Félix Guattari's treatment of machinic assemblages are, to use Gayatri Chakravorty Spivak's phrasing, "thought ... trying to touch the ontic."[10]

If, however, poststructuralism has ontological implications, it is not itself an ontology of presence. It rather problematizes an ontology of presence. It offers an ontological perspective, such that ontology is always haunted by what Derrida refers to as the "given."[11] "This extremely difficult perhaps impossible idea," as Derrida describes "the gift," forces ontology to "break off ... with all originary authenticity."[12] To be a pure gift, neither obligation nor debt can be induced; the gift cannot be returned or produce exchange. This impossible idea of the gift, of the given, therefore, ruins any presumption of origin or authenticity in Being. It is preontological, or what Derrida refers to as "hauntological."[13] It is in this sense that I want to suggest that the ontological implications of poststructuralism cross through the ontology of presence, put origins and authenticity under erasure, making ontology impossible or only impossibly so. The shift in ontological perspective that poststructuralism implies makes ontologizing impossible but imperative, necessary for thinking Being anew, that is, for bringing Being back to the opening of ontology, to the preontological, and thereby inviting a rethinking of technicity as well. Poststructuralism, I want to suggest, offers an ontological perspective in which nature and technology, the body and the machine, the real and the virtual, the living and the inert are given in *différantial* relationships, each inextricable from the other.

Mapping Unconscious Thought in the Age of Teletechnology

Of course, connections have already been drawn between the teletechnological and poststructuralism. The focus on writing and textuality produced in the deconstruction of the western modern discourse of Man has been recognized as an elaboration of a teletechnological aesthetic. The works of Derrida and Foucault, as well as Deleuze and Guattari, already have been treated in such terms.[14] But the scholars I want to consider are the cultural critics who have been engaged with poststructuralism, such as Fredric Jameson, Stuart Hall, Stanley Aronowitz, Richard Dienst, Michael Hardt, Paul Virilio, Kaja Silverman, Judith Butler, Elizabeth Grosz, James Clifford, Gayatri Chakravorty Spivak, Pheng Cheah, Trinh T. Minh-ha, Donna Haraway, Dorothy Smith, and Bruno Latour. I want to take up their works, tracing their engagement with poststructuralism, and thereby explore the reach of cultural criticism in the late twentieth century to the teletechnological. I want to make more explicit the ontological perspective to which teletechnology has drawn cultural criticism. The chapters that follow, then, represent a process of pursuing the unconscious thought of teletechnology in the cultural criticisms of the past three decades of the twentieth century.

A certain way of reading and writing is required, something like what Deleuze and Guattari refer to as rhizomatic reading and writing, which brings conceptualizations from various writings together, assembling them on the same plane so that these concepts can be made to provoke a problematization. In *What Is Philosophy?* Deleuze and Guattari argue that a concept is better understood as the construction of a question that urges one to adopt a new perspective. The concept in this sense has a "becoming" that refers to its relationship with other concepts on the same plane: "Here concepts link up with each other, support one another, coordinate their contours, articulate their respective problems."[15]

A concept is not connected to a problem in order to reformulate earlier concepts. Instead there is an assembling of concepts from across various problematizing series so that they can interfere with each other: as Deleuze and Guattari put it, "Each concept will branch off toward concepts that are differently composed but that constitute other regions of the same plane, answer to problems that can be connected to each other, and participate in a co-creation [of new thought]."[16] Concepts, therefore, are not referential; they do not explain by way of propositions. Being neither particularizations nor generalizations, concepts, Deleuze and Guattari suggest, are instead made of singularities. Concepts are to be treated in terms of "coincidence, condensation, or accumulation" of singularities as well as a shedding of singularities onto a "plane of consistency."[17] On such a plane concepts resonate with each other rather than cohere or correspond. They "vibrate" and give a new sense to thought.

It cannot go without saying, therefore, that a plane of consistency makes no reference to a unifying transcendental principle. The plane of consistency is a nontranscendental plane. It is a "machinic assemblage" — neither organism nor mechanistic. It is a composing apparatus, a composition of desire. The machinic assemblage is an unconscious surface upon which singularities move by desire that is neither individual nor personal, although a subject's desire can itself be part of a machinic assemblage. Whether it is really possible to conceptualize without presuming unities remains a question; nonetheless, the effort to do so, to do away with a pregiven transcendental principle of unity, is for Deleuze and Guattari, as well as for me, an effort to shift concepts from the regime of truth to that of desiring so that "categories like interesting, remarkable, or important ... determine success or failure."[18]

All of this, however, already resonates with teletechnology and is to be more closely explored in the chapters that follow, where I want to set off vibrations between the teletechnological and various cultural criticisms engaged with poststructuralism; that is, I do not want to argue that teletechnology is the condition of possibility of these cultural criticisms or that they are historical symptoms of technological changes. It is not possible to make these claims, because it is not possible to say exactly what teletechnology will become or what its conceptualizations can still make happen. Furthermore, it is not possible to say much or as much as I have already said about the teletechnological without the cultural criticisms I consider in this book. My point is that these cultural criticisms are following thought to the teletechnological; they are giving cultural criticism the presumption of the teletechnological. They are giving us the unconscious thought of the age of teletechnology.

Many of the scholars whose writings I treat in this book already have had considerable attention, especially in the United States academy, but not only there. Many of these scholars have international star status, a teletechnological effect itself. Although this has meant that their writings have received as much dismissing commentary as careful criticism, the chapters that follow are not aimed at engaging the reception of various scholars. Still, in treating their writings again I do mean to suggest why they are interesting, remarkable, and important in ways, it seems to me, that even their authors do not or do not always realize. That is, putting various cultural criticisms on the same plane along with teletechnology allows me to draw out concepts with which to follow unconscious thought to a new ontological perspective.

The following chapters are especially focused on Marxist treatments of capitalism and psychoanalytic treatments of the unconscious, which, after all, poststructuralism meant to make problematic and which also have been central to the cultural criticisms that have engaged poststructuralism. These chapters, therefore, are still moving against the traditions that inform Marxism and psychoanalysis; they drain the condensations around Marxism and psychoanalysis in order to follow unconscious thought beyond them. In what follows, then, I try to evoke the unconscious thought of teletechnology and follow its reach to the ontic by drawing a map through cultural criticisms that engaged Marxism and psychoanalysis just as poststructuralism was deconstructing the authority of both by opening wide a disjuncture between the individual subject and nation-centric collectivities.

To give a first mapping, the chapters that follow take up early feminist film criticism, that of the 1970s and early 1980s, which elaborated Lacanian psychoanalysis not merely as a way to "read" film texts, but more as a way to flesh out the phallocentric logic of unconscious desire deployed in the dominant oedipal narrative organizing the western modern discourse of Man. Meant to give a feminist turn to Marxist cultural studies that also had taken a cue from Louis Althusser's treatment of the subject's unconscious formation in ideological-textual forms of mediated cultures, feminist film theorists offered a discourse on the interpenetration of technology and the unconscious, pointing to what I refer to, following Derrida, as "the technical substrates of unconscious memory."[19] The discourse of feminist film theory gestured toward the relationship of technology to the psyche, space and time, Being and technicity, which was, however, never fully elaborated and never to fit comfortably in the then-burgeoning field of Marxist cultural studies of television.

At issue was the difficulty of treating television with the narrative approach of feminist film theory, not to mention the limitation of early feminist film theory in focusing on sexual difference to the exclusion of differences of race, class, ethnicity, sexuality, and nation, or in focusing on film texts to the exclusion of institutional analysis of industries and audiences. The drift of Marxist cultural studies from Althusser to Antonio Gramsci and Ernesto Laclau under the influence of Stuart Hall also allowed a shift of emphasis from text to audience and the multiple and locally situated audience responses to television viewing. Justifying this shift, however, had the effect of narrowing the meaning of text. Certainly Derrida's treatment of textuality became widely misrepresented in this process.

In this context the complexity that Fredric Jameson found in rereading the generic form of the novel from romanticism to realism to the cinematics of Joseph Conrad's high modernism no longer seemed compelling; instead he and other Marxist critics moved from a treatment of the literary evidence of a political unconscious into a protracted debate over "postmodernism as the cultural logic of late capitalism," in Jameson's influential phrasing. No matter what was argued in this debate, culture seemed to flatten out into a barrage of meaningless texts. Jameson, along with other Marxist cultural critics, all but sealed the fate of the concept of textuality. They both linked textuality to the superficial and connected it to the development of technology, which, however, was reduced to the capitalist organization of production. Although suffocated under a barrage of meaningless texts, History was thereby given one last chance — to overcome postmodernity through the dialectic logic of capital. But what now seems more important about the debate over postmodernism and textuality is that Marxist studies of television were once again revised, turned away from studying audiences to thinking about television as the machine central to the technology defining postmodernity.

Although I intend to recover Derrida's treatment of textuality, connecting it with unconscious thought's reach to the ontic, nonetheless, my mapping goes through Marxist cultural studies of television, because it opened a path to rethinking teletechnology by forcing a reconsideration of the dependency of Marxist cultural studies on a post–World War II welfare state capitalism, where a certain structural configuration of family and national ideologies, the state and civil society, and the public and private spheres was presumed and linked to a discourse on democracy as well as made to underwrite a post-Althusserian political economic analysis. That is to say, the treatment of television awakens Marxist cultural studies and cultural criticism generally to the "flexibilities" of a post–welfare state and a postmodern or late capitalism that is dependent on the globalization of teletechnology and the transnationalization of capital.

Ironically, it is this awakening that forces a view of late postmodern capitalism as inextricable from teletechnology, that is, that neither can be separated from the other so that neither can be the context for the other, the condition of possibility for the other. Reducing teletechnology to the capitalist organization of production becomes politically unexciting. Thinking of one capitalism, thinking of capitalism as totality, or thinking of the history of capitalist organization of production as becoming-universal — all are increasingly unremarkable. Rather, the political and cultural antagonisms of localized capitalisms would be better treated as irreducible to the economic and as pointing instead to what Pheng Cheah has called "a global miredness" in order to describe the complexities of the condition of agency in a transnational frame of glocalized cultures.[20] Although undoubtedly the complexity of agency finally yields states of power including identified subjects, institutions, and groups, it also releases uncharted resources for politics.

Thrown forward, therefore, to rethink capitalism as something other than totality, its history other than becoming-universal, it seems necessary to grasp the historicity that teletechnology gives with its technical substrate of unconscious memory. It also seems necessary to feel the pressure this historicity exerts on a new ontological perspective to reconfigure the opposition of Being and technicity, so that nature and technology, body and machine, the virtual and the real, and the living and the inert might be understood in terms of *différantial* relationships rather than oppositional or even dialectical ones. But in order to imagine that the body and the machine, the virtual and the real, and nature and technology are inextricably implicated, always already interlaced, which a *différantial* relationship is meant to suggest, it also is necessary to think of materiality and the unconscious differently.

In this sense, what has been thought to be the context constituting the unconscious, that is, the oedipal complex, is to be rethought, especially for the way it functions as the dominant narrative logic informing the construction of the subject's identity and social reality. If early feminist film theory was left without full extension into a Marxist cultural studies of television, it did, however, produce a legacy in what has become known as queer theory; at least Judith Butler and Elizabeth Grosz have drawn heavily on the revisions of Lacanian psychoanalysis that feminist film theory elaborated. For this, Butler's and Grosz's writings are part of my mapping of the unconscious thought of teletechnology in its reach to the ontic. Their writings, when taken together, have brought feminist theory to the ontological implications that have been folded within it from the start, that is, when feminist theory first undertook to rethink sexual difference and the "nature" of the woman's body. Butler and Grosz have provided treatments of bodies, images, and unconscious desire that aim to bring these and feminist theory beyond the oedipal complex, even beyond the human subject. Deconstructing the psychoanalytic configuration of the imaginary, the symbolic, and the real even beyond the efforts of feminist film theorists, Butler and Grosz have made it possible to think of bodies as intensities in a flow of electronic images, texts, and sounds, that is, as imagined materialities.

Against Butler's critics, who have accused her work of voluntarism or "ludic feminism,"[21] that is, of failing to deal with the material contingencies of political economy or the institutional arrangements of power, my mapping of her work is meant to take her focus on the imaginary construction of bodies as a gesture toward rethinking bodily matter in the age of teletechnology, to question what institutional or materiality means in relationship to teletechnological flows of capital, information, labor, and abstract knowledge. What kind of bodily matters are these? I follow Butler's theoretical focus, which, steadied on the limits of psychoanalytic discourse, has forced the treatment of the imaginary to move into a transnational frame where the oedipal narrative is not the only dominant narrative of desire and where it has also become much more important to think about the technical substrate that teletechnology gives unconscious memory. Although Spivak was first to set this agenda for feminist scholars and to show the relevancy of Derridean deconstruction for situating feminism and psychoanalysis in "an international frame,"[22] Butler more specifically has drawn the deconstruction of the oedipal narrative to the construction of bodily matter, and therefore to a shift in ontological perspective.

In her effort to reconfigure the symbolic, the imaginary, and the real in such a way as to deprive the phallus of its transcendental status, Butler not only has let loose the oedipal logic of desire that holds together the structural configuration of family and national ideologies, the state and civil society, and the private and public spheres, upon which Freudian and Lacanian psychoanalyses depend. Her work also has signaled the need to rethink the real and the nature of bodies, sexualities, and subjectivities in terms of the speeds of teletechnological flows and the vulnerabilities of exposure to media event-ness, which are linked to the glocalism of world cultures. For all this, Butler's work fits with the work of a number of postcolonial scholars, who have made it possible to read "the question of woman" into rethinking democratic politics in a transnational frame.

Although Butler's writings must be drawn to the ontological implications of the teletechnological, Grosz's writings have been more explicitly aimed at rethinking ontology. Although, like Butler, Grosz begins by treating the sexed human body in relationship to the limitations of psychoanalysis, she also treats bodies other than human bodies. Grosz, like Butler, means to take the unconscious beyond the oedipal narrative, and, like Butler, she has done so by queering sexual desire. But Grosz finally has turned from psychoanalysis toward Deleuze and Guattari's treatment of desire, the "body without organs" and machinic assemblages.

In Grosz's terms, bodies are "volatile." They are about connections, intensities of vibrations over a surface and its folds, where the difference between concepts, images, institutions, and discourse are indistinguishable from the perspective of desiring production. Bodies are what desire produces. Grosz's thought of volatile bodies, including but not privileging the human body, is an argument for a *différantial* relationship of nature and technology, body and machine, the virtual and the real. Rather than being in an oppositional or dialectical relationship, nature and technology, body and machine, the virtual and the real are interimplicated; culture is nature deferred, as is technology. Unlike Butler, Grosz does not treat the body in terms of a radicalized social construction. Rather, for Grosz, bodies are given, in the Derridean sense, in specific modalities of materiality, and it is as such that they are engaged with cultural inscription devices such as racism, sexism, heterosexism, and ethnocentrism, with varying political effects.

In this sense Grosz's treatment of bodies makes explicit the dynamism of matter that in Butler's treatment of bodies remains implicit. For Grosz, all bodies are virtualities of this dynamism; they are images in process, lines of flight to the future. This dynamism, which Cheah has referred to as the subindividual, finite forces of mattering, is for Grosz the desiring of postpersonal thought. Her work on bodies turns into the thought of bodies as desiring production, bodies as machinic assemblages, bodies as the movement of forces. For example, her treatment of architecture suggests a shift from the structural to the mobile, to the speed of flows of singularities into and through bodies, where desire is the movement.

Grosz's work, in my mapping of it, suggests a resolution of what began as the deconstruction of the oedipal narrative and the ungrounding of the structural configuration of family and national ideologies, the state and civil society, and the public and private spheres into a much more complicated but more flexible network for desiring production in the speeds and exposures of the teletechnological. Desire is delivered from the limitations of the historical and geopolitical specificities of the oedipal narrative, and the opposition of the real and the imaginary is displaced. Grosz's writings, along with Butler's, have made it seem that feminist theory was meant all along to deliver desire from its modern elaborations, to give the unconscious over to thought, to make thought and affect inseparable, to make all this palpable through the deconstruction of the psychoanalytic configuration in which the unconscious has been held—that is, the real, the symbolic, and the imaginary.

With thought becoming indistinguishable from affect, the unconscious, and desire, cultural criticism not surprisingly not only has focused on rethinking bodies. It also has turned reflexively upon knowledge and science, calling into question their epistemological underpinnings in rationality and its discursive forms of legitimizing authority. My mapping, therefore, goes from feminist theory, Marxist cultural studies, queer theory, and postcolonial theory to the cultural studies of science, especially the new sociology of science and the criticism of ethnographic writing as a textual production of the scientific authority of anthropology. Sociologists, historians, philosophers, and scientists who have been engaged in science studies, along with the cultural critics who have questioned the location of those producing knowledge in terms of identity politics on one hand and antiessentialist treatments of identity on the other—all have become ways for following poststructuralism and the unconscious thought of teletechnology in its reach to the ontic.

Arguing that power is internal rather than external to science, the field of cultural studies of science has rethought knowledge as power/knowledge and science as technoscience. As such, it has overseen the displacement of labor by abstract knowledge as central to production in the transnationalization of capital in neocolonialism. Deconstructing the ideology of scientificity and facing the commodification of abstract knowledge, cultural studies of science treats science as a doing, or as everyday practices elaborating an ability to do, where the ongoing results are directed to the making of "machines" or "centers of calculation," to use Bruno Latour's terms.[23] The machine functions as a black box, whether it be a vaccine or a measuring apparatus, and as such it is something like what I refer to in this book as a technical substrate of unconscious memory; the machine is productive of society—that is, all that is to matter socially must pass through it. Cultural studies of science has brought thought to technoscience as a primary agency of power/knowledge, the awareness of which gives shape to what have been called postmodern or "knowledge societies."[24]

The field of cultural studies of science also has taken writing and textuality as the machine metaphors or the vehicles of its criticism; even ethnographic studies of science has become focused on inscription devices — machines or centers of calculation. From graphs to cyborgs, how inscription devices come to be, how they are deployed, and with what effects, all have been made central to cultural studies of science, drawing it to the semiotic-material objects, or "knowledge objects" and their "interobjective sociality."[25] My mapping is meant to suggest that cultural studies of science has given human-machine attachments an ontological glow in the blue light of teletechnology.

Not surprisingly, the field of cultural studies of science also has become self-consciously aware of its own inscription devices, its own doing and practices of representation and methodology. In this sense, cultural studies of science has found itself engaged with what might be considered cultural studies of the social sciences, that is, the criticism of the production of anthropological authority in ethnographic writing. Although the criticism of ethnography began as a criticism of the authority of western anthropology in the context of colonization, it quickly provoked reflection on the transformation of world cultures in the context of the globalization of teletechnology and the transnationalization of capital in neocolonialism. The criticism of ethnographic writing turned cultural criticism to a reflection on the transformation of labor, identities, agencies, knowledges, sexualities, bodies, and practices of migration and immigration in diaspora. In the experimental writing of ethnography that followed the criticism of western anthropology, there was, therefore, an attempt to open up the ethnographic text to the multiple voices of those who have for so long been the object of anthropological study.

The criticism of ethnographic writing in anthropology also led to extended self-reflection on the production of authorized knowledge in the social sciences generally, and therefore to various experimental writings in which a self-conscious self-reflection on the part of the writer was performed in the text. Yet for some critics, among them Donna Haraway,[26] a performed self-conscious self-reflection already is insufficient to the task of a cultural criticism of technoscience. Haraway has suggested instead that a critical practice other than self-reflection has become necessary, a practice that, as I would put it, can treat the speed of territorializations and adjust to the vulnerabilities of exposure to media event-ness in the networks of teletechnology. Self-reflection as a practice of rethinking or as a thinking-over is, according to Haraway, to be displaced by critical direct action on semiotic-material objects across all the sciences. There already are examples of a cultural politics of nature, where cultural criticism takes the form of direct action aimed at biotechnology or technonature.

Yet autoethnography has remained one of the most common responses to the criticism of ethnographic writing; its aim is to give a personal accounting of the location of the observer, which is typically disavowed in traditional social science writing, traditional ethnography especially. It does this by making the ethnographer the subject-object of observation, exploring experience from the inside of the ethnographer's life, emphasizing emotions or feelings. Tragedies, including the tragedies of oppression, often are the focus, but autoethnography also has had a subtext that has functioned to produce a subject identity for its author. Autoethnography not only performs itself as "a politics of location," to use Adrienne Rich's term.[27] It also draws legitimacy from various standpoint epistemologies, especially those developed by feminist scholars such as Nancy Hartsock, Dorothy Smith, Patricia Hill Collins, and Gloria Anzaldua; these scholars have emphasized the situatedness of knowledge production and have privileged the increased capacity of the oppressed to yield, if not accurate knowledge of systems of oppression, domination, and exploitation, then surely oppositional consciousnesses.

With all this, the field of cultural studies of science, including the criticism of ethnographic writing, became enmeshed in arguments over identity politics and its antiessentialist reversals. Although trying to grapple with the issue of thought's and affect's becoming indistinguishable from each other, these debates have, nonetheless, reduced writing and textuality to too narrow a definition, as the whole effort of experimenting in autoethnographic writing often does. Therefore, the autobiographical revision of ethnography and of the social sciences has seemed both a symptom and a dull or slow response to teletechnology. It is a symptom in that it is a performative practice engaging the vulnerability of exposure to media event-ness; its melodramatic focus on the personal, especially the tragic, is televisual. But it is slow precisely because it repeats too closely levels and kinds of exposures common to television without much interfering with them or redirecting them.

In most cases of autoethnography, there also is the forgetting of nonknowing or of the unconscious and desire altogether, so that what began as a criticism of the authority produced in ethnographic writing comes back at times as a naive production of autobiographical authority, as if writing about what one knows about oneself can be so much fuller or more accurate or even more ethical than writing what one knows about others. The subindividual finite forces realized with the deconstruction of the subject have little play here; indeed, the subject of autoethnography often ends up full of its self-identity. Perhaps, then, what is more remarkable about the autoethnographic treatment of emotions from inside the experience of the autoethnographic subject is its difference from poststructuralism and deconstruction. That is to say, beyond the deconstruction of the subject, there has been a remarkably insistent effort to restore the subject expressed in the desire for voice or the voicing of one's experiences, claiming them as one's own. In Derrida's earliest writings, he referred to this desire as autoaffectionate.

Autoaffectionately Yours

Derrida uses the term "autoaffection" to mean "giving-oneself-a-presence or a pleasure," "hearing oneself speak" in the closed circuit of mouth and ear, voicing and hearing.[28] It is autoaffection, Derrida argues, that gives the natural grounds to the subject privileged in the western modern discourse of Man. It is autoaffection that allows the presumption of the unity of speech and precommunicated thought, giving the subject an inner presence, an inner voice, so that the subject, when it speaks, is presumed to speak its own voice, to speak its intention and to express its inner being.

Autoaffection, therefore, is laced through and through with the repression of *différance* and the disavowal of the unconscious as the mark of the noncoincidence of subjectivity with consciousness, that is, of the subject with its conscious self. Autoaffection is the resistance to recognize the technical substrates of unconscious memory, and therefore autoaffection is crucial to any refusal of an intimacy between the body and the machine, nature and technology, the virtual and the real, the living and the inert. It is, after all, against the natural circuitry of self-heard voice that Derrida places technicity, the machine, the text, writing — all as bearers of unconscious thought. Unconscious thought breaks the natural circuitry of autoaffection, but of course it also makes it possible, or impossibly so, by providing the fantasy of the natural voice of a self-same subject. Unconscious thought, that is, provides the mechanisms for repressing and disavowing the unconscious itself. So autoaffection remains around unconscious thought; a certain desire for voice reasserts itself after deconstruction as its *différance*.

In the chapters that follow, for which I have just given a first mapping, the concepts and the cultural criticisms that find their way onto the same plane with teletechnology are not only those that have had an enormous impact on intellectual and academic discourses over the last three decades of the twentieth century. They also are the writings and concepts that have shaped what I think, what I feel. They are the concepts and writings to which I am and have been drawn. My unconscious desire, therefore, also plays its part in constructing the plane of consistency given over to teletechnology.

I became involved with poststructuralism and with the cultural criticisms that have engaged it just when I was beginning what would end up being fifteen years of psychoanalysis; my own psychoanalysis plays an important part in what follows. True, the unconscious thought I am pursuing is beyond the oedipal narrative, and therefore beyond Freudian psychoanalysis; after all, Freud argued that the recognition of the oedipus complex is "the shibboleth that distinguishes the adherents of psychoanalysis from its opponents."[29] Yet, like Marxist criticism, psychoanalysis, my psychoanalysis especially, has been crucial in my understanding of what I am referring to as postpersonal thought. Understanding has come through the experience of the psychic domain and of the failure of that experience to ever give full understanding of self or other, and therefore the necessity of political strategies that know not fully their conditions of possibility or their full implications. Of course it is risky to too closely align poststructuralism or the cultural criticisms that have engaged it with Marxism and psychoanalysis, but this risk is worth what can be gained in working through psychoanalysis and Marxism rather than just going around them.

But I am also introducing my psychoanalytic experience as a guide for the reader. I am, after all, about to suggest that the thought that teletechnology draws from poststructuralism and various cultural criticisms gives itself over to an ontology that refuses to privilege the body over the machine, nature over technology, the real over the virtual. I am instead endorsing a *différantial* relationship between nature and technology, the body and the machine, and the real and the virtual as a nonoriginary origin. This shift in ontological perspective, of course, has implications for politics. Even though I am in no way sure of the full range of the political implications, I am especially aware that my focus on ontology leaves aside the histories of technological development that locate modern technology in close relationship to war and the military, usually under the direction of capitalist interests.[30] Although I am not unaware of the horrors resulting from the intimate relationship of war, capitalism, and modern technology, I do not begin with this relationship. I do not begin in the presumption of any critical position that is simply for culture by means of opposing it either to technology or to nature or to both, as so often cultural criticism has done. I am hoping that we can learn how to distinguish one deployment of technology from another within one machinic assemblage or another and how to treat these differences critically. This will not be easy for scholars and intellectuals, like myself, who have for so long taken as bottom line the capitalist mode of production or the oedipal complex or some tight fit between these and for which an ontology of Being is necessarily presumed.

Because I am proposing a certain acceptance of machines and technologies, although I am aware as much as anyone else of their terrible possibilities, I am offering my psychoanalytic experience as a hesitation for the reader, as my effort to locate myself in my unconscious thought, to locate myself in the chapters that follow. In my own psychoanalysis I came to realize that death, so often figured as a machine or as technology in the western modern discourse of Man, is internal to life. Yet death is different than life. The struggle for life must be fought to the end, but there is no overcoming finitude; death or finitude will not be overcome by life. In the first chapter that follows, therefore, I take up Derrida's treatment of the technical substrate that he argues is inextricable from unconscious memory and is, in this sense, like death or finitude, internal to life.

I begin here because Derrida's writings about Freud have been for me a burden and a gift throughout my psychoanalysis. Because Derrida suggests that the unconscious is shaped as much by a technical substrate as by the individual subject's history, he provides a way to think of unconscious memory beyond Freudian psychoanalysis. He thereby provides me with a way to think about the insights of my psychoanalysis beyond myself and to link them with the future of thought. This gift has also been a burden in that it also easily serves as a defense against my own psychoanalysis.[31] I therefore also want to frame the chapters that follow with a treatment of my own unconscious, with its fantasy of the machine, with the struggle of the forces of life and death within my being. I want to lay bare my leanings toward technology, my attachment to machines, as a way to remind the reader of the vulnerabilities involved in my proposal of a new ontological perspective and of an unconscious other than one organized by an oedipal narrative. To do this I have placed here and there between the chapters some few prose poems — autoaffections of sorts.

The prose poems are not so much about the experiences of my life, but rather are meant to reveal something of my unconscious, about the way in which it first took shape, at least as I came to understand it in psychoanalysis. The references to my life, therefore, are limited to my early childhood up to and through my early adult years, when, in the language still in use in psychoanalysis, the formation of the unconscious first begins to condense onto itself and project itself onto the world. The prose poems, therefore, are not just about my unconscious; they are about the unconscious of others, which has come to haunt mine. In this sense the prose poems are about the openness of the unconscious, or about the opening of my unconscious to encrypting the unconscious of others. In two of the prose poems I do not appear as myself so much as in and in between the sound and feel of others, others' fantasies taken up in me, worked through and abandoned, or too deeply embodied for an I to speak them, as such — overworked machines.

Finally, the language of the prose poems is sometimes patched with words and phrasings from the various writings that have engaged me over the last three decades. These words and phrasings appear in the prose poems as an enactment of the attraction I have had, and still do, to the abstract, postpersonal, but passionate language of poststructuralism and the cultural criticisms that have engaged it.

Television: A Sacred Machine

The title is meant to trick you.
It is meant to keep you,
perhaps to keep me too,
from being afraid of me
because I am drawn to the machine,
because I am drawn in by the machine
that draws me out,
that draws me apart.
I am afraid that you will see that it excites me
being drawn out,
being drawn apart,
being drawn out into parts.
It is an apparatus of display: the machine.
It holds me on display,
holds me to the display.
It is made of tacking devices
that sometimes attack me without pity,
like projectiles,
tacking me in parts to the display.

I am afraid you will be afraid when you see what comforts me.
It is a holding apparatus: the machine.
It holds me up,
cradles me.
It is made of framing devices that negate,
reverse, and enlarge —
to perfect and protect.
It makes me an ideal surface of projection and reception.

Is it already getting too difficult for you?
But you said you wanted to understand,
to understand what I was saying.
I am not saying.
I am desiring. The machine.
I am the machine's desire.
The desiring machine alone knows my desire.
It keeps it; it repeats it.
In the machining of my desire, I am.
Not located,
I am
arrested and displayed in arresting positions —
held and beheld.

Would it be easier if I said:
"In postmodernity, increasingly control is applied to people's routine
existence by the apparatuses of education and entertainment which
exact identification and consensus."[1]

It would be easier, but it would be different.
This is not only about ideology,
about arrested and arresting political positions.
Nor is it only about self-exposure,
an autobiographical antidote for a closeted all-knowing eye.
My machine has more parts; it has more action,
like the action of fingertips attached to ivory keys,
playing in between the beats of a metronome's patterning.

My first piano came from my grandparents' house.
A barrelled organ, it became a street piano as it transmigrated
from Brooklyn to Queens,
to the three-room apartment
where I lived
and where it was placed up against a wall.
From there it beckoned me, gently at first,
but then more and more insistently.
It was an upright and when being played,
the wooden covering over the strings and hammer devices
was to be left slightly ajar.
It was for better sound.
But it also was an exposing.
I could see the machine's action.
I could see the strip of red felt ribboning through the strings.
I could see the hammer devices which, when moving,

seemed like marching toys —
not so friendly,
but regular and regulating.
And there were the keys
meeting the eye
meeting the fingers all at once,
as the wooden hood over the keyboard was slid back
into the piano's insides.

I once heard a jazz musician say that reading music
is not as good as playing one's feelings,
playing with one's body.
But I do read music,
and I am ashamed that my body won't speak to the piano,
that my body wants to be so dumb.
I want the piano to speak to me.
I want it to draw me out, as it first drew me to it,
drew me apart from the three-room apartment.

If I were very still, the piano would grab me by the finger tips.
It would speed me up,
slow me down,
take me high,
take me low.
The piano bench would gently support my body's flight
into a trance of mobile immobility.

This is not simply sex.
I fear that in your ear my words are already simply sex.
A child becoming piano.
A body becoming machine.
Not yet a sex. It is desiring.
A thousand tiny sexes, in between the keys,
in the action of the hammering devices,
in the strings —
those wrapped thinly for startling and stinging vibration,
those more heavily wrapped for somber and sobbing sounds.
Not yet a sex, I tell you.
I tell you because what I must now tell you
is more difficult. The eyes,
the eyes of the piano are down below me,
looking up from where my feet attach to the pedals.
They see into and through the indifferent spacing
between piano and piano bench.
Backs of knees and elbows,

bits of arm and slices of thigh
are cut out by the piano's vision.
It is not yet a sex, this vision from below, shooting up into me.
It is not because it is sex that it excites me.
It is that I can't see myself like the piano sees me.
Its eye is the eye of the outside.
It has its own eye.
The piano lives,
and it gives
its music to me.

And when it does become a sex, one sex, my sex,
it is all but unbearable.
So, it is not until I am nineteen
and already for some time a Roman Catholic nun,
wrapped tight in black from head to toe
and coifed in starchy white,
that I dare approach the church organ for the first time.
I tremble before the doubled set of keys at my fingertips.
There also is a keyboard beneath my feet,
and pipes rise up into the heavens,
as if growing out of my backside.
Once the organ is turned on,
I can hardly move without making music.
Even the sound of my breathing is in harmony with the aves
ringing through the organ's pipes.
I have turned my sex,
having been made one, back into a bisexuality
of the black and white of cloth and keys alike.

Would it all make more sense if I reminded you that sociologists say that
by the early 1960s the television had replaced the piano that was once in
every lower-middle-class home, an object of entertainment and a mark
of upward mobility?
I would have told you that right from the start.
But by the early 1960s,
I was already gone from my home,
in flight,
a musical organ,
a sacred machine.
Actually it was before the early 1960s that I first took flight from my
home. It was in the early 1950s, when I had my first piano lesson. I was
six. My teacher, Sister Bernadine, met me at the back door of the
convent. Following her, I passed a long row of wooden rockers up
against the porch wall.

Suddenly, we were inside.
I was told to walk on my tippytoes,
to be quiet, as she led me through the refectory.
It startled me.
Cream-colored dishes and cups, all in a row,
turned down on the tabletops,
under which were tucked small wooden stools.
The air pressed its lightness heavily against my young body.
I felt faint with the pleasure of the secret life
I was glimpsing,
although it did not seem to be for the first time.

We moved quickly down a long hall to a small room with a piano
up against a wall.
The piano was much like mine at home,
but it seemed much more tired and somehow proud.
Sitting before it and next to Sister Bernadine,
my back was up against a wall.
There was no room in that room
that wasn't taken up with the music-making machine.
The lesson began with Sister Bernadine holding my ten fingers
between her two hands, saying:
"Don't just play it; pray it."

I wanted to think,
but I couldn't,
that I was an anointed one. Destined.
But thinking that I wasn't worthy was a way of being
I already knew at six.
Still, I was destined by that piano,
destined to find myself in attachment to machines.
To find comfort in difficult compositions of sounds and images —
drawn less to meanings
and more to ordered angles,
corners of frames, coordinates of moving surfaces.
Drawn to the machine, drawn in desire
to the machine of desire.

I am certain that it would have been easier if I had said: "Television is
part of the way in which exchange value is constructed, distributed,
and attached to bodies formed in the general circulation of labor,
commodities, and money. It has expanded the zones of value by
changing, mediating, that is to say, mechanizing the imaginary of
social relations."[2]

It would have been easier, but different.
This is not only about a political economy of value
attached to exchanges between body and machine.
It is about the attachment itself.
It is about the tacking devices that put up the display —
the framing devices that are given to hold and behold.
It is to wonder about their location,
and the micro-movement of the singular forces of music,
accompanying the machine's vision.
Ah, the wonder of it!
To wonder how it is that the machine's vision is
not secondary to my vision.
An auto-tele-vision.

It is the flickering up
and the passing away
of conscious contact.
It is to be zapped in and out of a rush of images and sounds:
A mother who only loves just enough — not quite the wire mommy of
the rhesus monkey experiments of an earlier scientific research
agenda — the cold machine.
A mother's sensual singing accompanied by a red and white kitchen
radio — the music machine.
The fat of a childhood body that would be without organs,
without a sex, all filled up,
all closed in — the dumbing machine.
A World War II father, needy, limited and unfaithful,
shot in his army fatigues, late in the afternoon with a Brownie camera —
the war machine.
A father's lap, like a couch to sink into while he reads nursery rhymes —
the babysitting machine.
A 1939 photograph of a bride and groom, young, handsome, red lips and
white gardenias — the wedding machine.

I Remember Mama,
Make Room for Daddy,
Kukla, Fran, and Ollie.

A set of *Encyclopedia Britannica* bought at the beginning of years of
disciplinary study meant to rationalize the rhythm of images, the
cadence of words.

What's My Line? the McCarthy hearings, *The X-Files, Aliens 3,*
Terminator 2, cybernetics, The Genome Project.
The wound culture of late afternoon talk shows.

Machine anthropology,
cultural studies, African American studies, women's studies, queer studies,
science studies.

Sheet music keeping safe a strange language
between eye, fingers, mind, body, and machine.

Bright flashes against thousands of lines of pulsing light —
a veritable vision of flows
of energy and matter.
A blue white halo all around,
surrounding me,
holding my face up close,
touching the screen,
hard but never too hot, just cool enough.

Stories of techno-organic kinships,
stories that cannot be passed on only as history.

"The singularities of a life, which when mined for their richness, should
not be made to encourage a swapping of memories, a textual game of
'Oh that happened to me too' that stalls the movement of chance,
disarms pivots of unpredictable necessity in the relations of bodies and
machines."[3]

Notes

1. Alberto Melucci.
2. Richard Dienst.
3. Elsbeth Probyn.

CHAPTER ONE

The Technical Substrates of Unconscious Memory

Deconstruction and the Freudian Unconscious

In *Archive Fever*[1] Jacques Derrida returns to an earlier essay where he first traced Freud's steps from treating unconscious memory in terms of neurology to when, in 1925, Freud finally treated the unconscious in the metaphor of a writing machine, a child's toy that Freud referred to as the "mystic writing-pad." In this earlier essay, "Freud and the Scene of Writing" (1978),[2] Derrida points to Freud's failure to recognize the existence of archiving machines or technologies that are surely more sophisticated than the toy mystic writing-pad. Derrida goes on to argue that the metaphor of the mystic writing-pad, which Freud claimed to be the best rhetorical device for treating unconscious memory, is made possible "only through the solid metaphor, the 'unnatural,' historical production of a *supplementary* machine, *added* to the psychical organization in order to supplement its finitude."[3] Not only does Derrida suggest that there is a relationship between unconscious memory and historically specific machine metaphors or that unconscious memory is inextricable from the various "technical substrates" given it with historically specific technologies, to use the bolder formulation of *Archive Fever*. He also suggests that from the start a certain technology oversaw Freud's treatment of unconscious memory; a certain technology drew Freud to the metaphor of the writing machine.

If, as Derrida would have it, Freud did not recognize the technology that oversaw his project, the same might be said about Derrida, at least in his earlier rereading of Freud. But in *Archive Fever,* where Derrida returns to "Freud and the Scene of Writing" in the context of interrogating the relationship of unconscious memory and teletechnology, what is suggested is that it is teletechnology that allowed for the connections Derrida first elaborated between his own project and Freud's. Or, as I would like to put it, in "Freud and the Scene of Writing" Derrida begins to complete Freud's project in the machine metaphors given with teletechnology and suggests, therefore, that teletechnology always already drew the Freudian unconscious to it and to the future.

To propose that teletechnology oversees the Freudian and Derridean treatments of unconscious memory is, of course, to raise a question about history. What kind of history would place teletechnology at the scenes of both Freud's writing and Derrida's writing? What is history if there is some relationship between unconscious memory and historically specific technical substrates? Surely history cannot simply be linear or homogeneous if technologies give unconscious memory historically specific technical substrates that are the condition of possibility of various historicities or of various relationships of temporality and spatiality. It would seem that there is an "aporia of time," to use Derridean terminology: a history of technological development that undermines history. History, even the history of technological development, can be only impossibly so. This impossibility is, nevertheless, productive. It is the condition of possibility of more than one historicity. It allows for the anticipation of various historicities. It also permits an understanding of Freud's treatment of unconscious memory as anticipatory or, better, compensatory, that is, as compensating for what could not be thought without the machine metaphors yet to come in the future, but which future, nonetheless, drew Freud's treatment of unconscious memory to it—from neurology to writing machine.

And if Freud's treatment of unconscious memory may be grasped in this way, what of recent cultural criticism, specifically film criticism of the 1970s and the 1980s? After all, among contemporary cultural critics it was film theorists, most notably feminist film theorists, who were persistent in engaging the relationship of unconscious memory and the machine metaphors of the cinematic apparatus. If feminist film theory is a compensatory treatment of unconscious memory in relationship to teletechnology, television especially, then rereading feminist film theory may show what is no longer necessary for an understanding of the unconscious in the age of teletechnology.

In the second part of this chapter I take up the relationship of feminist film theory and television, but first I want to turn to Derrida's rereading of Freud's treatment of unconscious memory in order to show how Derridean deconstruction problematizes the history of technological development so profoundly that it returns thought to ontology. Although not itself an ontology, Derridean deconstruction undermines the ontology of presence or, as Richard Beardsworth has suggested, it draws an "originary Being" down into an "originary technicity."[4] Derridean deconstruction thereby reconfigures the oppositions that an ontology of presence grounds, such as the opposition of nature and culture, body and machine, the real and the virtual, the living and the inert; it displaces these oppositions with *différantial* relationships. In this sense Derrida's project is to be understood in terms other than those that restrict it to the linguistic turn; *différance,* whether as textuality or writing, is to be understood instead as thought reaching to the finite forces of mattering or the dynamism of matter.[5]

Although Derrida, therefore, seems to approve of Freud's steps away from neurology to the writing machine, Derrida does not mean simply to dismiss nature, neurology, or biology. He does not mean to turn nature into cultural text or machine writing, which has so often been (mis)understood to be the aim of Derridean deconstruction. For example, in her reading of "Freud and the Scene of Writing" Elizabeth Wilson argues that Derrida endorses Freud's move away from neurology and in so doing misses the productive link that might have been made between Freud's neurological treatment of unconscious memory and various new models of cognition, such as "connectionism."[6] The irony, Wilson proposes, is that connectionism has enabled researchers in the fields of artificial intelligence and psychology to rethink cognition in terms of neural nets or "the effect of relational differences in the activation between units and across a network (of neurons)."[7]

In treating neurology as a matter of movement over spatial and temporal differences in a network of neurons that is without origin or ends, template or stored rules, connectionists, Wilson proposes, use terms that show a strong likeness to Derrida's, especially when he is treating *différance,* whether as writing or textuality. She therefore proposes that Derrida's rereading of Freud can be deployed not only to link Freud's neurological treatment of unconscious memory to connectionism, but thereby to reinforce the effort of connectionists to think of neurology as operating "in excess of the limits of presence, location, and stasis."[8]

Although I find Wilson's take on Derrida and Freud provocative, I want to offer a reading of "Freud and the Scene of Writing" that suggests that Derrida does not dismiss neurology, biology, or nature, but rather refuses to oppose these to culture. He therefore refuses to oppose the unconscious to the machine. In following Freud's steps from neurology to the writing machine, Derrida wants to pose certain questions, such as: What is the machine that it lends itself as a metaphor for unconscious memory? What is the inside and outside of the machine? What is the inside and outside of unconscious memory? In posing these questions Derrida not only means to treat nature or biology as inextricably interimplicated with culture or the machine; he means to do so in relationship to a historically specific technology.

What Wilson misses is that Derrida reconfigures the relationship of nature and culture in terms of the solid metaphor or supplementary machinery of a historically specific technology. She therefore also fails to appreciate that the object that connectionists treat as a neural net is a knowledge object, inseparable from its technological enframement. Furthermore, the very terms with which connectionists treat the neurological are terms not only befitting Derridean deconstruction, but also teletechnology, upon which connectionism depends. It is in the machine metaphors of teletechnology, I want to suggest, that Derrida draws Freud's treatment of unconscious memory to the future to register the dynamism of matter out of which nature and culture are given, always already interimplicated. In other words, I want to suggest that in following Freud's steps, Derrida has a tele-vision.

Step by Step to a Tele-vision at the Scene of Writing

The mystic writing-pad, although a child's toy, is a writing machine. It is made of a wax slab to which is attached, on one end, a sheet made of two layers; one layer is celluloid, and it protects the other layer, a waxed paper. The device is worked by lifting the sheet at the side where it is not attached. This completely clears the writing while leaving traces only on the deepest layer, the wax slab, which Freud proposed might be compared to the unconscious "behind" perception. The mystic writing-pad has the metaphorical capacity that Freud had been seeking in order to properly represent the functioning of unconscious memory. As Derrida puts it, the mystic writing-pad has "the potential for indefinite preservation and an unlimited capacity for reception."[9] The device can turn one surface out to the world, remaining open to every excitation because the traces of excitation can be stored elsewhere than on the writing surface. But when the traces are stored — or, better, when an impression is made on the wax slab beneath — the impression entirely changes the network of traces that makes up what is below or what is taken to be the unconscious. Although the example of the mystic writing-pad proposes that unconscious memory allows the perceiving surface above it to remain open to the world, it also suggests that there is no presence present beneath, in the unconscious. The unconscious has no place; it is a space that is temporally dynamic, a spacing of ungraspable traces.

It is Freud's notion of the ungraspable trace that interests Derrida. Earlier, in the *Project for a Scientific Psychology* (1895),[10] Freud had introduced the notion of a trace as a kind of writing of forces in relationship to the accumulation or the discharge of energy in the nervous system. As Freud explained it, the primary function of the neurons is to receive excitation and discharge energy. But Freud also argued that there is a secondary function of the neurons that, on the other hand, operates simultaneously with the primary function. This secondary function, which might be better referred to as the deferral of the primary function, is to resist the discharge of energy. Instead, energy is accumulated so as to allow the neurological system to face what Freud described as "the exigencies of life" — that is, to enable the activity of living.

Freud went on to argue that the resistance to discharge occurs at the "contact barriers" between neurons, so that when the discharge of energy is inhibited, the accumulated energy forces open a trace or a path at the contact barriers. In Derrida's terms, Freud suggested that, against resistance, a "path of facilitation" is opened or "breached"; "the tracing of a trail opens up a conducting path."[11] The contact barriers between neurons thereby become variably capable or incapable of repeated conduction of energy, and some contact barriers offer no resistance at all. Unconscious functioning is a matter of the different paths of facilitation in a network of neurons and the variation in the conduciveness to repetition thereby allowed. But Freud further suggested that neuronal networks reconfigure themselves with each excitation, endlessly changing, and as such remain fully open to excitation. In this, "the first representation" or "the first staging of memory (Darstellung)," as Derrida puts it,[12] Freud refused to describe the nervous system as compartments for storing memories; instead his description proposed that the nervous system is a substrate in motion, which allows the unconscious to function as a memory-making — not a memory-keeping — apparatus.

Derrida emphasizes that it is the difference in the breaching, the difference in the spacing and timing of the traces, that makes unconscious memory possible. It is not, then, that there are paths or connections present in the nervous system. As Derrida puts it, "it must be stipulated that there is no pure breaching without difference. Trace as memory is not a pure breaching that might be reappropriated at any time as simple presence; it is rather the ungraspable and invisible difference between breaches."[13] For Derrida, Freud's neurology suggests that "psychic life is neither the transparency of meaning nor the opacity of force but the difference within the exertion of forces."[14]

There is therefore no memorized content in the nervous system. Although there is repetition, it is not remembered content that is repeated. Instead the repetition is of an impression or a trace that is only a repetition of the difference in the exertion of forces. Derrida argues that this repetition is an "originary" repetition; it is not the repetition of an original. That is to say, the "orginary" of originary repetition is always already crossed through or put "under erasure," as Derrida puts it. Repetition is labeled originary only to undermine the idea of an origin: "It is a non-origin which is originary."[15] In this sense, Derrida brings Freud's treatment of repetition closer to Gilles Deleuze's treatment of repetition as "pure repetition."[16] For Deleuze, pure repetition is repetition without an originary essence or a transcendental principle. It is neither a oneness turning into multiplicity, nor is it a matter of different elements of a concept that itself remains the same. Repetition is thought meant to grasp the irreducibility and contingency of singular forces. Only pure repetition releases the possibility of pure difference.

For Derrida, to think of Freud's neurology as a matter of originary repetition suggests that, although the resistance to the discharge of energy makes repetition possible so that the exigencies of life might be met, life is, nonetheless, not originary. It is not life already present that is protected by resistance to the discharge of energy. Rather, life is made possible in the repetition of the protective resistance. But if life is not an originary presence, life also is not-life. In this sense and only in this sense, Derrida argues, life is death, just as memory is forgetting or repression or the force of impression. Or to put this another way, this life-giving repetition, like the compulsion to repeat a trauma in order to forget and master it, which Freud called the death drive, is internal to unconscious memory and to life as well.

Although Freud finally opposed the death drive to the life force, Derrida argues for a *différantial* relationship between life and death. Therefore, Derrida's aim in following Freud from neurology to the writing machine is made clearer; it is not to dismiss neurology, biology, or nature as inert or dead matter, but rather to bring neurology, biology, and nature closer to technology, even to suggest an ontological perspective that allows for a *différantial* relationship rather than an oppositional or dialectical relationship between body and machine, nature and technology, the virtual and the real. Not only does Derridean deconstruction suggest that nature and culture are deferrals of each other; it also proposes that nature and culture are given out of *différance,* or the dynamism of the singular, subindividual, finite forces of mattering. At least this is what Derrida's treatment of *différance* seems to imply.

Différance refers to a weave or network of differences that are nonlocatable, ungraspable; *différance* refers therefore to the impossibility of presence or identity, except in the disavowal of *différance.* As Derrida suggests, *différance* refers to a pure interval:

> An interval must separate the present from what it is not for the present to be itself, but this interval that constitutes it as a present must, by the same token, divide the present in and of itself, thereby also dividing, along with the present, everything that is thought on the basis of the present, that is, in our metaphysical language, every being and singularly substance or the subject. In constituting itself in dividing itself dynamically, this interval is what might becalled *spacing,* the becoming-space of time or the becoming-time of space (temporization). And it is this constitution of the present, as an "originary" and irreducibly nonsimple (and therefore, *stricto sensu* nonoriginary) synthesis of marks, or traces of retentions and protentions . . . that I propose to call archi-writing, archi-trace, or *différance. . . .*[17]

These remarks appeared some years after the publication of "Freud and the Scene of Writing" in an essay in which Derrida offers his most extensive treatment of *différance.* But already in "Freud and the Scene of Writing," in its very first pages, Derrida describes his primary concern: there is autoaffection, presence, logocentrism to be put into play with *différance* as "the pre-opening of the ontic-ontological difference."[18] Derrida is proposing that *différance,* so often understood as linguistic undecidability, moral relativism, or political indifference, is not simply any of these. *Différance* rather points to thought reaching to touch the ontic. It is meant to give an ontological perspective.

By deconstructing autoaffection, logocentricism, and presence, Derrida's treatment of *différance* draws Being back into preontological forces—the subindividual, finite forces of mattering. In this sense Derrida's treatment of *différance* is close to Foucault's treatment of power as a "moving substrate of force relationships which by virtue of their inequality constantly engender states of power but the latter are always local and unstable."[19] Like Foucault's treatment of power, Derrida's treatment of *différance* gives nature and culture over to unstable, unlocatable networks of differences. But Derrida's treatment of *différance* further proposes that the thought of *différance,* the thought of the *différantial* relationship of nature and culture, is to be grasped through machine metaphors, the technical substrates of unconscious memory. It would seem that it is the metaphors given with teletechnology that especially enabled Derrida to arrive at his thought of *différance* in following Freud's steps from neurology to the writing machine.

But Freud had not yet himself thought of the unconscious in terms of the mystic writing-pad. Although in the *Project for a Scientific Psychology* he treated the nervous system as a network of differences or traces, he did not consider the unconscious apparatus as itself operating as a network of differences or functioning as a writing machine. Before he came to do so, Freud turned from neurology to the question of unconscious memory, asking how it functions or reaches to and through conscious perception or cognition. He treated this question as a question about translation, which he took up in relationship to dream texts and their interpretation. Derrida follows Freud to his treatment of the dream text as a writing of hieroglyphics and to the question Freud thereby raised: If the text of unconscious memory is written in hieroglyphics, what kind of translation is possible?

The answer given by Freud was that the hieroglyphics of the dream text are not translatable in the usual or narrow senses of the term; dreams have a materiality, "a scenic quality," that cannot be translated. The hieroglyphics of the dream text are not meant to be meaningful, and in this sense there is no dream text present in unconscious memory. Derrida notices that, in the case of the Wolf Man, Freud had already argued that the interpretation of unconscious material, such as fantasy or the dream text, is not a matter of returning or referring to an originary moment; interpretation is rather a matter of *Nachtraglichkeit,* or deferral. Freud argued that it was only through the scenic production of the unconscious, the unconscious production of a "screen memory," that the Wolf Man could experience the primal scene — his parents engaged in *coitus à tergo* — long after the "event." In fact, the Wolf Man never had experienced the event at an earlier moment, at least not consciously; the event may even have been nothing more than an infantile fantasy. No matter; the screen memory, in this case a fantasized pack of wolves on a tree outside the Wolf Man's bedroom window, allowed the primal scene to be experienced, albeit only as deferred traumatic effects.

For Freud, then, the untranslatable hieroglyphics of the dream text suggested that the dream is an "originary" production that gives its own grammar. This grammar is irreducible to any other code, foreclosing any thought of translation as a matter of re-presentation. Freud therefore proposed that the "dream thoughts," that is, the free associations to the dream's content, can only be read back into the dream content — not as a reconstruction, but as a deconstruction. After all, the dream content, as Freud saw it, already is a construction, indeed, a repressed or defensive one. The grammar of the dream content is singular — not simply because it refers to the individual subject, but because it refers to subindividual, finite forces that are singular. The forces of repression, which make the translation of the dream text impossible, have their own singular vicissitudes. It is the singularity of the forces of repression that makes possible individual subjectivity rather than the other way around. For Freud, then, the dream content and the dream thoughts remained "in two different languages"; they were "two different modes of expression."[20]

Freud's treatment of the interpretation of dreams leads Derrida to conclude that the movement from the unconscious to conscious perception is not a matter of translating a text present in unconscious memory. The unconscious is not a presence. Derrida puts it this way:

> There is then no unconscious truth to be rediscovered by virtue of having been written elsewhere. . . . There is no present text in general, and there is not even a past present text, a text which is past as having been present. The text is not conceivable in an originary or modified

form of presence. The unconscious text is already a weave of pure traces, differences in which meaning and force are united — a text nowhere present, consisting of archives which are always already transcriptions..., whose signified presence is always reconstituted by deferral, *nachtraglich*, belatedly, *supplementarily.*[21]

It is in similar terms that Freud would treat the apparatus of unconscious memory when finally he approached it through the metaphor of a writing machine or the mystic writing-pad. No possibility of translation would be posited between the systems of the psychic apparatus — from preconsciousness to the unconscious, from the unconscious to conscious perception. There would only be, as Derrida puts it, "original prints," "archives," "always already transcriptions." Not only is unconscious memory a movement of traces and erasures, but each of the systems of the psychic apparatus is also only this. Once Freud treated the unconscious in the metaphor of a writing machine, the psychic apparatus became what Derrida describes as "a depth without bottom, an infinite allusion, and a perfectly superficial exteriority: a stratification of surfaces, each of whose relationship to itself, each of whose interior, is but the implication of another similarly exposed surface."[22]

It surely is Freud's treatment of the psychic apparatus as a machine production of an infinite depth of meaning without foundation that is inextricably linked to Derrida's own treatment of the text. Derrida may have already written, "There is nothing outside the text (there is no outside-text)," the infamous sentence appearing in the *Grammatology*,[23] first published in the same year "Freud and the Scene of Writing" was published. Although the statement "There is nothing outside the text" has been so often (mis)understood to mean that there is no reality or even any materiality that much matters or that there is no meaning but what is given in written texts, the statement instead must be understood as "There is no present text" — "A text is nowhere present" in the unconscious. Or, as Derrida puts it in "Freud and the Scene of Writing": "What is a text, and what must the psyche be if it can be represented by a text? For if there is neither machine nor text without psychical origin, there is no domain of the psychic without text."[24] Nor without the machine.

It is therefore the literary text, when narrowly conceived as written text, that Derrida proposes is a disavowal of the impossibility of an unconscious text or a general grammar for translating dream texts. The production of a literary text is the production of an identity; it is "the becoming literary" of *differantial* traces (traces of the timing and spacing of *différance*), which always implies repression, forgetting or the disavowal of *différance*. The deconstruction of the text as "book" or "finished corpus of writing" opens up the text or returns it to "a differential network, a fabric of traces, referring endlessly to something other than itself, to other differential traces."[25]

The production of a text and the possibility of its deconstruction, therefore, cannot be disconnected from the unconscious, where there is only production without an outside, without an urtext, but where there also is disavowal, repression, and the death drive. In their effects disavowal, repression, and the death drive produce a text and give an outside to production; they make the outside into a transcendental figure of the origins and ends of thought so that outside-ness loses its heterogeneity, its *différance*, its virtuality, its futurity. Derrida gives a list of such figures, which have operated in western thought to produce a text and give origins and ends to thought: "*eidos, archē, telos, energeia, ousia* (essence, existence, substance, subject) *alētheia*, transcendentality, consciousness, God, man, and so forth."[26] The deconstruction of the text and of the origins and ends of thought returns the text to the thought of *différance*, to the thought of production without beginning or end, that is, to a writing machine that is an apparatus of originary repetition.

In insisting that the psychic apparatus is a matter of originary repetition, Derrida turns Freud's mystic writing-pad into a perpetual motion machine. It is no surprise that after the publication of "Freud and the Scene of Writing," when, in "Signature Event Context," Derrida returns to treat writing and communication as part of a criticism of speech act theory, Freud's mystic writing-pad has become a distributed network of transmissions without beginning or end, which functions only to permit the pure repetition of unconscious memory. Against the privilege that speech act theory grants the speaking subject as the origin and end of communication, Derrida instead refers to communication as a writing machine, for which the software of the program and the hardware of the apparatus are indistinguishable, so that the distinction of form and content are inoperative and there is no central executor or stored rules. It is here, in elaborating a criticism of speech act theory, that Derrida describes the writing machine of unconscious memory as "telecommunication," when every communication is "being sent" without a sender, when the machine is internal to every communication—"a machine that is in turn productive" and "which a subject's future disappearance in principle will not prevent from functioning and from yielding and yielding itself to reading and rewriting."[27]

Surely a machine other than Freud's mystic writing-pad seems to be making itself available as a metaphor for unconscious memory. It would seem that Derrida is having a tele-vision. But rather than recognize the technology that draws him to follow Freud step by step to the future, Derrida instead worries about the future of Freud's treatment of unconscious memory. Since Freud came to treat the unconscious as a writing machine, what will become of psychoanalysis? What will be Freud's legacy to psychoanalysis, to its authority? These are questions to which Derrida will return, especially in *The Post Card*.[28] But now, for the first time following Freud, Derrida notices that Freud suddenly experienced a letdown. The mystic writing-pad has its limits. Freud was disappointed that the mystic writing-pad cannot go on its own. Freud complained that once the writing has been left on the wax slab beneath the surface layer, the mystic writing-pad cannot "reproduce it from within." The mystic writing-pad fails to mimic unconscious memory perfectly. Someone's hands are necessary—writing hands.

Derrida also is disappointed. He is disappointed in Freud. When the limits of the mystic writing-pad became apparent to Freud, he retreated and privileged the "organ," the unconscious that can do what it does on its own or can do it naturally. The toy writing pad that Freud deployed to supplement unconscious memory, making its capacity for limitless receptivity seem a natural matter, was itself devalued for its limitations, for being "unnatural." The technical substrate that supported unconscious memory was cast off and would be forgotten. Freud refused to think that unconscious memory and the machine are inextricably inter-implicated. He refused to think what Derrida dares to: "The machine—and consequently, representation—is death and finitude *within* the psyche."[29]

But Derrida already has gone beyond Freud in proposing that the machine does not "surprise" memory from the outside. The machine is not only metaphor, outside the unconscious. Unconscious memory is inextricable from its technical substrates. The opposition of the unconscious and the machine is to be deconstructed. There is to be no dismissal of nature or biology, no opposition between nature and culture, biology and technology, the unconscious and the machine. Rather, for Derrida the machine is the unconscious deferred, as culture and technology are nature deferred. As he puts it: "[A]ll the others of *physis*—*technē, nomos, thesis,* society, freedom, history, mind, etc. [are to be thought] as *physis* different and deferred, or as *physis* differing and deferring. *Physis* in *différance.*"[30]

Surely the thought of *différance* has ontological implications. The ontology of Being by which nature and culture are opposed is undermined; culture and nature instead are drawn back into the play of the differences of preontological forces—the singular, subindividual, finite, forces of mattering, which subtend and yet are immanent to the *différantial* relation of body and machine, nature and technology, the virtual and the real. But if there are ontological implications arising from the thought of *différance,* there is something that prevents Derrida from fully articulating them here, when first following Freud. Derrida hesitates. He turns back from ontology or turns ontology to the historical production of technology. It is here that Derrida wonders how it is that Freud did not notice that besides the child's toy machine, there already are machines "in the world" that more closely resemble memory—"machines for storing archives."

Derrida goes on to propose that Freud failed to address the question his treatment of unconscious memory raised; he failed to ask about the analogy between the psychic apparatus and the machine in the context of what Derrida describes as the "historico-technical production" of technology. In addressing the question, which Freud did not, Derrida posits an unconscious memory beyond the individual's psychic organization, which therefore calls forth its own method of study—a discipline other than psychoanalysis or the "sociology of literature"—a discipline that can treat the "sociality of writing as drama."[31] This discipline, Derrida proposes, must take up the question of *technē* and technology once again; however, he says that "technology may not be derived from an assumed opposition between the psychical and the nonpsychical, life and death."[32] For Derrida the drama of writing in the unconscious is the scenography of a "cruel theater," to use Antonin Artaud's terminology. It is the timing and spacing of life through the permeation of Being with the deferred trauma of death and finitude.[33]

Derrida thereby comes to the end of his reading of "Freud and the Scene of Writing," having brought ontology as close as possible to the historico-technical development of technology—crossing one through the other. There are no other steps to be taken here by Derrida, beyond registering his suspicion about psychoanalysis and a sociology of literature—the suspicion that psychoanalysis and sociology cancel each other out when it comes to treating unconscious memory in terms of the historico-technical production of technology. Perhaps what stalls Derrida is that he is unable to embrace the technology that has given deconstruction its machine metaphors. If Freud had the disappointing toy writing machine, Derrida has the much-maligned television, the exemplary machine of teletechnology. Derrida cannot go all the way and fully articulate an ontological perspective befitting the technology that has been drawing deconstruction to it all along.

Derrida, of course, does return again and again to treat teletechnology; there are references to it in *Grammatology, Limited Inc,* and *The Post Card.* Teletechnology also makes a star appearance in *Specters of Marx,* where Derrida concludes:

> Techno-science or tele-technology. . . obliges us more than ever to think the virtualization of space and time, the possibility of virtual events whose movement and speed prohibit us more than ever (more and otherwise than ever, for this is not absolutely and thoroughly new) from opposing presence to its representation, "real time" to "deferred time," effectivity to its simulacrum, the living to the non-living, in short, the living to the living-dead of its ghosts. It requires, then, what we call . . . hauntology. We will take this category to be irreducible, and first of all to everything it makes possible: ontology, theology, positive or negative ontotheology.[34]

Teletechnology obliges us more than ever to think what Derrida has been thinking when he has been thinking beyond Freud and raising questions about the technical substrates of unconscious memory. These are questions of preontology or hauntology that put ontology close to what Derrida describes as the shared "history of psyche, text, and technology."[35] What is this shared history that Derrida takes up instead of ontology, a history about which he nonetheless equivocates, suggesting that what the history produces is neither "absolutely" nor "thoroughly new"? What can be made of this pull toward and away from history, toward and away from ontology—this "aporia of time," which is produced when the thought of the historico-technical production of technology crosses through ontology?[36]

What I think can be proposed is that the shared history of text, psyche, and technology historicizes ontology, making an ontology of Being impossible or impossibly so. To put this another way, the historico-technical production of technology gives different technical substrates to unconscious memory and thereby produces different historicities or different relations of temporality and spatiality, the thought of which displaces the Being of an ontology of presence. The historico-technical production of technology pulls "originary Being" down into an "originary technicity" — finitude and its different historicities. Teletechnology, therefore, not only offers a different historicity specific to it, but registers and oversees the drawing of ontology into preontological finite forces of mattering. Teletechnology oversees the becoming dynamic of matter that undoes the opposition of nature and culture, body and machine, the real and the virtual. Derrida proposes that in the age of teletechnology we must think "another historicity — not a new history or still less a 'new historicism,' but another opening of event-ness as historicity . . . as promise and not as onto-theological or teleo-eschatological program or design."[37]

Although Derrida insistently refuses to elaborate an ontology, he does, however, profoundly problematize an ontology of presence. He does so by taking unconscious memory beyond Freud's treatment of it; he offers a generalized unconscious, an unconscious other than that of the individual human subject; he offers a more generalized forgetting, disavowal, and repression. Provoking a move from treating unconscious memory in the metaphor of the mystic writing-pad to treating it in the metaphors of teletechnology, Derridean deconstruction makes it possible to think of the unconscious as a matter of thought's movement through a network of differences, of which subject identity is neither origin nor end. This does not mean that the subject's identity or the subject's unconscious are made irrelevant. Rather, it means that the confinement of thought and unconscious memory to a certain narrative fiction of the subject is resisted, thereby providing a chance for thought to return to its unconscious, to the unthought, in order to escape to the future.

Although the subject's identity or the subject's unconscious memory is not made irrelevant in Derrida's deconstructive problematization of an ontology of presence, nonetheless it is suggested that unconscious memory no longer may need certain machine metaphors with which to supplement its productivity. The unconscious may not need the narrative fiction of the subject as its origin and end. If there is disavowal of *différance*, repression, and forgetting, these may operate through something other than fictions of origins and ends, something other than the narrative fiction of subject identity. Derridean deconstruction proposes that, as a certain technology draws the unconscious to it, the metaphors with which Freud gave the unconscious, and gave it over to the individual subject's identity, are no longer all that is necessary to an understanding of the unconscious.

What, then, of Freud's insistence on oedipus as the narrative logic of unconscious memory? What of the oedipal narrativization of the subject's identity, its sexuality and unconscious fantasy? After all, the oedipal narrative is central not only to the Freudian treatment of unconscious memory; it is also central to Jacques Lacan's rereading of Freud. What of Derrida's deconstruction of Freud's treatment of unconscious memory in relationship to Lacan's rereading of Freud?

Derrida makes no mention of Lacan in "Freud and The Scene of Writing"; but there is little doubt that Lacan's rereading of Freud is already at play in the essay—enabling Derrida's deconstruction while at the same time being its target. Is it not Lacan's rereading of Freud that Derrida wishes to go beyond as he follows after Freud? After all, Lacan proposed that the unconscious is structured like a language and thereby shifted the focus of psychoanalysis to the autoaffecting speech of the subject—the circuit between speaking and hearing oneself speak, which is meant to disavow the writing machine or the technical substrate of unconscious memory. However, in turning psychoanalysis to the analysis of the subject's speech and to the treatment of its disturbances, Lacan was not merely proposing to restore to the speaking subject a unified identity or a self-same presence. For Lacan, autoaffection is possible only as a fantasy disavowing the Other and denying the unconscious altogether. In the unconscious, Lacan proposes, the subject speaks, but with "the voice of no one."[38] Lacanian psychoanalysis, therefore, shows that the unconscious is a resource both for producing and for breaking into the autoaffecting circuit of the subject; through psychoanalysis the unconscious is shown to provide the mechanisms for disavowing the Other, and, as such, psychoanalysis also is a way to recover (from) the unconscious disavowal.

Derrida recognizes a connection between deconstruction and Lacan's rereading of Freud; he finds the connecting point at the repetition compulsion of the death drive, or what Freud also refers to as "the drive for mastery." Commenting on Michel Foucault's *History of Sexuality,* where finally Foucault rejects Freud and psychoanalysis generally, Derrida differs with Foucault, arguing that the "French heritage of Freud would not only not let itself be objectified by the Foucauldian problematization but would actually contribute to it in the most determinate and efficient way . . . beginning with everything in Lacan that takes its point of departure in the repetition compulsion. . . ."[39]

Derrida proposes that Lacan makes clearer that Freud's treatment of the death drive as a repetition of what is painful in order to master it severely problematizes the agency of power and undermines mastery "with the greatest radicality." But for Derrida, when the death drive goes into overdrive the authority of the narrative of mastery, the oedipal narrative, is undermined. Derrida closes his comments on Foucault suggesting, "It is very difficult to know if this drive for power is still dependent upon the pleasure principle, indeed, upon sexuality as such, upon the austere monarchy of sex that Foucault speaks of. . . ."[40] Has not the history of sexuality been opened up to the shared history of text, psyche, and technology, opened to the historico-technical production of technology, so that the death drive breaks its connection to an oedipalized sexuality or an oedipal narrativity? Has not the history of technological development opened Freud's treatment of the repetition compulsion to the thought of "pure repetition" or "originary repetition," thereby taking the unconscious even beyond Lacan's rereading of Freud.

Derrida's difference with Foucault, therefore, also refers to his differences with Lacan. Although he recognizes that Lacanian psychoanalysis problematizes the oedipal narrative for the analysand, Derrida nonetheless complains that Lacan makes the oedipal narrative a transcendental framing for the analyst, a transcendental enframing of psychoanalysis's will to speak the unconscious in the discourse of truth. It is against this transcendental framing of the truth of the unconscious that Derrida aims his criticism of Lacan, most notably in the essay "Le Facteur de la Vérité."[41] But even before this essay, in "Freud and the Scene of Writing," Derrida begins the deconstruction of the transcendental framing of the truth of the unconscious by proposing that the unconscious is inextricably related to historically specific technical substrates or writing machines.

In "Freud and the Scene of Writing," therefore, Derrida gestures to but does not fully elaborate a deconstructive criticism of the oedipal narrative. But around the time of the English publication of "Freud and the Scene of Writing," the detailed work of deconstructing the authority of the oedipal narrative and of rethinking the relationship of oedipal narrativity to the psychic apparatus and its historically specific technical substrates would be initiated elsewhere by feminist theorists who, seemingly indifferent to Derridean deconstruction, nonetheless would be drawn to Lacanian psychoanalysis in order to critically engage the relationship of the cinematic apparatus and unconscious fantasy. Although at first television would be denied the attention of feminist film theorists, the draw of teletechnology on feminist film theory eventually would be recognized.

The Lacanian Unconscious and the Historicities of the Gaze

In the latter half of the 1970s, when feminist theorists first focused on the relationship of unconscious fantasy and the cinematic apparatus, they seemed to pay little attention to the historical specificity of the relationship; they surely were indifferent to television, which by the 1970s already had eclipsed the cinema as the dominant mass medium. It was, however, not until 1990 that Patricia Mellencamp, in her contribution to a collection of essays on television, would propose that television requires a shift in feminist film theory away "from theories of pleasure," away "from desire, lack, castration, Oedipus, the unconscious" — all of which had been central to feminist film theory since the mid-1970s.[42] Yet in 1992 Kaja Silverman would publish a grand theoretical synthesis of two decades of feminist film theory. She not only would ignore television; she also would insist that an understanding of the cinematic apparatus remains central to treating "a society's mode of production and its symbolic order."[43]

Although Louis Althusser's treatment of ideology had by then met with strong criticism, Silverman argued for Althusser's engagement with Lacanian psychoanalysis as the starting point for treating the cinematic apparatus and thereby getting hold of "the ideological belief" through which "a society's reality is constituted" and "a subject lays claim to a normative identity."[44] Offering a systematic review of the treatment of unconscious fantasy that feminist theorists had contributed to Marxist cultural studies, Silverman proposed that unconscious fantasy is "the ultimate sense of reality for the subject" and that it is organized through a "dominant narrative fiction," that is, the oedipal narrative.

Borrowing the notion of dominant narrative fiction from Jacques Rancière and taking up his treatment of it as an "image of social consensus" offered to members of society, Silverman argued that the oedipal logic of narrativity organizes the system of representations through which the subject is subject-ed to the symbolic order and through which the symbolic order is aligned with the capitalist mode of production, along with a kinship structure and family and national ideologies. The oedipal narrative, Silverman argued, not only offers gendered subject identities; it also "forms the stable core around which a nation's and a period's 'reality' coheres."[45]

Not so surprisingly, Silverman's grand theorization of the subject and social reality was met with a lack of enthusiasm by readers who already were uncomfortable with totalizing theories, such as Lacan's and Athusser's proposed to be, especially when taken together. On one hand, there was a growing suspicion that postmodern capitalism could not be reduced to one cultural reality informed by a dominant narrative, and on the other hand, psychoanalysis was suspected of not being capable of treating differences other than sexual difference, such as differences of race, ethnicity, class, nation, or sexual orientation, all of which had become central to questions of subject identity in postmodernity.

Perhaps it was these same discomforts and suspicions that led Silverman to return in 1993 to an earlier essay in which she had treated unconscious fantasy while paying some attention to differences other than sexual difference.[46] In this 1993 essay Silverman raised a question that had not been asked in feminist film theory. She questioned the historicity of the camera, and in doing so she brought feminist film theory up against the historical specificity of the relationship of technology and unconscious fantasy. Although Silverman makes no mention of teletechnology, it seems to me that it is the machine metaphors of teletechnology, television especially, that draw her to rethink the camera's historical specificity and thereby to recognize the possibility of different cameras, screens, subject identities, and social realities.

In what follows I want to offer a reading of feminist film theory and Silverman's grand summation of it. I want to propose that in rethinking the camera, Silverman, like Derrida, has a tele-vision and that it allows her to show, albeit inadvertently, that feminist film theory has all along been drawn to the future by teletechnology. Teletechnology, television especially, has overseen the elaboration of feminist film theory, first drawing it to frame the problematic of unconscious fantasy in terms of a dominant narrative fiction, and then to deconstruct the narrative that no longer seemed essential to television criticism. Feminist film theory, therefore, finally severed the link between unconscious fantasy and the oedipal narrative, which it had first made central to the project of turning film criticism to feminist ends.

In her 1975 publication "Visual Pleasure and Narrative Cinema," Laura Mulvey offered a convincing argument for engaging psychoanalysis in a feminist treatment of the way film, especially classic Hollywood film, elicits the viewer's unconscious identifications.[47] She argued that psychoanalysis could be used to show how film reinforces the "unconscious of patriarchy"; Mulvey suggested that the film identifications offered the viewer are organized in terms of a sexual difference, such that the masculine figure is identified with the gaze of the camera, while the feminine figure is identified with the spectacle or the screened image, for which the female character's "to-be-looked-at-ness" is emblematic. Treating the fetishism and the narcissism deployed in projecting the screened image of the woman, Mulvey turned feminist theory to an analysis of unconscious desire that drew on Lacan's rereading of Freud, especially Lacan's elaboration of Freud's treatment of the "mirror stage."

As Lacan had argued, the mirror stage occurs sometime after an infant-child is six months old and before the infant-child is eighteen months old. During this time the mirror image displaces the look of the mother, which has stood in for the "gaze of the Other," or what might better be thought of as a culturally authorized visual regime. The mother not only assists the infant-child in separating from her look, but encourages the infant-child to join with his own image as part of a series of images in a regime of images. The mirror stage allows the infant-child to find in the mirror image the bodily form of his ego; a "gestalt" is offered the infant-child in contrast to his experience of perceptual incapacity and motor immaturity, part of being a "body-in-bits-and-pieces." For Lacan the mirror image also provides a frame of reality for the ego, a grid of cultural intelligibility. As Lacan puts it, the infant-child "anticipates in a mirage the maturation of his powers."[48] The image also will allow the ego to protect itself in projecting itself, perfected with

"the phantoms that dominate [the infant-child], or with the automaton in which, in an ambiguous relationship, the world of his own making tends to find its completion."[49]

According to Lacan, the infant-child's attachment to the image is
narcissistic and necessarily constitutes a misrecognition or an idealization of the ego, which comes to characterize the infant-child's preoedipal imaginary. It is Lacan's treatment of the notion of misrecognition
that is deployed in Althusser's study of ideology and that becomes central to feminist film theorists' earliest treatments of the viewer's unconscious identification with film images, which are thought to promote
the society's ideology. In what way a preoedipal imaginary functions in
relationship to the oedipal narrative, however, would become a pressing question for feminist film theorists.

In her treatment of the viewer's identification with the film image,
Mulvey had made mention of narrative, even pointing to the importance of it in the title of her essay. But she had not fully elaborated the
function of narrative in relationship to film identification or to the mirror stage. It was in reworking Mulvey's argument that Teresa De Lauretis
moved narrativity to the center of feminist film criticism. She proposed that Mulvey had not fully displaced Christian Metz's argument,
that is, that the viewer "identifies with himself, with himself as an act of
pure perception" and observes "a story from nowhere, that nobody tells,
but which, nevertheless, somebody receives."[50] Mulvey had only begun
to undermine Metz's argument by treating film imaging in terms of a
sexual difference; she pointed to the way films, dominant Hollywood
films, privilege the masculine over the feminine such that it is the masculine subject who is presumed to have the capacity to identify with a
pure act of perception and to take up a view from nowhere. But as De
Lauretis sees it, not only had Metz uncritically presumed that the masculine subject is the subject of perception; his treatment of viewing de-
pended on an analogy with the preoedipal infant's relationship to the
mirror image as Lacan had described it. Questioning the analogy and the
possibility of a film viewer's returning to the condition of the mirror
stage, that is, to preoedipality, De Lauretis revised Mulvey's argument,
proposing that the viewer's identification with the film image is possible
only through a "prior, narrative identification with the figure of narrative movement."[51]

According to De Lauretis, it is an oedipal logic of narrativity that promotes the viewer's unconscious identification with film images. The oedipal logic of narrativity deploys a rhetoric of sexual difference to represent the subject and elaborate his development through suffering various experiences of separation and identification. The oedipal logic is one of development toward a full subject identity that is struggled for throughout the story and realized only in the end. This final realization, however, retroactively makes sense of the subject's experiences and of his struggle to separate from the feminine other, thereby constituting the subject's identity and authorizing the social reality of his experiences.

In De Lauretis's terms, the oedipal logic of narrativity deploys a rhetoric of sexual difference that ideologically constructs a subject position for the viewer by linking the activity of perception to the engendered plot spaces of the narrative:

> Much as social formations and representations appeal to and position the individual as subject in the process to which we give the name of ideology, the movement of narrative discourse shifts and places the reader, viewer, or listener in certain portions of the plot space. Therefore, to say that narrative is the production of oedipus is to say that each reader — male or female — is constrained and defined within the two positions of a sexual difference thus conceived: male hero human, on the side of the subject; and female obstacle-boundary-space, on the other.[52]

De Lauretis's argument not only redirected the focus of feminist film theory toward oedipal narrativity; it also raised a question as to whether the mirror stage might itself be a reconstruction of preoedipality as part of the resolution of the oedipal complex. Sometime later Jane Gallop would in fact make this argument.[53] In her rereading of Lacan, Gallop would propose that the mirror stage be understood as coming into play sometime after the initiation of the oedipal complex, thereby suggesting that every aspect of the subject's identity formation is under the sway of a cultural norm of intelligibility that is given in the dominant oedipal narrative. Feminist film theorists, therefore, were led to rethink the preoedipal imaginary and the mirror stage in the suspicion that it is impossible to return to preoedipality except in fantasy or through a fantasmatic construction of preoedipality. This suspicion would become central to the antiessentialism of a later feminist theory in its refusal of any reference to biology or prediscursive reality in the construction of subject identity, the body or social reality; by then, however, the link of antiessentialism with narrative and film technology would be all but forgotten, whereas the link of antiessentialism to unconscious fantasy would become troubled with the thought of differences other than sexual difference.

But for early feminist film theorists, rethinking preoedipality was not only a resource for their film criticism. It also allowed them to propose feminist film theory as a model for a general treatment of the socially or culturally informed fantasmatic construction of the subject and social reality. For example, De Lauretis would follow up her treatment of the cinematic apparatus, the oedipal logic of narrativity and unconscious desire, with an elaboration of what she described as "gender technologies." Borrowing from Michel Foucault's treatment of sexuality, she would suggest that gender "is not a property of bodies or something originally existent in human beings," but rather an effect of "a complex political technology," including "social technologies such as cinema," along with "institutionalized discourses, epistemologies, critical practices as well as practices of daily life."[54]

When, however, feminist film theorists first began to rethink preoedipality as a fantasmatic reconstruction shaped by the resolution of the oedipal complex, it was not only to think about subject identity as a fantasmatic construction. It was also to take up the matter of unconscious desire or unconscious fantasy without submitting uncritically to the privilege Lacanian psychoanalysis seemed to afford masculinity in linking the masculine subject to phallicity. Feminist film theorists would instead emphasize the fantasmatic construction of the masculine subject's phallicity and its required narration of the devaluation and negation of the feminine Other. Feminist theorists would emphasize that the loss that Lacan described the infant-child as experiencing in preoedipal separation and individuation is given its meaning only in the resolution of the oedipal complex, when "the law of the father" commands the infant-child to accept his identity as subject in terms of an opposition of masculine and feminine, phallic or castrated. It is then that the loss experienced in preoedipality is given the name "mother," as the "feminine" is made to figure castration, or what Lacan refers to as "lack." It is then that the oedipalizing law of the father symbolically figures the subject as masculine, as "having" the phallus, whereas the feminine is made the figure of the Other, who seemingly *is* the phallus, at least for the subject who is yet able to be threatened with castration.

In arguing that the symbolic castration of the oedipal complex retroactively forces the assignment of a name for preoedipal loss, feminist film theorists intended to make clear that there is no essential link between lack and femininity or between phallicity and masculinity; there surely is no essential biological link. Rather, these links are fantasmatic constructions, and therefore require the ideological support of technologies, such as cinema, that reproduce this fantasmatic structure in organizing cultural imagery in terms of the dominant narrative fiction — the oedipal narrative. But as feminist film theorists suggested, cinema not only reproduces the dominant narrative fiction; it also depends on it. That is to say, the loss provoked by film viewing for the viewer in experiencing both the absence of the actual object and the erasure of the actual production of the image can be displaced onto the feminine figure of the film due to the ongoing function of the dominant narrative fiction to project lack onto the feminine figure.

But for all this, feminist film theorists still were faced with the conflation of the feminine figure with lack and castration, if only at the level of narrativity; they were challenged by the question of how to change a dominant narrative fiction. In facing this challenge they were forced to rethink the link of unconscious fantasy and the oedipal narrative. The efforts of feminist film theorists to rethink unconscious fantasy and the oedipal narrative were initiated with a close rereading of Lacan's rereading of Freud.

Feminist film theorists noticed that in treating the oedipal complex Lacan had insisted that the oedipal narrative always fails in its imposition of the "law of the father"; it fails, that is, to fix subject identity in terms of an opposition of masculine and feminine, phallic and castrated. As Jacqueline Rose put it:

> The unconscious constantly reveals the failure of identity. Because there is no continuity of psychic life, so there is no stability of sexual identity, no position for women (or for men) which is ever simply achieved. Nor does psychoanalysis see such "failure" as a special case inability or an individual deviancy from the norm. Failure is not a moment to be regretted in a process of adaptation, or development into normality. . . . Failure is something endlessly repeated and relived moment by moment throughout our individual histories. It appears not only in the symptom,

but also in dreams, in slips of tongue and in forms of sexual pleasure
which are pushed to the sidelines of the norms. . . . There is a resistance
to identity at the very heart of psychic life.[55]

It was this thought of the failure of identity that in fact enabled feminist film theorists to grasp how the subject or the viewer can return in
fantasy to preoedipality. What feminist film theorists proposed is that
in the resolution of the oedipal complex, subject identity is never fixed,
and its instability is played out in the structure of a set of unconscious
fantasies. There is, they argued, a certain mobility of subject positions
in unconscious fantasy that is born of resistance to the law of the father
in the resolution of the oedipal complex. Or, to put it another way, in
the resolution of the oedipal complex unconscious fantasy both informs
the ego's subject identity and gives it over to various subject positions
in scenarios of unconscious desire. In these fantasies symbolic castration is both accepted and refused, that is, internalized, disavowed, and
displaced. It is in this sense that these fantasies can be said to "return"
to or to persist in preoedipality.

According to Silverman, all of this means that the subject repeats or
replays both the negative and the positive resolutions of the oedipal
complex in fantasies such as those Freud catalogued: the fantasy of primal scene, the fantasy of parental seduction, the fantasy of castration,
and the fantasy of the child's being beaten. Drawing on Jean Laplanche's
and J.-B. Pontalis's treatment of the "fantasmatic," Silverman proposed:

> The fantasmatic generates erotic tableaux or *combinatoires* in which the
> subject is arrestingly positioned — whose function is, in fact, precisely to
> display the subject in a given place. Its original cast of characters would
> seem to be drawn from the familial reserve, but in the endless secondary
> productions to which the fantasmatic gives rise, all actors but one are
> frequently recast. And even that one constant player may assume
> different roles on different occasions.[56]

Although the fantasmatic allows the ego more or less to consolidate
a subject identity in the repetition of certain subject positionalities, the
fantasmatic nonetheless holds the ego open to unconscious identifications with various other subject positions, including those that figure
resistance to the law of the father. This is important, as Silverman would
come to see it, because it suggests that unconscious fantasy may inform
political resistance, beginning in the resistance to a positive resolution
of the oedipal complex. Since the positive resolution of the oedipal com-

plex aligns the subject not only with a privileged masculinity, but with the family and national ideologies of the symbolic order and its mode of production, the negative resolution of the oedipal complex, Silverman would argue, carries potential for political resistance. As she puts it:

> There are subjectivities which have established a different relationship to the family — and, in some cases, even to the laws of language and kinship structure — than those valorized by the dominant fiction. For these subjectivities . . . psychic reality has a different consistency than that dictated by the dominant fiction. The desire and identifications through which they are constituted may even sustain a disjunctive or oppositional relation to the *vraisemblance*.[57]

According to Silverman, it is the subject's fantasmatic that not only makes film identifications possible, but also gives film viewing its political implications. The viewer's fantasmatic, Silverman proposes, is projected onto the film, allowing the viewer to unconsciously identify with the figures offered in the film narrative; it allows the viewer to grasp those figures in terms of its unconscious scenarios of desire. Therefore, although the viewer's fantasmatic identification makes adherence to the dominant oedipal narrative possible, it also makes it possible to resist the dominant narrative, thereby permitting ideological change.

But in arguing that the unconscious exceeds the dominant fiction and that all individuals probably escape the positive oedipal narrative to some degree, Silverman begins to think about the possibility of generalizing the disjuncture between the dominant narrative, the incest taboo, the psyche, and the symbolic order. She proposes that each of these must not be reduced to the other, at least not at a theoretical level. Silverman does not yet imagine taking account of the actual existence of different situations that do not fit or refuse to fit the configuration of social spaces that she assumes to be organized through an oedipal narrative — that is, the certain configuration of family and national ideologies, the state and civil society, the public and private spheres presumed in subject-centered, nation-centric, modern western discourse. As a nationalist discourse cinema does not seem to urge rethinking the configuration of social spaces or the environment of the viewing subject. But television, in its global reach, surely does.

Still presuming the cinema as her starting point, Silverman points her disagreement with Freud and Lacan at their conflation of the symbolic order, the incest taboo, the law of the father, and the psyche. She refuses Freud's and Lacan's equation of an oedipalized castration with the loss suffered by any ego that must take its subject identity from an exterior image and its frame of reality from the grid of cultural intelligibility given with a symbolic order. Silverman argues instead that although every ego must suffer a lack of being in taking its meaning — that is, its subject identity and its social reality — from a symbolic order, it is not the case that this occurs in every place and at every time through the imposition of the oedipal narrative. Silverman's difference with Lacan and Freud turns out to be a large one. Although she does not question the centrality of narrative to subject and national identities, Silverman does undermine the universality of the oedipal narrative; she suggests that the dominant narrative by which an ego becomes and lives its subject identity and its social reality is culturally and historically specific.

It would seem that Silverman's rethinking of the oedipal narrative as part of her grand summation of two decades of feminist film theory might well have had profound implications for the future of film criticism, as well as for an Althusserian treatment of ideology. But these implications were never fully elaborated among feminist film theorists, and although Althusser's treatment of ideology would be rethought, it would be in relationship to the cultural studies of television when feminist film theory already had lost its central place in cultural studies.

Except for those feminist theorists who contributed to what would be referred to as queer theory, many cultural critics had already by the late 1980s shifted their theoretical focus from the psychoanalytic treatment of the subject to the analysis of identity in terms of the intersection of race, class, gender, sexuality, and ethnicity. Although not all of these cultural critics refused psychoanalysis, many criticized it[58] for being a white, middle-class practice and a nineteenth-century Eurocentric theory. The refusal of psychoanalysis was at least implicit in the demand that attention be paid to the intersection of race, class, gender, sexuality, and ethnicity, especially when it was made through an autobiographic writing in which the writer would claim authority for speaking in his or her own voice attuned to localized conditions. The refusal of psychoanalysis on historical, geopolitical grounds, however, left off any systematic treatment of the unconscious in historical, geopolitical specificity. The question of the relationship of the unconscious to historically specific technologies also was put aside. Such questions were not asked, even though the globalization of teletechnology surely was implicated even in the earliest shift of theoretical attention to the intersection of race, class, gender, sexuality, and ethnicity in the construction of identity in cultures around the world.[59]

Yet Silverman persisted in thinking that the fantasmatic, as feminist film theorists had revised it, allows for a treatment of the intersection of race, class, gender, sexuality, and ethnicity in the construction of the subject and filmic identifications. It was in the early 1990s that, when elaborating this possibility, Silverman was led to rethink the camera and the gaze as well as their usual apprehension only in the terms of the oedipal narrativization of sexual difference. In two remarkable essays Silverman took up Lacan's diagrammatic treatment of the gaze from which feminist film theory had first drawn its founding presumption, that is, that the gaze is masculine whereas the feminine figures the spectacle. Just as the interest in early feminist film theory was waning, Silverman's attempt to rethink it produced an argument about the historical specificity of the camera and the gaze in which, it seems to me, the machine metaphors of teletechnology are drawing feminist film theory to the future.

In her first essay,[60] which treats a film by Rainer Werner Fassbinder in which a black male body appears as spectacle, Silverman sets out to articulate the operation of the cinematic apparatus in relation to the differences of race, class, gender, sexuality, and ethnicity. In doing so Silverman corrects the presumption of early feminist film theory that the gaze is a transcendental position of mastery that is identified with the look of the camera and projected through the eye of a male character; she rethinks this presumption in terms of feminist film theorists' own revisions of fantasy and preoedipality. Returning to the three diagrams with which Lacan treats the gaze,[61] Silverman suggests that what they propose is that the gaze, like the phallus, cannot be appropriated. No subject can possess the gaze and thereby make it one with the look of the eye. The gaze always exceeds the look of the eye of both male and female characters; the gaze rather produces the field of vision in which the subject is screened. Silverman now reads Lacan's three diagrams to underscore his treatment of the gaze as "that 'unapprehensible' agency through which we are ratified or negated as spectacle."[62]

Lacan's three diagrams map the relays among the screen, the image, the camera, the point of light, and the subject of representation, such that the gaze is nowhere identified with the subject nor with objects;

nor is it even a characteristic of light. Nonetheless, the diagrams were first read by feminist film theorists to propose that a point of light is offered by the camera that seems a representative of the gaze. The gaze and the eye are seemingly conflated when the eye takes the point of view of the camera projected from behind the viewer. When, however, the diagrams are reconsidered by Silverman, she is able to show what is much more interesting to her, that is, that Lacan does not locate the gaze in a transcendental position, behind the viewer, nearer the camera. Instead he proposes that the gaze is opposite the viewer. The gaze is located by implication at the site of what was thought to be the feminized spectacle, at the "lit-up" screen. That is to say, the lit-up screen shows what Lacan describes as the "pulsatile, dazzling and spread out function of the gaze."[63] The ungraspable gaze is nearer the lit-up screen where the subject is pictured or "photo-graphed," as Lacan puts it.

Silverman emphasizes that although the gaze puts the subject in the field of vision, it does not give the subject its form or its various subject positions. It is the screen that does; indeed, Silverman argues that the subject must take the form given by the screen. She argues this on her way to insisting on the "ideological status" of the screen, which provides a "repertoire of images through which subjects are not only constituted, but differentiated in relation to class, race, sexuality, age and nationality."[64] All this, of course, resonates with the thought of the fantasmatic, its displaying of the subject in various positions elaborated in scenarios of unconscious desire. But it adds other differences than sexual difference, such as differences of race, ethnicity, age, and sexuality.

In shifting the location of the gaze nearer to the screen, Silverman also has made it possible to think about the way the screen looks back at the viewer not only with ideologically specific images, but with its own eye or its own definition of visuality. It is as if the machined screen has its own way of seeing, upon which the human eye is dependent and inseparably connected. At least this is the point of Silverman's second essay. In this essay Silverman argues that "the human subject's experience of the gaze may vary markedly from one period to another, and that different optical apparatuses may play a key role in determining this variation."[65]

In this essay, "What Is a Camera?, or: History in the Field of Vision," Silverman repeats her earlier rereading of Lacan's diagrams, but she confesses that in the earlier treatment she was concerned to show only how the subject is seen in the screened gaze and that the screen, "through which a given society articulates authoritative vision," is historically and culturally specific. In the earlier treatment, she admits, she did not see that the diagrams raised a question about the relationship of the camera and the gaze. Now Silverman proposes that the gaze is differently related to different screens that are given through historically specific cameras or technologies.

No doubt Silverman is attempting to adjust feminist film theory, however belatedly, to the work of Foucault and others who have proposed that perceptual apparatuses are historically specific. Especially important for Silverman is Jonathan Crary's *Techniques of the Observer*,[66] where the camera obscura and then the stereoscope are treated in terms of their historical specificities. Silverman especially notices Crary's suggestion that the stereoscope and, by extension, the moving camera are not tools; they are rather machines. The moving camera is a machine because the moving camera compensates for the insufficiency of the human eye and makes the human being dependent on or part of the apparatus. As such, the moving camera and the human eye are like "contiguous instruments on the same plane of operation,"[67] as Crary puts it. They are linked and inseparable. Silverman concludes that if the gaze is differently related to historically specific cameras, then in each case, the eye also is "visualized" differently with each camera, mediated differently by what she now refers to as the "image/screen."[68]

In all this it might be argued that Silverman moves the unconscious closer to Derrida's treatment of unconscious memory in terms of historically specific technical substrates. At least she suggests that there is a history of technology that gives historically specific optical machine metaphors to the gaze. Silverman tries to get ahold of the nature of such a history:

> Those optical metaphors through which the gaze manifests itself most emphatically at a given moment of time will always be those which are most technologically, psychically, discursively, economically, politically and culturally overdetermined and specified. However, as should be apparent by now, each of those metaphors will also articulate the field of visual relations according to the representational logic of a specific apparatus. The meaning of a device like the camera is consequently both extrinsic and intrinsic—a consequence both of its placement within a larger social and historical field and of a particular representational logic.[69]

Silverman's remarks suggest that there is an aporia of time that informs the historical specificity of the machine metaphors informing the gaze. Technologies that give these machine metaphors are constituted within a larger social and historical field, but the particular representational logics of these specific technologies offer particular historicities. In the case of cinema, for example, that representational logic is the oedipal logic of narrativity, which informs a certain historicity. The machine metaphors given with specific technologies, therefore, problematize the possibility of a linear, homogeneous history, including a history of technological development. History is put out of joint, a point that might have been further elaborated by Silverman.

Although Silverman historicizes and therefore problematizes the relationship of the camera and the gaze, she never fully questions what is to be made of the unconscious if the cinema is no longer taken as its machine metaphor or as its only machine metaphor. She does not notice that in historicizing the relationship of the camera and the gaze, she may have left behind thinking of the unconscious in terms of the cinematic apparatus and its oedipal narrative. She does not imagine that it may be television that gives the metaphors for her own understanding of the differing of the psyche from the symbolic order and the symbolic order from the oedipal narrative and the law of the father. Silverman does not imagine that it might be television that has drawn her to rethink the gaze and move it from a transcendental position of mastery somewhere near the camera behind the viewer's back and locate it instead across from the viewer at the lit-up image/screen. She does not imagine that the emergence of television metaphors not only gives a new sense of the gaze, but gives the very idea that visuality can be changed with different technologies, and that therefore the unconscious is historically specific.

Yet if Silverman's rereading of Lacan's diagrams were to be superimposed on Derrida's treatment of Freud's mystic writing-pad, it might be possible to see a television. The superimposition gives the picture back to Derrida's transmitting writing machine. After all, Derrida lost sight of the picture when he grew confident that Freud was no longer interested in optical machines to metaphorize unconscious memory after 1925, when he turned to the mystical writing-pad.[70] But the sophisticated archiving machines, which Derrida proposes that Freud ignored when he turned to the mystic writing-pad, surely included the motion picture camera. It is Lacan's rereading of Freud that brings back the picture and lights up the screen of Derrida's transmitting writing machine. As a result, the screened gaze can look back at the viewer through a pulsating light in which a subject and his or her reality cannot but be deconstructed again and again.

But how can it be that television informs Derrida's, Lacan's, and even Freud's treatments of the unconscious, as if television had come before cinema? Silverman notices only that rereading Lacan's diagrams requires a rearticulation of the relationship of the gaze, the camera, and the look of the eye. Given that the gaze, which the camera makes visible, nonetheless cannot be appropriated by the human eye, the camera may be said to be a "compensation" for the human eye. Yet Silverman does not imagine that the relationship between different technologies might also be one of compensation. For instance, it might be thought that cinema is a compensation for the not yet fully developed television — that cinema came before television only as its compensation. Richard Dienst has suggested something like this in posing a question as to "whether cinema has not always been compensating for its incapacity to transmit images — if, in other words, the dream of television as simultaneous inscription and diffusion has not haunted all cinematic forms from the beginning."[71]

But Dienst suggests that television not only came before cinema. Television also came after cinema because television leaves behind what has been crucial to the cinematic apparatus: that is, the linking of images by means of relays through subject positions. The cinematic apparatus, after all, works by machining the cut-up components of subjectivity into the narrated movement of the image. Indeed, as feminist film theorists have made clear, the important function of the oedipal narrative in film is to suture what has been cut up in order to give the machine vision back to the subject, as if it were a human vision, as if it were the viewer's vision.

In giving up this concern for relaying images through the subject, perhaps suspending any linkage between images altogether, television presents itself as a machine apparatus for the nonsubjective movement of images; it presents itself as a machine apparatus for an unconscious without the oedipal narrative. In this sense it might be better to argue that television, rather than coming before or after cinema, supercedes it. It does so by evoking the unconscious of nonsubjective images. In this sense, television in general evokes the thought of the interimplication of nature and culture and makes it possible to think of the relationship of the unconscious and its technical substrates as a nonmetaphorical one. It might be possible then to think of the unconscious and its technical substrates on the same plane. In all of this, television evokes the conceptualization that Gilles Deleuze and Félix Guattari refer to as machinic assemblage.

As they describe it, the concept of machinic assemblage is for thought when it is a movement that aligns human and machine, nature and technology, the virtual and the real, even the living and the inert — all on the same plane of consistency. The machinic assemblage is neither organic nor mechanical; rather, it refuses to presume an opposition between organic and nonorganic life, given its allowance of the dynamism of matter or the self-organization of subindividual, singular, finite forces of matter. Assemblages refer to pure or originary repetition or, as Deleuze and Guattari describe it, "a movement capable of affecting the mind outside of all representation . . . of inventing vibrations, rotations, whirlings, gravitations, dances or leaps which directly touch the mind."[72] It is this movement, which evokes the movement of television's electronic image on one hand and the movement of the unconscious resisting an oedipal narrativization on the other, that gives the possibility for the revelation of the different temporal and spatial relationships of teletechnology, that is, its historicity, its unconscious.

If Deleuze and Guattari have identified this unconscious as antioedipal, it is to insist that the unconscious need not be characterized by an originary lack, as they read Lacan to have proposed. But feminist film theorists make clear that the Lacanian unconscious is never successfully oedipalized, and that therefore the unconscious is always without recognition of lack. It is the failure of the oedipal narrative and the disavowal of lack that, however, psychoanalysis means to bring to truth. And in its deconstruction of the oedipal narrative, feminist film theory is implicated in the discourse of truth as well. No matter if its aim to deconstruct the rhetoric of sexual difference as it is deployed in the oedipal narrative; feminist film theory returns the unconscious again and again to the oedipal narrative in order to reveal the truth of the family and national drama.

But Deleuze and Guattari argue that unconscious desire has no need of a narrative of truth. The unconscious should not be thought of in the terms of narrative at all. Instead it should be thought of as desiring production, an assembling that is grasped in its effects. In this sense, the oedipal narrative might be reconceived as a line of flight, a line of escape. Rather than an order or command to overcome desire, the oedipal narrative might be conceived of as a line to set off from or as an abstract program by which to start up desiring production. It is this reconception of the oedipal narrative, begun with feminist film theory and further elaborated in Deleuze and Guattari's treatment of the unconscious, that is connected to teletechnology and to the teletechnological smoothing out and reconfiguration of the arrangement of family and national ideologies, the state and civil society, the private and public spheres presumed in subject-centered, nation-centric modern western discourse.

Certain questions are raised. One that l want to take up in the next chapter is the question of political economic analysis. What analysis of political economy is possible when the oedipal narrative no longer is the dominant narrative, when economy may no longer be contained within a subject-centered and nation-centric narrative discourse? Surely these questions refer back to the problem of history and the history of technology, which has been closely connected to the history of capitalist production, especially in Marxist analysis of political economy. What becomes of a history of capitalist production if history has been put out of joint by the historicities given with the specific technical substrates of unconscious memory?

It is to the Marxist cultural studies of television that I want next to turn, because it is these studies in which Marxist analysis of political economy first showed itself to be drawn to the future by teletechnology. It is these studies that first raised questions such as: What is television's place in political economy? If television has no interest in relaying images through subject positions, does television fail to do what cinema promised to do? Does television ruin the promise of a mass mediation of the individual and collective identities through ideological interpellation to subject-centered and nation-centric narrative representations, and therefore profoundly trouble the analysis of ideology through cultural criticism? What is the nature of the cultural in the age of teletechnology?

There Is a Story; or, A Home Movie Rerun

There is a story.
It is the story I am going to tell you.
I don't know if it is true.
I don't know if anyone does.
But it was the story told me
about my mother's father,
whom she adored
and whom she often said I was so much like.

Nothing could keep the delight from her eyes
when she talked about him —
a child's delight still caught in her eyes
when she described him.
Handsome, charming, and smart,
always the center of attention
with family and friends,
at the beach,
at the racetrack,
at parties,
playing cards.
Handsome, charming, and smart.
A man about town
dragging her home from the fruit store
when she had stolen an apple.
The storekeeper had called him.
So he came and took her home
and put straight pins in each of her ten fingers.

He drank; I knew that. But when he drank,
she told me once, he was brutal,
brutal to her mother.
To your mother?
To my mother and my brother.
To your mother and your brother? But not you?
But not me.

This is not the story my mother would write.
But I have never seen my mother write
anything but notes on the bottom of Hallmark cards.

When she was nine, her father took her out of school.
She was the oldest of five.
Her mother was then twenty-five, and he said
that my mother needed to help her mother
take care of the other kids.
She resented it, but she did it,
and she never undid it.
She never went back to school.

Still, nothing could keep the delight from her eyes
when she talked about him.
Her eyes a mirror image of his beautiful, dancing, Italian eyes
that I had seen in pictures
and once when he held me,
just after I was born,
just weeks before he died,
at an age close to the age I am now.

"His liver exploded," my mother said.

Everyone knew; he drank.
My father always said that it wasn't so odd, not in those days.
My mother's father owned the barbershop,
and the customers did not want to tip him.
So after a haircut, a customer brought him across the street
to the bar for a shot. For a shot, for a shot and a beer.
My father always said that my mother's father was
handsome, charming, and smart.
A man about town.
My father adored him, as he adored my mother.
Nothing could keep the delight from his eyes
when he talked about her —
a young man's delight still caught in his eyes

when he described her.
Glamorous, charming, and smart
always the center of attention
with family and friends,
at the beach,
at the racetrack,
at parties,
playing cards.
Glamorous, charming, and smart.
The girl of every guy's dream,
refusing his kisses,
screaming cruel words and hateful reproaches,
slammed doors and bitter silences,
living unhappily everafter.

Still, nothing could keep the delight from his eyes
when he talked about her —
a young man's delight still caught in his eyes.
His beautiful, dreamy blue eyes,
gone vacant and cold, just once when he told me,
I have never been faithful.
There was always some other.
Than my mother?
Than your mother.
But you always said that you loved and adored her?
I loved and adored her.
I did.

She walks toward the movie camera as her raven-black hair dances
around her beautiful heart-shaped face. Her ruby-red lips slowly part
and burst into laughter, flashing her milky-white teeth. She reaches
down out of sight and back up with a bundle of baby caught in her
arms. She kisses me on the tip of my nose, and then she points her
finger, out toward the camera. But my eyes do not follow. They stay fixed
on her eyes, deeply carved in her face, just below perfectly arched brows.
The thick, long lashes flutter like butterfly wings over the warm velvet-
brown orbs that rest on the rising curve of movie-star cheekbones.
Glamorous, charming, and smart. She looks back at me with a gorgeous
smile and then again out, beyond her finger, toward the camera, which
suddenly jerks and twitches out of control, the image coming undone,
and caught in that moment, a face gone hard, quickly swollen with
annoyance and anger and then just as quickly drained white, leaving
only a fear before the impending disaster,
the unbearable ruined moment.

I Grind

After his mother died, my father and I stood together
before her laid-out corpse.
She was still beautiful
in a gown of velvet and lace
that was the violet color of her eyes,
now covered over with dried hard lids.
A sweet, sickening smell of roses was everywhere around her.
At work, they had called her Rose,
although that was not her name.
She had been a dress designer and talented—
   caught up more in her work,
   caught up more in herself
   than she was in him.
"I still hate her," I thought I heard my father whisper.
But his lips were shut closed tight, when I looked up,
and saw him,
just as he turned,
and walked past the coffin
to far, far away.

When he was a boy, my father's mother sent him
to visit with his grandparents.
She sent him to Italy when he was five,
and left him there until he was ten.
It must have been then that the little boy's rage
was smothered under a silencing shame.
His anger never translated
from Italian to English.

He never again spoke without cutting off the words,
drawing back the sentences, keeping the meaning
yet to come, never ever to arrive.

I watched his mouth become the vanishing point of
dream after dream never ever to be lived,
sucked back in through a roundness
that became teeth clenched,
lips tightened and thinned.
I watched his mouth up close
until it became mine.

Alone, I stand perfectly still.
Only my mouth moves.
My tongue swells and pushes hard against my teeth,
trying to save itself from swallowing itself.
My teeth clamp and grind,
a moving fence against every line of flight.
I grind and choke on tongue swollen and imprisoned.

When he was a student, my father fancied himself a poet,
even though teachers told him
that he was really good at math.
He wrote love poems for my mother,
pages of words, made into rhymes, just for her.
I kept and cherished each page
that she tossed away,
not believing much in his love.
So his efforts soon let up,
and he turned his pen back to numbers. Hundreds of them
meticulously placed in the tiny boxes
of an accountant's balance sheet. I tried hard to go as fast
as his adding machine, when he challenged me
with multiplication and division, addition and subtraction.
But mostly, he was alone with his numbers.
He wanted to be alone with their secrets wrapping around him.
I thought that they must keep him safe —
lines in order one after another,
always adding up.

My mouth became a shallow grave for his buried dream of poetry.
I grind away, damning up the flow of words,
decomposing the music of every love song,
in his name.
I grind away, making it impossible for my lips to rest,

to be at rest,
mauved, full of kisses
and that light touch of a lover's finger
tracing the curve of a peaceful landscape.

My father worked hours without end bent over his desk,
like his mother,
bent over her workbench
with so many pins, held tight between her lips.
She fixed one at a time along the lines thrown up on silk
with a designer's chalk.
Her work stopped only on Sundays, when she sat
at the head of the dining-room table, fifteen of us all around,
drinking sparkling wine and eating
grand meals that my grandfather prepared,
fit for kings and queens
on lazy afternoons
of laughter and seamless conversation.
And my father ate too.
With little laughter and hardly a word spoken,
he ate much more than the others.
I watched him and tasted a hunger that outlasted every bite.
I felt a thirst that could not be quenched.

My mouth became an empty well,
a lenten fast with no hope of Easter.
I grind away,
chewing nothing but air.
Even the memory of hunger and thirst is all but erased
with the pleasures of crusty bread soaked in olive oil,
sweet lamb, and polenta,
plum flesh and purple-stained pits left behind,
all gone, one by one, into thin air.

When my father was a young man,
something caused him to lose all his teeth.
I tried not to stare at his mouth,
collapsed in pain and embarrassment,
while my mother remembered
that his mother had lost her teeth, too, when she was young.
My mother seemed cruelly indifferent to my father's fury,
which I watched creep up from the broken chin line
over the caved-in mouth,
up into the sunken cheeks,
flashing out from his narrowed eyes and then as quickly gone,

leaving nothing but resentment in the embers of a burnt-out rage.
My mouth became a store of lost affections,
mistrustful of the simple expression of emotion,
guarded against the return of a withdrawn anger.
I grind away, drawing back every utterance
into the contradictory meanings of forgotten memories,
retracing each story line,
reaching for the one word that releases understanding,
the lost kiss that frees poetry,
the loving touch that eases complexity,
the dream that offers another way
I await.
Alone, I stand perfectly still.
Only my mouth moves.

CHAPTER TWO

The Generalized Unconscious of Desiring Production

Marxist Cultural Studies: Televising the Political Unconscious

Jacques Lacan's book *Television*[1] is not about television. It owes its title rather to the fact that the book contains an interview with Lacan that was aired on French television in 1973. Taking note of this fact, Richard Dienst begins his book about television suggesting that, although not intended to be, the opening remarks of Lacan's interview turn out to be a good description of television, demonstrating television's capacity to turn everything into the televisual.[2] This is what Lacan said in his opening remarks to his television audience: "I always speak the truth; not all of it, because there's no way to say it all. Saying it all is materially impossible: the words are lacking. It's even through this impossibility that the truth holds on to the real."[3] Dienst goes on to treat Lacan's remarks in terms of television's "drive to transmission" — its dream of receiving and sending every message from and to everyone, everywhere, all of the time.

It may very well be its drive to transmission that has made television both the cultural apparatus through which Marxist cultural critics have focused their criticism of postmodern culture and the machine that has made it so difficult for them to reduce teletechnology to a capitalist organization of production, as they are wont to do. It turns out that Marxist cultural critics have had as much trouble just watching television as they have had seeing it as a machine. What makes Dienst's book about television remarkable is his effort to figure out what kind of machine television is. He seems able to keep his eyes steady and just look back at the machine without being distracted by the culture it transmits. Well, almost. Dienst is at times distracted by the debt he owes Fredric Jameson, from whom he has inherited the idea of television as the icon of late capitalist, postmodern culture.

Not that I do not owe a debt to Marxist cultural criticism; I do. Not only is there no way to get to unconscious thought in the age of teletechnology without raising the question of the relationship of television and capitalism. But there is no easy way to answer the question without taking up Jameson's essay on video art, "Surrealism without the Unconscious."[4] If nothing else, I am challenged by what seems to be Jameson's proposal that video art, and television by implication, can go on without the unconscious. But perhaps it is only the political unconscious of a dominant hegemonic cultural narrative which Jameson cannot find operating in television.

It would seem that although television first draws Marxist cultural criticism, most notably Birmingham cultural studies, to an engagement with Louis Althusser's treatment of the unconscious interpellation of the subject by the ideological state apparatuses, television finally leads Marxist cultural criticism beyond an Althusserean treatment of ideology to the deconstruction of both the subject and narrativity; television finally draws Marxist cultural criticism to face the challenge of postmodernity. In pointing to the transnationalization of capital and the globalization of teletechnology in the late twentieth century, television will trouble Marxist cultural criticism and its treatment of the economic in terms of the subject-centered, nation-centric discourse of the ideological apparatuses. That is to say, television undermines the presumed relationship of narrative, national and family ideologies and the unconscious construction of subject identity presumed in an Althusserian treatment of ideology.

Surely television's unfailing effort to appear as if trying to tell all and show all, no matter that it fails, suggests that television not only wants to be on everywhere, but also wants to be on all the time. So rather than calling forth the subject's unconscious identification through a narrative re-presentation, television hopes for a continuous body-machine attachment. Television is mechanizing the autoaffective circuit; it is displacing the sound of hearing oneself speak with the sound of television going on and on, cutting the pleasure of hearing one's own voice with the pleasure of television's just being on. Television operates on the unconscious of the circuit, befitting the notion of machinic assemblage. Television gives the thought of an unconscious that is irreducible to human subjectivity.

As such, television makes it difficult to remain indebted not only to Marxism, but to psychoanalysis as well, and even more difficult to remain indebted to both at once. But surely many Marxist cultural critics have tried. If I am still trying to do so, it is not without Jacques Derrida's remarks in mind. Commenting on his own debt to Marx and Freud, Derrida proposed, "Inheritance is a task, not a given."[5] It involves trying to get to "the most thinking thought" of the inheritance, but it also involves mourning no longer being able simply to be either a Marxist or a Freudian. Along with mourning, then, there is the task of finding the most thinking thought about television in Marxist cultural criticism. And surely Marxist cultural critics have had thoughts about television.

To get at the thought of television in Marxist cultural criticism, in the second part of this chapter I turn to Jameson's essay on video art and Dienst's elaboration of it in an analysis of television that takes the field of Marxist cultural studies beyond its fixation on the textual analysis of a dominant hegemonic cultural narrative to the globalization of teletechnology and the transnationalization of capital in the late twentieth century. But first I want to treat the earlier engagement of Marxist cultural criticism with Althusser's rereading of Marx; I want to turn to Birmingham cultural studies of television and then to Jameson's literary criticism of the novel's narrative logic in order to show how both made the text into literary evidence of a political unconscious such that criticism became a matter of "reading," whereas the notion of text was reduced to its narrower definition. Meaning was thereby located beyond or outside the text in the social context, saving Marxist cultural criticism from the effects of the ontological implications of poststructuralism. It was in these terms that seeing television as a machine became a difficulty for Marxist cultural critics.

Cultural Studies/Television Studies

In 1973, the same year Lacan appeared on television, Stuart Hall, the director of the Birmingham Center for Cultural Studies, published his essay "Encoding/decoding."[6] Destined to become a canonical text, "Encoding/decoding" turned Marxist cultural studies to textual criticism that would depend more on Louis Althusser's treatment of ideology and Antonio Gramsci's treatment of hegemony than on the cultural approach that E. P. Thompson and Raymond Williams already had established at the Center. But Hall's aim in "Encoding/decoding" was to complicate the established models of communication employed in the social sciences generally. To do so, he took television as his example. Yet Hall did not ask, Why television? He did not ask why television, rather than cinema or the novel, better demonstrated the argument he offered — that is, that in communication there is no transparent medium carrying the message from sender to receiver, and therefore there will be more than one reading of a text. The message encoded will not necessarily be the message decoded. Although Hall referred communication to television's "passage of forms" through "a continuous circuit," it is not the idea of the circuit that was the focus of Hall's treatment of television; rather it was the meaning of the text.

Without addressing the specificity of television as a medium, Hall turned television criticism into an analysis of texts that all but loses sight of what Raymond Williams would seem to have been after when he argued that television refers communication to "a planned flow."[7] Instead, Hall treated the television text as a discrete cultural production that, for him, raised the question of how the circulated text is consumed, that is, how it is read. For Hall, however, the question of how a text is read was a question about ideology. Abandoning the idea of class-specific ideologies, Hall preferred instead to focus on the political action of hegemony. He therefore assumed that television texts carry "the dominant hegemonic discourse" and are aimed to win "active consent" for a symbolic order or the belief in the way things are. The television text, therefore, is to be treated in terms of the larger ideological struggle between "the people" and the dominant discourse of a symbolic order. Although Hall assumed television's will to universalize an encoded ideological message, such that the television text is presumed to elicit unconscious identifications with what Althusser refers to as the ideological apparatuses of the state, Hall also insisted on the possibility of multiple readings of the television text.

But if the text allows for multiple readings, this possibility is located at the connotative level of the message. The possibility of multiple readings is located at the decoding end, at the point of reception. So although television texts mean to elicit a "preferred reading" of the text, readers, Hall argued, may engage in "negotiated readings" whereby pre-ferred readings are given local interpretation. Readers also may engage in "oppositional readings" whereby the encoded message is transformed in terms of a frame of reference that is an alternative to the one implied in the dominant discourse encoded in the text.

Although "Encoding/decoding" long would remain an influence in Marxist cultural studies, the notion of text that it elaborated would be criticized and revised by Marxist cultural critics. Treating the text as a discrete cultural production would be displaced by the notion of what Tony Bennett and Janet Woollacott called "inter-textuality." Yet, even the notion of inter-textuality serves only to refer the text to what Bennett and Woollacott[8] called "reading formations," referring to a particular reading context that activates a given body of inter-texts. For Bennett and Woollacott there is no "text outside a reading formation." Although inter-textuality implies that the text is always part of a network of insti-tutional arrangements, and therefore the notion of the text can be ap-plied to practices such as stylizing appearance, going to school, or shop-ping at the mall, nonetheless the evidence for the variety of readings always is found in the audience, the readers or users of practices.

Although Bennett and Woollacott's notion of inter-textuality has often been mistaken for Derrida's notion of intertextuality, the former is re-lated to the latter, but is also different. Bennett and Woollacott locate meaning in the social context in terms of which reading formations are defined. Hegemony and its formation still are what matter, and these still are thought to operate at the point of the reception or reading; in-deed Bennett and Woollacott argue that they are not offering the no-tion of inter-texuality "in the interest of a fashionable 'anything goes, everything is permissible' relativism tacked onto the coattails of Der-rida's project of deconstruction."[9]

Linking deconstruction with relativism in order to police textual analy-sis and define its political limits in terms of hegemony would become the rule in much of Marxist cultural studies, serving to reduce decon-struction to relativism wherever Marxist cultural studies would be taken seriously. Yet what Derrida proposes in deconstructing the border be-tween text and context is a revision of the notion of social context. Derrida's argument is that it is through the disavowal of *différance* that a text is constructed and given an identity that thereby produces a context — that is, produces the origin or end of meaning in all that is usually opposed to the text, such as speech, life, history, world, body, mind, consciousness, economics, and the real.

It is *différance* that gives the possibility that allows the text to "engender infinitely new contexts in an absolutely nonsaturable fashion."[10] Derrida argues further that this does not mean that the text "is valid outside its context, but on the contrary that there are only contexts without any center of absolute anchoring."[11] The reading context cannot ground the multiple readings of the text, because the text and its context are given together in the disavowal of *différance*. So to seek out evidence for the multiple readings of a text in the reading context is a matter of creating a context, not just finding one. For Derrida it is not possible to presume that there is a text and that it can arrive as such at the site of reception; nor is it any more possible to do so when the text is understood as inter-texts that refer to a dominant discourse.

Furthermore, it is not the case that Derridean deconstruction aims to take cultural criticism in the direction of relativism or, for that matter, the polysemy of multiple readings. Derrida is explicit in arguing that "the semantic horizon that habitually governs the notion of communication is exceeded ... by a *dissemination* (or *différance*) irreducible to polysemy."[12] This exceeding of the semantic horizon in *différance* is not, therefore, the "semiotic excess" that John Fiske argued characterizes television and constitutes its pleasures. As Fiske would have it, "television's openness, its textual contradictions and instability," allows readers "to construct subject positions that are theirs (at least in part) ... to make meanings that embody strategies of resistance to the dominant, or negotiate locally relevant inflections of it."[13] In making this argument Fiske made the appropriation of television discourse by audiences in behalf of their own pleasure central to his treatment of popular culture and the reading of its texts. Yet not only did Fiske still confine television's openness to a politics of hegemony that is at play at the point of reception; Fiske's language to describe television's excess of meaning and pleasure also shows the strong hold on Marxist cultural studies of Hall's "Encoding/decoding."

In Marxist cultural studies what has mattered most about television is located in the social context of reading. The focus is steadied on the audience as a potential of political resistance to the dominant discourse. The audience, therefore, figures "the people," who take the place of the

working class in figuring the agency of change. All of this allows Marxist cultural critics to be assured of being on the side of resistance to the
dominant discourse and of being able to offer their criticisms in the name
of the people. Meaghan Morris's complaint about Fiske's television
studies might be extended to most Marxist cultural studies. As she puts
it: "The people are also the textually delegated, allegorical emblem of
the critic's own activity. Their *ethos* may be constructed as other, but it
is used as the ethnographer's mask."[14]

That ethnography has been the preferred methodology for Marxist
cultural studies, even for its television studies, reveals its deep roots in
the cultural approach that both Thompson and Williams had proposed
as a way to champion the culture of the working class. Even when Hall
turns to Althusser's structural approach to ideology and to Gramsci's
and then Ernest Laclau's treatments of hegemony, he never abandons
the humanism and experiential base of the culturalist approach. With
its emphasis on the politics of hegemony embedded in the text and context of reading, Hall's "Encoding/decoding" might even be understood
not as a displacement of the culturalist approach, but rather as an adjustment of it to the mass consumption and the mass media of a Fordist
advanced western capitalism.

The irony of Hall's refitting the cultural approach with a structural,
textual approach to hegemonic discourse is that mass-mediated culture,
including watching television, become a matter of reading. Richard Johnson, when he was director of the Birmingham Center, even argued that
"the best studies of lived culture are also necessarily, studies of reading."[15] There is little thinking here about television as a machine that is
watched; for that matter, there is little thinking about television as a
machine at all. It is simply assumed that television is a machine befitting a Fordist advanced capitalism; it is assumed that television is a vehicle of advertising, a conveyor of an ideology of mass consumption.

What is missed in deploying the notion of "reading" television texts,
especially when the text is expected to carry the message of the dominant discourse, is the possibility that television's drive to transmission
is more about what Stephen Heath describes as television's "universalization of reception and the circulation of capital, not in particular
meanings — or not in the first instance in meanings other than those
of that circulation."[16] This understanding of television, however, requires recognizing that television is not primarily about texts; it does

not allow any distinction between the production of the text and its re-production in reading; it does not allow a distinction between text and reading context. Rather than being about a text that is offered to be read or even watched, television, Heath's remarks suggest, is more about a movement in a *différantial* network of traces referring endlessly to some-thing other than itself. Television is something more like what Derrida's treatment of textuality suggests, where the distinction between text and context is indistinguishable or always in the process of its construction, yet to be made and always deferred.

Television makes instantaneous transmission the limit of communi-cation. In these terms television is to be thought of as marking the ar-rival of the circuit as the machine metaphor meant to overwrite the dif-ference between production and reproduction, as well as production and circulation, thereby putting the text and all that has been opposed to it in a *différantial* rather than a dialectical or oppositional relationship. Then the questions that television raises displace the questions of mean-ing and interpretation; television becomes visible as a different cultural apparatus than the one it was presumed to be in the early Birmingham Marxist cultural studies of television. Television, that is, raises questions concerning the expanded circuit — What does it make and what does it make happen in collapsing the difference between production and re-production, production and circulation, text and context?

To think about television in this way, however, is to think of tele-vision as a machine of postmodern or late capitalism rather than as a machine of Fordist advanced capitalism. Ironically 1973, the year "En-coding/decoding" first was published, also would be claimed by Marx-ist cultural critics, most notably David Harvey, to be the year when the problems of Fordism-Keynesianism "erupted into open crisis," and thereby when the the "sea change" in the organization of capitalist pro-duction could be linked to an eruption and extension of postmodern culture.[17] But by the late 1980s, when Marxist cultural critics would focus on postmodern culture, Hall too had begun to treat culture in terms of the "new times" of postmodernism, connected to the post-Fordism of late capitalism in postwelfare and neocolonial states.[18] For Hall, culture then becomes a matter of differences — racial, gender, sexual, and na-tional differences — where the notion of a unified identity is decon-structed into a cultural hybridity and where Derrida's treatment of *dif-férance* is given a more serious reconsideration. It is in these new times of Marxist cultural studies that television as a cultural apparatus becomes increasingly indistinguishable from television as a technical apparatus or technical substrate. Television, that is, is seen as a machine, a productive/reproductive/circulating machine.

Surely by the late 1980s Marxist cultural critics were becoming aware that there is no culture that is not technologically mediated. It is this realization that Jameson best registers in his treatment of television as the icon of postmodern culture, informed with the economic logic of late capitalism. Jameson proposes that television even makes it possible to see that culture always has been mediated: "The older forms or genres, or indeed the older spiritual exercises and mediations, thoughts and expressions were also in their very different way media products."[19] For Jameson, television revises the history of technological development as a history of literary form, as a play of culture's machine metaphors or technical substrates.

In response to his tele-vision, Jameson rethinks literary form in terms of media technologies. He defines a medium as "an artistic mode or specific form of aesthetic production" that becomes "social institution" and belongs to "a specific technology, generally organized around a central apparatus or machine."[20] But Jameson's reformulation of literary form in terms of media technologies is not without his mourning what seems to him to be teletechnology's dismissal of narrative, which in his influential study of the novel he had found to be the vehicle of a political unconscious. The future to which Jameson's tele-vision will draw Marxist cultural criticism, therefore, would be difficult to realize, detoured as it is through Jameson's own treatment of the political unconscious of the dominant narrative of the state apparatuses.

Like Hall's 1973 "Encoding/decoding," Jameson's 1981 analysis of the novel is remarkable for its lack of any explicit treatment of postmodernism or late capitalism. But more than "Encoding/decoding," *The Political Unconscious: Narrative as a Socially Symbolic Act*[21] seems to have been motivated by the political economic changes in postmodernity that are registered in Althusser's rereading of Marx, even though Jameson does not explicitly address these. Instead Jameson claims that the aim of *The Political Unconscious* is to restore to Marxist cultural criticism the Hegelianism that Althusser's rereading of Marx seems to dismiss. In seemingly pursuing this aim, Jameson turns Althusser's rereading of Marx into a criticism of poststructuralism, thereby enabling Marxist literary criticism to become an authoritative interpretive diagnostic of postmodern culture.

Denarrativizing and Renarrativizing the Political Unconscious

In *The Political Unconscious* Jameson set himself a difficult task: to appropriate and contain the poststructural criticism of narrative and history by treating both narrative and history in terms of what Althusser referred to as a "structure-in-effect," or becoming structured around "an absent cause." Jameson notices that Althusser's rereading of Marx raises the question of representation in a way that is relevant to narrative and history. Not only does Althusser propose that economy and culture are relatively autonomous, both overdetermined levels of the mode of production; he therefore also suggests that the totality of the mode of production "is no where empirically present." How, then, can one represent it? How is there to be representation of a structure in the absence of the empirical presence of it as a totality?

Althusser's answer focuses on Marx's notion of *Darstellung,* or representation in the weaker sense of staging. Jameson quotes the following from *Reading Capital*:

> Structural causality can be entirely summed up in the concept of "Darstellung," the key epistemological concept of the whole Marxist theory of value, the concept whose object is precisely to designate the mode of *presence* of the structural in its *effects,* and therefore to designate structural causality itself. . . . The structure is not an essence outside the economic phenomena which comes and alters their aspect, forms and relations and which is effective on them as an absent cause, absent because it is outside of them. The absence of the cause in the structure's "metonymic causality" on its effects is not the fault of the exteriority of the structure with respect to the economic phenomena; on the contrary, it is the very form of the interiority of the structure, as a structure, in its effects. . . . the structure which is . . . nothing outside its effects.[22]

Althusser's remarks seem to suggest that the structuration of capitalism can be represented only by a rhetorical mechanism or a representing machinery (a *Darstellung* machine) that stages and thereby produces the effects of capitalism as a whole, that is, as an economy or a system for the realization of value. If so, what is implied is the possibility of thinking that the machine for production of value and the rhetorical mechanism or machinery for staging the effects of capitalism as a whole are one and the same machine. Or, to put this another way, the mechanism for the reproduction of the subject in ideology has become the machine for the production of value. Some transformation of capitalist production — some transformation of machinery or technology — is the historical condition of possibility for Althusser's rereading of Marx. Yet Jameson, like Hall, misses seeing the historically and technologically specific machine implied in Althusser's rereading of Marx; like Hall, Jameson focuses more on the rhetorical devices for representing capitalism to the subject in order to induce in the subject an unconscious identification with the dominant discourse or the cultural hegemony. Jameson especially focuses on the novel and the relationship between the novel's narrative logics and the history of the capitalist organization of production.

Going back to Lacan's treatment of the "Real" as that which "resists symbolization absolutely," Jameson insists that Lacan's formulation does not mean that there is no referent to which History refers. Jameson argues instead that History, like the Real, cannot be known directly; but History can be apprehended in its textualization or narrativization. As Jameson puts it: "History is not a text, not a narrative, master or otherwise, but . . . , as an absent cause, it is inaccessible to us except in textual form, and . . . our approach to it and to the Real itself necessarily passes through its prior textualization, its narrativization in the political unconscious."[23] Although Jameson seemingly gives up on a notion of History as origin or cause, he nonetheless proposes that History can be found through its various narrativizations. History finally is realized as "the collective struggle to wrest a realm of Freedom from a realm of Necessity"; History, therefore, is told "within the unity of a single great collective story," which, however, is moved by the antagonisms between the social classes.[24]

It is in these terms that Jameson's analysis of the novel means to show how its narrative logics displace or defer the single great collective story yet to be realized in a revolutionary change of the capitalist organization of production. He proposes that the novel's narrative logics — from romanticism to realism, from modernism to high modernism — each prematurely totalizes a symbolic order by covering over the contradictions or differences between the levels of a mode of production, thereby preventing the possibility of revolutionary change. For Jameson, a Marxist literary criticism must recover the contradictions by drawing on the dialectic and its provision of the notion of mediation that allows for relating levels of a mode of production without reducing them one to the other.

In this sense a Marxist literary criticism involves deconstructing totalizing narrative logics. It seeks "rifts and discontinuities." It looks for the "strategies of containment" in forms that gloss over the *non dit,* the *impense,* that is, the political unconscious. But a Marxist literary criticism also must go beyond deconstructing a narrative logic. For Jameson, deconstruction is: "only an initial moment in Althusserian exegesis, which then requires the fragments, the incommensurable levels, the heterogeneous impulses, of the text to be once again related, but in the mode of structural difference and determinate contradiction."[25]

In order both to reveal hidden contradictions of a narrative logic and to begin to reconstruct the unity of the single great collective story, Jameson reads or deconstructs the novel's narrativity against three horizons. First the narrative is read as an "imaginary resolution of a real contradiction" that cannot be resolved. The narrative is apprehended at first as "a symbolic act," an aesthetization of a contradiction. Then the novel can be read against a second horizon, that is, in terms of the "essentially antagonistic collective discourse of social classes,"[26] whereby "individual phenomena are revealed as social facts and institutions."[27] Finally the text is to be read against a third horizon — the "overlay and structural coexistence of several modes of production all at once."[28] Here Jameson follows Nicos Poulantzas's argument that in a mode of production there are "vestiges and survivals of older modes of production, now relegated to structurally dependent positions within the new, as well as anticipatory tendencies, which are potentially inconsistent with the existing system but have not yet generated an autonomous space of their own";[29] it is when these vestiges and anticipations become "visibly antagonistic, their contradictions moving to the very center of political, social and historical life,"[30] that revolution of the mode of production is made possible.

At the heart of Jameson's reading strategy is the deconstruction of the subject form given with the novel's narrative logics. Unlike Derridean deconstruction, however, Jameson's deconstruction of the subject is not in the direction of the subindividual, that is, through the individual subject's unconscious to the subindividual, finite forces or singularities of mattering. Instead Jameson aims the deconstruction of the subject in the direction of the collectivity and thereby insists that only a collective unconscious is relevant to revolutionary change. Only a political unconscious is relevant. It is in this sense that Jameson attributes only a limited importance to Lacan's rereading of Freud, proposing instead a more generalized unconscious that is trans-subjective but, nonetheless, insistently humanistic.

Jameson, therefore, recognizes that Lacan especially problematizes the category of the subject by turning psychoanalysis against Freud's "notion of individual wish fulfillment" with "its buttressing ideologies and illusions (the feeling of personal identity, the myth of the ego or the self, and so forth)."[31] Yet as Jameson sees it, Lacan's emphasis on unconscious desire in the displacement of wish fulfillment does not go far enough; Jameson complains that "desire, like its paler and more well behaved predecessor, wish-fulfillment, remains locked into the category of the individual subject, even if the form taken by the individual in it is no longer the ego or self, but rather the individual body."[32]

Seeking an "ultimate Utopian vision of the liberation of desire and of libidinal transfiguration" that is more congenial to a Marxist perspective, Jameson turns from Lacanian psychoanalysis to the myth-criticism elaborated by Northrope Frye; it is Frye's myth-criticism that informs Jameson's elaboration of the three horizons against which to read the novel's narrative logics in terms of a political unconscious. As Jameson sees it, Frye allows for a link between unconscious desire and community in the interpretation of collective representation. Whereas for Frye collective representation is to be interpreted for its religious implications, for Jameson the interpretation of collective representation is a matter of ideology. Frye's understanding of religious myth as figuring "the symbolic space in which the collectivity thinks itself and celebrates its own unity" is thereby drawn to literary criticism where "all literature must be read as a symbolic meditation on the destiny of community."[33] Turning from Lacanian psychoanalysis to Frye's myth-criticism allows Jameson to conclude both that literature is "a weaker form of myth or a later stage of ritual . . . informed by . . . a political unconscious"[34] and that "only the community . . . can dramatize that self-sufficient intelligible unity (or 'structure') of which the individual body, like the individual 'subject' is a decentered effect."[35]

In deconstructing the individual subject's body in the direction of the collectivity, Jameson effectively forecloses any treatment of the rhetoric of sexual difference at play in the unconscious identifications that the novel's narrative logics engender. Jameson's literary criticism, unlike feminist film theory, does not, therefore, open itself to questions of differences other than sexual difference, such as differences of race, ethnicity, sexuality, or nation. Although Jameson resists the normative and restrictive aspects of reducing the unconscious to the individual subject, he does not recognize the particularity that the individual subject's body registers and the subindividual finite forces of matter to which that particularity refers. Instead Jameson would draw the unconscious — the political unconscious — from individual to collectivity in order to make the unconscious function in centering the structure of a mode of production to be its absent cause and the difference of its levels.

Jameson's literary criticism of the novel's narrative logics troubles but leaves in place the arrangement of family and national ideologies, state and civil society, and the private and public spheres presumed in the subject-centered, nation-centric discourse of ideology in terms of which he would treat the capitalist mode of production. Therefore, Jameson elaborates a literary criticism that seems indifferent to the effects of technological development on the political unconscious and the arrangement of social spaces, family and national ideologies, the state and civil society, and the private and public spheres, in which the political unconscious is situated.

Although Jameson proposes that the narrative logics of the novel are productive, their relationship to historically specific machines of production is not elaborated. Jameson's literary criticism maintains a distance between the metaphoric machinery of ideology and the machinery of production. It thereby leaves in place the reduction of technological development to the history of the capitalist organization of production implicit in Jameson's Marxist perspective. However, it is this reduction of technology to capital logic that will become more explicit and troubled as Jameson turns to treat late capitalism. Then the narrativity upon which Jameson's literary criticism is focused will be opened up to the historically specific technical substrate of the political unconscious of teletechnology.

What Jameson argues in *The Political Unconscious* — that literary forms have ideological content, that they resonate with the struggle of social classes and the contradictions of modes of production, that they provide a subject form for the cultures of various modes of production — seems to fit easily his study of the generic changes of the novel's narrative logics through romance, realism, modernism, and high modernism. Linking the changes of the novel's narrative logics to the rise of market capitalism, followed by industrial capitalism and finally by advanced or monopoly capitalism seems easy. But when Jameson seeks the generic form of the subject given to the culture of late capitalism in order to study its ideological content in relationship to the struggle of social classes, he finds it necessary to shift his focus from the study of the narrative logics of the novel to the study of machines, primarily the machines of teletechnology. It is then that Jameson comes to recognize that all cultural forms are technologically mediated. It is also when he begins to struggle with the machine face on, when he begins to struggle with teletechnology, especially television. It is then that Jameson's effort to reduce technological development to the history of the capitalist organization of production becomes problematic and seems overly strained.

No doubt Jameson first looks at teletechnology in order to find its political unconscious, in order to subject teletechnology to a cultural criticism informed with an Althusserian treatment of the capitalist mode of production. He therefore asks whether the culture of teletechnology erases the difference between the levels of the mode of production, whether it makes the contradictions between the levels illegible, so as to make the final realization of a single great collective history all but impossible. In asking these questions Jameson means to propose that culture in the age of teletechnology, what he refers to as postmodernism, can yet be transcended by the dialectic logic of capital. It is therefore to Jameson's treatment of postmoderism, and the video art that he takes as its example, that I want to turn next, along with Dienst's revision of Jameson's treatment of television.

The Time-Image and Machinic Assemblages

Having considerable influence throughout the late 1980s and the early 1990s, Fredric Jameson's treatment of postmodernism served to powerfully contain a certain elaboration of poststructuralism, especially Derridean deconstruction. One of the effects of Jameson's treatment of poststructuralism along with postmodernism as symptomatic of the late capitalist mode of production was the widespread failure to even imagine treating the late twentieth-century development of technology along the lines suggested by poststructuralism, that is, to treat technology in its own terms or to give thought to an ontological perspective that allows for the various historicities given with technological development—an ontological perspective whose elaboration is only made more pressing with the development of teletechnology.

Yet Jameson's effort to treat postmodernism and poststructuralism as expressions of the cultural logic of late capitalism finally does draw him to the temporal/spatial relationships given with television, which Jameson finds elaborated in video art. In "Surrealism without the Unconscious" Jameson offers a cultural criticism of video art as an example of postmodernism that leads him to argue that technologies are best characterized for the way they machine time. His argument even seems to raise the question of whether the development of technology can be reduced to the history of the capitalist organization of production, and therefore whether cultural criticism can subject capitalism to a teleology aimed at the revolution of the mode of production conceived as a totalized structure, if only in its effects. Although Jameson's treatment of video art troubles his effort to reduce teletechnology to the late capitalist mode of production, his insistence on an Althusserian treatment of structure prevents him from rethinking a cultural criticism of teletechnology by rethinking the reduction of the development of technology to the dialectic of capital. It does, however, lead him to rethink the relationship of narrative logic, the subject, the unconscious, and ideology, taking Marxist cultural criticism toward realizing in television the becoming indistinguishable of productive machines and the machine metaphors of literary forms.

Although Jameson nonetheless shrinks from fully recognizing that without a dependency on narrativity, television forces mediation to break loose from the dialectic logic of capital, Richard Dienst is less timid. Focusing on the postmodern transnationalization of capital, Dienst struggles to elaborate a Marxist cultural criticism for a capitalism that has become inextricable from the globalization of teletechnology when neither capitalism nor technology can be reduced to the other, when neither can be the condition of possibility of the other. Therefore, Dienst follows teletechnology, television especially, to Gilles Deleuze's treatment of the time-image, as well as to his treatment of "control societies," where the arrangement of family and national ideologies, the state and civil society, and the public and private spheres presumed in subject-centered, nation-centric modern western discourse of ideology is deconstructed.

Although Dienst passes over rethinking teletechnology in terms of the ontological perspective suggested by Deleuze, he does make it clearer that a Marxist cultural criticism can no longer be a matter of treating the narrative logic of a political unconscious of a single great collective story. In doing so, Dienst draws Marxist cultural criticism to the irony of the postmodern transnationalization of capital: that is, given the world-wide reach of postmodern capitalism, it therefore seems necessary to follow the lines of flight of localized criticisms of globalized cultures. In this Dienst gives Marxist cultural criticism over to the unconscious of machinic assemblages, where the opposition of nature and culture is deeply troubled, making it impossible to take use value as originary in the elaboration of cultural criticism in the age of teletechnology.

Jameson's Postmodernism

Jameson treats postmodern culture in terms of a set of symptoms, now well known, which, he proposes, are intimately connected to the technological development of the late twentieth century. As Jameson sees it, postmodern culture is characterized by the "fragmented" subject, the "waning of affect," the collapse of the difference between pleasure and pain, the intensification of emotionalism, the severing of the image from its material basis, the snap of signification into "a rubble of distinct and unrelated signifiers."[36] All of these symptoms, Jameson concludes, demonstrate the challenge that postmodern culture poses to the hermeneutics of a depth model of cultural criticism. With depth gone, postmodernism allows for nothing but pastiche—a quoting or citing without end(s), beyond which there is only nostalgia, save a meaninglessness, in the turning of every work of art into "nothing but texts."[37]

Although Jameson worries that postmodernism blinds us to the contradictions between the levels of the late capitalist mode of production, refusing mediation, and therefore undermining the very possibility of History, he also argues that postmodernism is a cultural enactment of capital logic or a reflection of the history of the capitalist organization of production. Reduced to a capital logic, postmodernism can still carry the promise of the possibility of a historical transcendence through the revolution of the late capitalist mode of production.

But to secure the possibility of transcendence in the reduction of postmodernism to a capital logic, there needs be a prior reduction of technological development to the history of the capitalist organization of production. Drawing on Ernest Mandel's *Late Capitalism*, Jameson argues:

> Technological development is however on the Marxist view the result of the development of capital rather than some ultimately determining instance in its own right. It will therefore be appropriate to distinguish several generations of machine power, several stages of technological revolution within capital itself.... there have been three fundamental moments in capitalism, each one marking a dialectical expansion over the previous stage. These are market capitalism, the monopoly stage or the stage of imperialism, and our own ... multinational capital.[38]

Reducing technological development to capital's interest allows Jameson to treat postmodern culture in the historical terms of a single History, even if it is the historical terms of History's threatened end in the development of technology. Jameson refuses to think of the possibility that teletechnology informs unconscious memory with a technical substrate, and therefore gives a specific historicity; he cannot think that the history of technological development crosses through or crosses out the universal history of the capitalist organization of production. He cannot think of the aporia of time that allows for the intimate connection of teletechnology and the transnationalization of capital in late twentieth-century capitalism, but that does not permit the reduction of one to the other.

For Jameson, teletechnology does not give a specific historicity. Instead, teletechnology, like postmodernism, seems simply to challenge all possibility of History; this is because teletechnology, like postmodernism, raises a question about History in terms of its relationship to nature, especially as it is inscribed in the Marxist notion of use value. Indeed, as Jameson sees it, postmodern culture is both a treatment and an effect of a "society where exchange value has been generalized to the point at which the very memory of use value is effaced" and when—and here Jameson quotes Guy Debord—"the image has become the final form of commodity reification."[39]

Jameson gives a sense of such a society by way of a reading of Vincent Van Gogh's painting *A Pair of Boots*, contrasting it with Andy Warhol's painting *Diamond Dust Shoes*. Taking up Van Gogh's painting allows Jameson to engage Martin Heidegger's reading of the same painting, whereas treating Warhol's painting allows Jameson to reinforce his criticism of postmodernism with Heidegger's treatment of modern technology. The reference to Heidegger's treatment of technology is meant to underscore Jameson's sense of modern technology as that "anti-natural power of dead human labor stored up in our machinery" that, nonetheless, "constitutes the massive dystopian horizon of our collective as well as individual praxis," what Jameson might have once engaged in terms of a repressed political unconscious.[40] The reference to Heidegger, therefore, is meant to bolster Jameson's effort to return to nature figured in the use value of labor, thereby restoring use value, if only in its effacement, as the originary reference of cultural criticism. It is in this appropriation of Heidegger's treatment of technology that Jameson differs with poststructuralism, providing him a way to contain the ontological implications of poststructuralism.

For Heidegger,[41] technology is to be understood in terms of the distinction of *techne* from *physis* or nature, although both *techne* and *physis* refer to *poiesis,* and therefore are under the compass of Being; that is, both are "destined" to bring forth the truth of Being, to reveal Being in a framing that makes the real possible. For Heidegger, *techne* is not merely technological or mechanical. Like *physis, techne* brings forth; but it does so differently than does *physis. Physis* brings forth of itself, whereas *techne* makes use of another, the craftsman or artist, the human laborer.

Although *techne,* like *physis,* is coupled with *poiesis, physis* is the highest form of *poiesis* because the revealing of *physis* is immanent to it. The privilege afforded *physis* is underlined, however, in Heidegger's further distinction of modern *techne* from premodern *techne.* It is in terms of this distinction that Heidegger refers to *Ge-stell,* a *techne,* or framing that is productive of the real, which Heidegger links to modern technology, that is, machine technology in the age of science. If premodern *techne* brings forth through a caring or taking care of, as in craft or premodern agriculture, modern technology does not; it uses up living labor. If premodern *techne* does not threaten nature with its obliteration, *Ge-stell* does. Nonetheless, like *poiesis, Ge-stell* still brings forth or reveals; but it reveals only by ordering, normalizing, objectifying, or reifying—an unending putting in place or an emplotment that is unendingly undone. *Ge-stell,* it would seem, all but goes beyond the compass of Being.

That is to say, *Ge-stell* is an enframing or an emplotting that means to give what is real by concealing its mode of revealing or its mode of bringing forth. It is an enframing that conceals so that both the revealing and the concealing become all but indiscernible from each other. Or, to put this another way, *Ge-stell* is driven to bring forth, so that any reality it enframes, and therefore brings forth, is readily displaced by another framing. The cycle of revealing and concealing, of placing and displacing every enframement of the real, is repeated with such rapidity that what is real seems to be nothing but framing.

There is nothing left but frames and frames of nothing — nothing but reproductive technology of framing. The linkages from the real to nature to Being are deeply disturbed, and this is disturbing to Heidegger, so that the conclusion that is to be drawn from his discussion of modern technology can be only this: although Being is nothing but the beings in which Being leaves a trace as it retracts from them, nonetheless Being is self-same, a purity; it is so at least in the opposition of nature, not to *technē* in general, but to *Ge-stell* or modern technology in particular. In this sense Being is an originary presence, a self-same identity.

Although it seems it would be easy to fit teletechnology, especially television, to Heidegger's treatment of *Ge-stell,* Jameson does not do so directly; instead he follows Heidegger to Van Gogh's painting of the peasant's shoes. Jameson, therefore, reads the painting as "a disclosure of what the equipment, the pair of peasant shoes, is in truth . . . the uncon-cealment of its being by way of the work of art."[42] As for Heidegger, for whom the truth of the equipment — the pair of peasant shoes — is that it "belongs to the earth," for Jameson, too, the truth of the pair of shoes is revealed along the route from use value to a peasant/laborer; the boots still resonate with nature (a use value) worked into a second nature of a caring agri-culture, a *technē* not yet attached to modern technology and modern labor. As such, the painting also refers to the future, to the possibility of transcending the present; the painting offers a horizon against which to overcome the exhaustion in laboring. Jameson proposes that the Van Gogh painting gives, by means of vivid color, a utopian wash to the drab world of the broken, exhausted peasant who haunts the painting, traced in the shoes presented.

In contrast, the Warhol painting, as Jameson sees it, offers no return to the use value of labor. It instead reveals *Ge-stell,* without labor. War-hol's painting, therefore, also discloses. But in contrast to the van Gogh painting, the Warhol painting seems to strip away "the colored surface of things," revealing "the deathly black and white substratum of the photographic negative."[43] Jameson concludes that the Warhol painting points not to a content at all, but rather to "some more fundamental mutation both in the object world itself—now become a set of texts or simulacra—and in the disposition of the subject."[44] The world seems to have "lost its depth and threatens to become a glossy skin a stereoscopic illusion, a rush of filmic images without density."[45] The equipment that the Warhol painting discloses, therefore, is reproductive rather than productive—or, better, a reproductive technology that seems to deny completely the productivity of human labor. The Warhol painting points to what Jameson laments, that is, the impossibility of getting back to use value or to the use value of human labor. Whereas the Van Gogh painting still speaks of the peasant's labor, the Warhol painting "no longer speaks to us with any of the immediacy of van Gogh's footgear; indeed, [it] does not really speak to us at all."[46]

For Jameson, the comparison of the two paintings demonstrates that the relationship of nature and culture that existed in precapitalist society, such that nature was "the other" of culture, seems no longer to obtain. Since the displacement of nature by reproductive technology in the late twentieth century, postmodern culture no longer figures human labor, not even human labor displaced by productive machines. But for Jameson this seeming displacement of nature by technology, specifically teletechnology, is often misrepresented as the effect of technological development in the late twentieth century; indeed, it is this misrepresentation, Jameson argues, that must be the subject of a Marxist criticism of postmodern culture. As he puts it:

> Our faulty representations of some immense communicational and computer network are themselves but a distorted figuration of something even deeper, namely, the whole world system of a present-day multinational capitalism. The technology of contemporary society is therefore mesmerizing and fascinating not so much in its own right but because it seems to offer some privileged representational shorthand for grasping a network of power and control even more difficult for our minds and imaginations to grasp: the whole new decentered global network of the third stage of capital itself.[47]

Jameson has returned technology to the logic of capital and to the possibility of the transcendence of late capitalism through human agency collectively realized. In doing so, however, Jameson restores the privilege given to nature in the opposition of it to *Ge-stell* or modern technology, which allows only a narrow and terrifying sense of modern technology. Jameson refuses to think the possibility that Derrida proposes, that is, that culture and technology are nature deferred, or that nature and technology, like nature and culture, are in *différantial* relationships, interimplicated one with the other all the way down.

But then Derrida finds Marx's notion of use value often deployed in a faulty search for a natural origin, ignoring the interimplication of use value and exchange value from the start and leaving in place an ontology of presence, of originary Being. Nature is thereby privileged by Marx, and by extension so is human nature in the form of human labor. In this light it is interesting to note that Derrida also gives a reading of the Van Gogh painting — a reading that Jameson only mentions in passing. Derrida focuses on the differences between Heidegger's reading of the painting and Meyer Schapiro's reading some thirteen years after Heidegger offered his reading. Derrida discusses the dispute over whether the shoes belong to a peasant, as Heidegger (and Jameson) assumes, or whether, as Schapiro claims, they belong to the "artist, by that time a man of town and city."[48] But for Derrida, what is noteworthy about this dispute is the critics' need to locate the person who wears the shoes and the experience that the shoes present: "the desire . . . to make them [the shoes] find their feet again on the ground of the fundamental experience."[49] Derrida proposes that the "detached" shoes can be made to speak only if the subject who is imagined to wear the shoes is reattached to them and to the fundamental experience of laboring: "a general reattachment as truth in painting."[50]

Derrida's reading of the Van Gogh painting is emblematic of Derrida's resistance to treat technology in terms that privilege nature for its capacity to bring forth of itself or from within itself. Derrida rather suggests that *technē* should be allowed to contaminate Being, letting an "originary technicity" (or textuality) cross through an "originary Being." Technicity is made to mark Being with finitude. It brings Being down into finitude. It brings immanence into transcendence. It thereby makes a finite technicity the transcendental condition of possibility of Being when, of course, transcendence "only mimics a phantom of classical transcendental seriousness" — "a quasi-transcendental," as Derrida puts it.[51]

If Derrida's treatment of an "originary technicity" suggests a change in ontological perspective, the reach of thought to the *différantial* relationship of human and nonhuman, body and machine, nature and technology, the living and the inert, it is teletechnology that registers this shift, especially because teletechnology makes more apparent agencies other than human agency. It is these agencies that Jameson refuses, and with that refusal, it would seem, he also avoids rethinking the history of technological development in terms of the *aporia* of time, failing, therefore, to recognize the specific historicity given with teletechnology. All this leaves Jameson with an understanding of teletechnology given in terms of postmodernism that ceaselessly frames the real, and therefore empties it of every meaning other than that of the revealing and the concealing of its framing.

To confirm this understanding, Jameson finally approaches television, but still not quite directly. He approaches television by way of video art as an example of postmodern culture, which surely makes his argument about teletechnology easier to make. Still, through his treatment of video art Jameson is able to think of television as technology. Not overly anxious about the mass audience or the social context of reading, as those engaged in earlier Birmingham cultural studies of television had been, Jameson in his treatment of video art is able to get closer to television as the machine of the immense communicational and computer network of teletechnology. As such, Jameson's treatment of video art is drawn back to the relationships of time and space that television gives and is thereby drawn to the future of thought in its reach to the ontic. Jameson opens up the possibility of seeing in television the becoming indistinguishable of producing machines and the machine metaphors for representing capital's effects, which is the starting point of Dienst's treatment of television. It is also the end point of a Marxist cultural criticism that insists on the structured mode of production as totality, that is, as a structure to be overcome in the dialectic logic of capital.

The Darstellung Machine of Postmodernity

To get to Jameson's treatment of television as machine, it is necessary to work through his criticism of postmodern video art. Like Jean Baudrillard's treatment of postmodern "hypertelic" simulation, upon which Jameson draws, the rhetoric of Jameson's treatment of video is excessive, as if to make visible by mimicking to excess the threat to narrative and to history that Jameson and Baudrillard seem sure postmodernism brings. Not surprisingly, then, Jameson argues that the textuality of video art "resists meaning." Its "fundamental inner logic is the exclusion of the emergence of themes as such," and therefore, video short-circuits narrative closure.[52] There is instead "the capture of one narrative signal by another: the rewriting of one form of narrativization in terms of a different momentarily more powerful one, the ceaseless renarrativization of already existent narrative elements by each other."[53] In video the move seems to be to a ceaseless flow of information and images, where "the situation in which one sign functions as the interpretant of another is more than provisional. . . . Signs occupy each other's positions in a bewildering and well-nigh permanent exchange."[54] Jameson concludes that video flattens or empties History because reference is "systematically processed, dismantled, textualized and volatized."[55] But beneath his excessive rhetoric and despite his limited reading of video art itself, Jameson's attention to the deconstruction of narrative in video art draws him to the relationships of time and space that television gives.

When video undermines History and narrative, Jameson argues, it also gives an experience of machine time, showing time to be a matter of a measuring machine. As he puts it, "Measurable time becomes a reality on account of the emergence of measurement itself . . . ; clock time presupposes a peculiar spatial machine — it is the time of a machine, or better still, the time of the machine itself."[56] Video art does this revealing of machine time by delivering images from "fictive time." Fictive time is the foreshortening of time by way of an editing narration that nonetheless goes unnoticed so that fictive time can be taken to be real time.[57] But since video art is in "real" time, time is not its fiction. Instead, video is the "only medium in which this ultimate seam between space and time is the very locus of the form."[58]

In subtracting fictive time from images, video points to teletechnology's potential to sever representation from a narrative logic. Paul Virilio also has suggested that teletechnology shifts the focus of aesthetics from the narratological to something he refers to as the "chronoscopical." As he puts it:

Henceforth, the "real" time of telecommunications will probably refer no longer solely to "deferred" time, to feedback, or to time lags, but also to an outer chronology. Whence my constantly reiterated point about replacing what is chronological (before, during, after) with what is dromological or, if another formula fits better, the chronoscopical (underexposed, exposed, overexposed). In effect, ... the notion of exposure replaces, in its turn, that of succession in terms of present duration and that of extension in immediate space.[59]

The chronoscopical characterizes aesthetics not only when teletechnology subtracts fictive time from imagery, but when it puts perception beyond "the sphere of influence of the human body and its behavioral biotechnology."[60] As Virilio sees it, the chronoscopical is part of the displacement of time as duration and space as extension in their interface with speed when, however, speed is no longer solely about travel. As it becomes more apparent that speed is instead a relationship among phenomena, it also becomes more obvious that speed is used "to see, to hear, to perceive, and thus to conceive more intensely the present world."[61] In the context of teletechnology, the logic of the image, Virilio argues, is no longer dialectic; it is rather "paradoxical" because the image is as real as or more real than the thing represented.

It is all of this that Jameson glimpses in video art. He not only suggests that "the deepest 'subject' of all video art, and even of all postmodernism, is very precisely reproductive technology itself."[62] He also proposes that reproductive technology, first seen by him in the Warhol painting and then in video art, more clearly, is a matter of machine time, or the machining of time. Yet, having seen in video art the becoming of reproductive technology as productive technology and that, therefore, video art makes more apparent that time is machined and is inextricable from different technical substrates, Jameson nonetheless does not go further. He does not think that video art registers a shift in thought, a shift to thinking an "originary technicity" as the impossible condition of the possibility of time and, therefore, of Being. For Jameson, video art rather registers only the effects of the history of capitalist organization of production, which is to be held responsible for flattening History or compressing time, first by separating referents from signs and then signifiers from signifieds, resulting finally in a free play of signifiers. Although Jameson does not say so, perhaps it is this wild play of signifiers that is like surrealism but without the unconscious.

Yet it is in the way it produces an image for display that surrealism presents the unconscious in terms of a mechanical seeing very much like that of television. Focusing on Max Ernst's "The Master's Bedroom,"

Rosalind Krauss[63] has suggested that the surrealist image is not produced by putting pieces together on a blank surface, as in a collage, but
rather by a method of subtraction, that is, by overpainting or painting
out elements in an already printed set of images. A different ordering
of images is thereby displayed, as if it came up out of the painting and
had always already been there.

Krauss shows that the painting was produced by using gouache to
cover over a number of the elements on a Lehrmittel sheet, that is, a
printed sheet with rows and rows of objects — animals, vegetables, trees,
tables, windows, and beds. In surrealism, then, the material ground out
of which vision is to be produced is a "space of inventory" — or what
the surrealists themselves called "the readymade." What Krauss wants
to emphasize here is that the material ground of surrealism is always
already filled, unlike a modernist visuality, where the blankness of the
canvas allows an image to appear as if a projected picture of the preexist-
ing external world, thereby encouraging an understanding of perception
as a matter of the human eye's opening onto the external world as if
that world were simply there as such. This — the modernist cinematic
elaboration of vision — is turned down in surrealism.

The surrealist painting does not allow an understanding of perception such that the human eye sees an external world as if that world
were simply there as such. The surrealist painting rather gives an understanding of perception that points to the repeated return to "a structure of vision"; it means to bring to the surface this mechanical seeing
in the reordering of always already-given images. It is this repetitious
return to a structure of vision that is displayed, like an eye looking back
at the viewer, an eye that has been overtaken by an automaton. Krauss
argues that surrealism, "with a prescience that is amazing for 1920," gives
a paradigm for a mechanical seeing, "the automatist motor turning over
within the very field of the visual."[64]

All this leads Krauss to propose that the surrealist readymade be compared to Freud's mystic writing-pad — the wax slab always already
filled up with a network of traces, covered over by a filmic sheet more
like a hardened skin such as the gouache produces on the surface of the
painting. Given that the mystic writing-pad is a machine metaphor for
unconscious memory, the comparison of surrealism to it suggests that
the unconscious of surrealism is informed by a technical substrate, also
like that given with teletechnology, which is productive in endlessly re-

configuring elements in the already saturated field of the gaze. Surrealism points to the working of unconscious memory that shows itself, at the surface, as surfaces repeatedly appearing and disappearing without narrative links being necessary. If there is continuity, it is a nonnarrative one; the logic is not a narrative one, but one of exposure, over- and underexposure.

That "The Master's Bedroom" can be interpreted as a primal scene fantasy also suggests, however, that surrealism is connected to a trauma that does not know its cause, but is repeated as if to find its cause. In this sense "The Master's Bedroom" can be linked to the oedipal complex. Krauss gives evidence of the probability that Ernst was producing not only his primal scene fantasy; he also may have been reproducing Freud's oedipalizing description of the primal scene fantasy in the case study of the Wolf Man, with which Ernst was engrossed. In other words, the oedipal complex is quoted and doubled and dispersed in the readymade. As such, the oedipal complex is only another element of the readymade, and thereby is crossed through as an urnarrative or an originary narrative. In this sense, too, the readymade is closer to television than to cinema, where the oedipal logic of narrativity has functioned to produce—even has been central to the production of—fictive time. The readymade points instead to the coming of a tele-vision on a surface that is never blank, but always already filled, and where narrative is displaced, no longer central to seeing and to unconscious memory.

Yet Jameson suggests that teletechnology may do away with the unconscious altogether, most likely because there is no dominant hegemonic narrative. Even though there are narratives scattered throughout television programming, television does not seem to offer a dominant hegemonic narrative as a means to treat a repressed political unconscious. Jameson cannot make class antagonisms visible, nor can he prepare for their overcoming. Although Jameson ends his treatment of video art giving a brief sketch of the history of the capitalist organization of production, it is only to uncover the destiny that history has given to the sign, revealing the itinerary of its reification and dematerialization. There is no return, however, to the machining of time, which video art registers; there is no attempt to treat its effects on viewers. But, then, what about the viewer? What about the audience, which, in Birmingham cultural studies of television, was made to bear the weight of social class, not only in being figured as the subject-ed to ideological interpellation, but also in being imagined as the agency of revolutionary change? Following Jameson's treatment of video art, Dienst begins his treatment of television by dismissing these questions; for him these are not the questions by means of which to grasp the technical production of temporality that television suggests.

In extending Jameson's treatment of video art to a treatment of television, Dienst presumes from the start that the idea of a viewer's "reading" television texts or an audience's consuming ideological images is not a good idea with which to begin when trying to understand television as a machine. Dienst's understanding of television starts, instead, with the spectacle of teletechnology's recent extension and intensification, whereby cable services, satellite systems, interactive CDs, video games, VCR innovations, and camcorders all have moved the apparatus of television beyond a broadcast model. Dienst thinks television, along with zapping, time-shifting, and engaging in multiple forms of storage and replay, has become a reference point of a vision to interface television and the computer, making use of what is described as push-pull programming, which occurs when the operation of browsing the Internet is drawn into the machine further from the user's consciousness and is offered instead as part of the program, so that what is offered is beyond the viewer's choice. Transmitting both entertainment and information, television will always be on.

Dienst argues that television, as part of an expanded and intensified teletechnology, is not to be treated as a vehicle of ideology in the domain of consumption. Television does not just support a worldwide market economy. It brings the world market wherever it goes; therefore, television represents the transnationalization of capital and the globalization of teletechnology. As Dienst puts it, "Television captures distance and defines its social territory by grounding itself as a set of material objects: it exists as a vast number of scattered machines, connected by the diffusion of a production occurring elsewhere and everywhere at once."[65]

No matter whether a transnationalized capital is described in terms of a flexible accumulation or a flexible specialization, a matter of neo-Fordism or neo-Keynesianism, the centralization of financial services and their centrality to the accumulation of wealth, as well as the displacement of human labor by technoscience as central to capitalist production, means that a transnationalized capital works on the fast, nearly instantaneous, circulation of information, money, and abstract knowledge of a globalized teletechnology. Television not only brings the market wherever it goes; it brings the market in information, money, and abstract knowledge wherever it goes.

For Dienst, television's representation of a transnationalized capital is not, therefore, a faulty representation or distorted figuration of the "third stage" of the capitalist organization of production, as Jameson would have it. Dienst is arguing that television registers the indistinguishability of a globalized teletechnology and capitalist production in late capitalism. Television is the "*Darstellung* machine" of postmodernity. Television offers itself as the machine metaphor for representing the structuration of capitalism in its effects; it also is the machine that produces the effects — that is, value. Television not only represents a transnationalized capital; it does so as a machine of production. What it produces, Dienst argues, is socialized time for exchange.

Going beyond the argument that television produces value through the sale of advertising time, Dienst argues that the time bought by advertisers is socialized time that television produces. His argument is notable for its closeness to Marx's treatment of the labor theory of value in the first book of *Capital*. As Dienst puts it:

> If the machine system of large-scale industry radically collectivized and redistributed social labor time according to capitalist imperatives, the television system now performs the same function for other segments of time: pleasure time, public or community time, household time, parenting time, childhood time, even animal and vegetable time.... Certainly, advertisers buy time, but it is socialized time. Just as the capitalist buys labor power rather than an individual's labor, so the advertiser buys a unit of social time-power — the hypothetical fusion of "free" time and "free" images calibrated in price according to estimates and averages of productivity and potential return. Television, in its fundamental commercial function, socializes time by sending images of quantifiable duration, range, and according to its own cultural coordinates.... Everybody is free to spend time in their own way only because, on another level, the time is gathered elsewhere, no longer figured as individual.[66]

It is here that Dienst returns to the viewer whom Jameson had left aside and whom Birmingham cultural critics of television had made a figure of the working class, and then the people. Although Dienst does not imagine the viewer in terms of class antagonisms, he does imagine the viewer as a worker. But the viewer is not a worker who sits before television as a mechanism of ideological instruction. Dienst imagines the watching viewer as a worker at work, the work of watching. It is not, therefore, in reading images and then consuming advertised commodities that the viewer produces surplus value. The viewer produces surplus value when he or she watches, that is, when a unit of viewing time and television image, having already been capitalized, is used up. Noting that this production of surplus value seems without effort, Dienst suggests that "the peculiar property of watching television is that time (the socialized 'free' time of viewers) enters into a cycle of value without being treated as a commodity by those (viewers) who spend it."[67] It therefore appears that television networks make value out of nothing, when in fact they " 'buy' (with images) and 'sell' (as ratings) this socialized time."[68] In such a situation, Jonathan Beller argues, labor becomes "a subset of attention, one of the many kinds of possible attention potentially productive of value"; the labor theory of value is thereby displaced by what Beller refers to as "the attention theory of value."[69]

Turning the television viewer into the watching worker, of course, raises questions about human labor and work in postmodernity. Does it not become possible to think about the television watcher as working because increasingly workers are machine watchers, or the connecting links in the machinery, as Marx described laborers? Or is it possible to think of laborers as watching television because the flows of information and images, having been machined or computerized, have displaced human labor as central to capitalist production? Has not technoscience become central to production, as a number of Marxist critics have argued?[70] If increasingly laborers are technoscientific workers, are not the workers' subjectivities the point at which capitalist production engages them?[71] Does capital need any longer to depend on the state to organize workers into a laboring collectivity, socializing them into the nation through state ideological apparatuses? Surely, in discussions about work under post-Fordist, post-Keynesian conditions of transnationalized capital, questions have been raised about the necessity of work or the meaning of work. Is it a meaningful social request to ask for more and more jobs or more and more work when machines can do the work?

Must the distribution of wealth and well-being be attached to jobs or work? Should we not be left with more time for our pleasures?[72]

What, then, of the pleasure of television watching? Dienst, for one, is not much concerned. Having set out to show television as a machine of capitalist production involved in the transfer of value to the commodified time-image unit, Dienst argues for giving up on trying to connect the pleasures of television watching with Althusser's treatment of ideology, in which the individual is hailed or unconsciously interpellated into a narrated subject position by the television message. Drawing on Derrida instead, Dienst argues that the aesthetic appreciation or cultural criticism of television can start only in recognizing that messages are not centrally disseminated or broadcast. No message is carefully targeted, so no message ever hits right on target.

Since images are the unit of value, neither narrative nor stories are necessarily or primarily the way in which the viewer and television are attached to each other. Television aims primarily to capture attention and modulate affect through a logic of exposure, over- and underexposure; television works more directly than cinema in attaching the screen/image and the body. To borrow from Beller's description, television is able "to burrow into the flesh."[73] But, as Stephen Heath sees it, this is because television is not "a subject-system," that is, a technological system understood to be perfecting the human being, serving as an extension of the human body, while maintaining the intentional knowing subject at its center and as its agency. Instead, television makes the subject only one element in a "network imagination" of teletechnology.[74]

As such, television points to and produces itself in a network of a vast number of machinic assemblages, crisscrossing bodies — not just human bodies — producing surplus value, pleasures, and signs all on one plane. As Heath puts it, television "negotiates the breakdown of the subject-system unity through the assembling of meanings, voices, sights, viewer-moments into the continuum of its functioning."[75] Television especially makes visible a certain movement in and of images that belongs to the machine's functioning, where the subject is neither origin nor end. Releasing the image from narrative, television makes it necessary to think of the image outside the subject-system.

Perhaps this is why Dienst turns to Gilles Deleuze's study of cinema, where Deleuze treats the image as nonsubjective, a matter of machinic assemblages, where there is "no representation, only images in conjunction at different angles and speeds, intersecting aspects of bodies in motion"[76] and where the viewers "must always be considered images on the same plane with the filmic ones."[77] It is Deleuze's treatment of "the time-image"[78] that Dienst finds especially interesting, because it seems to fit television's electronic images even better than it fits filmic images. Dienst goes so far as to argue that Deleuze's treatment of the time-image is drawn from outside cinematic thought by television. He thereby makes it possible to think the thought of television in all that Deleuze describes when treating the time-image, the virtual, and nonsubjective memory.

Nonetheless, for Deleuze the time-image is realized in film, especially the films of De Sica, Passolini, Rossellini, and Godard, but also in the thinking of Nietzsche, Peirce, and Bergson. Taken together, all of these works lead Deleuze to argue that the time-image is not a matter of representation, but rather a matter of the relations of visibility, of time and space in a particular technical substrate. Deleuze contrasts the time-image of post–World War II avant garde film with "the movement-image" of pre–World War II film. The most important variant of the movement-image is the action-image. It is in the action-image that the time-image is contained; it is fixed to the movement of a human "sensory-motor schema," which, as Deleuze argues, fixes time to the unfolding movement of a linear narrative. Although Deleuze takes no interest in the work of feminist film theorists, there is a similarity between their description of the oedipal narrative logic of classical Hollywood cinema and Deleuze's description of the narrated action-image. But unlike feminist film theorists, Deleuze emphasizes the way narrative modulates, even domesticates, the movement of images or even the movement of machine time. He contrasts the action-image of pre–World War II film and classical Hollywood film to the time-image of post–World War II avant-garde film; he proposes that the time-image first presented in post–World War II avant-garde film is an image released from narrative.

As such, the time-image gives a direct image of time. No longer deployed to make something seen or to make a viewer see something, the image makes time visible in its own movement and without appearing as a movement aberrant to narrative. As Deleuze puts it: "Movement is no longer simply aberrant, aberration is now valid in itself and designates time as its direct cause. 'Time is out of joint': it is off the hinges assigned to it by behavior in the world. It is no longer time that depends on movement; it is aberrant movement that depends on time."[79]

For Deleuze, what is remarkable about the time-image is that it shows that time does not simply belong to the subject any more than thought, desire, or the unconscious does. Deleuze instead follows Henri Bergson, arguing that "the only subjectivity is time, non-chronological time grasped in its foundation, and it is we who are internal to time, not the other way round. . . . Subjectivity is never ours, it is time, that is, . . . the virtual."[80] Deleuze also follows Bergson in his treatment of the virtual. For Deleuze, as for Bergson, the virtual is to be contrasted with the actual rather than the real. The virtual is real; the virtual coexists with the real. The virtual is never realized; instead it calls forth actualization, but the actualized has no resemblance to the virtual. Actualization out of virtuality is creation out of heterogeneity or pure difference. The virtual-actual circuit, therefore, is different from the possible-real circuit. The real is related to the possible by resemblance. The possible anticipates the real, or, as Deleuze suggests, the real "projects backwards" to its possibility as if always having been.

Unlike the real, the actual is invention. Actualization is not a realization of possibilities. Actualization is not a specification of a prior generality. Actualization is an experiment in virtuality, an effecting or materializing of a virtual series. It is a divergence to the new or the future. The thought of the virtual-actual circuit makes it possible for Deleuze to elaborate the relationship of images and memory in terms of what he calls "the crystal image," where the past and the present of temporality are visible outside the subject's consciousness. Just as the time-image shows time in its own movement, the crystal image points to a memory store outside the subject's consciousness. Deleuze's description of the crystal image is striking for its positing series or channels of images out of which each image surfaces.

In contrast to "the organic image," the crystal image, Deleuze argues, has two sides at once. It turns on itself, divides in two: "it is a perpetual self-distinguishing, a distinction in the process of being produced."[81] The distinction always being produced is between the present and the past, the actual and the virtual. That is to say, for there to be past, the image must be actually present and virtually past all at once. The past and present are not moments such that the latter follows the former. They coexist: the present does not cease even as it passes and the past never ceases to be, even as presents pass through it. The present is constituted as past when it is constituted as present. As Deleuze puts it: "For every present there corresponds a vertical line which unites it at a deep level with its own past, as well as to the past of the other presents, constituting between them all one and the same coexistence, one and the same contemporaniety, the 'in-ternal' rather than the eternal."[82]

Here Deleuze refers to "the past in general" — not as a psychological matter, but as having ontological significance. The past in general makes all pasts possible; it makes possible the passing of all pasts into the present by memory. Deleuze goes on to argue that the past in general is where consciousness goes to look for "recollection images or this reverie that it evokes according to its states."[83] Recollection images, like dream images, are virtual images; to actualize recollection images, consciousness leaps to one of the levels of the past, all of which includes the past in general or all the pasts of other presents.

Although each level psychically repeats the past in general, each level is contracted around "variable dominant recollections." So when actualizing a recollection image, its level is actualized along with it because it is the level that is explored for a recollection image. Deleuze refers to Bergson's notion of "sheets of past" to describe where one finds recollection images; sheets of past support the invention of memory. Deleuze also refers to Bergson's notion of "peaks of present," which are found rolled up in an event as its event-ness — that is, its present-presentness and its past-presentness; event-ness allows every new present also to be an image belonging to a sheet of past.

If Deleuze's treatment of the time-image, virtuality, memory, and the past in general is meant to get at what is internal to time, it also seems to be drawn to the thought of teletechnology or to the technical substrate that teletechnology gives to unconscious memory. This surely is what Dienst proposes. If Deleuze's return to Bergson is remarkable for the way it is able to make Bergson's thought about image and time visible as the time-image of post–World War II avant-garde films, Dienst's return to Deleuze's treatment of the time-image is just as remarkable for the way it shows the time-image to be televisual, indeed, to be common to television. Dienst makes Deleuze's thought of the crystalline time-image and the virtual-actual circuit into a zapping revelation of television's deep flat images, made up of what Dienst calls "a cleaving force."

This force refers every image "not only to the innumerable points of visibility called viewers but also to other streams of images unseen, which nevertheless share the same moment and which always stand ready to emerge into a new present."[84]

Dienst draws both the notion of "sheets of the past" and the notion of "peaks of the present" into a description of the two predominant movements of television's imaging, what he refers to as "still" imaging and "automatic" imaging.[85] The former is used when television switches from one image to the next, each turning over and disappearing from view, slicing off images that not only designate the past, but give a sense of the past in general, that is, as an endless resource of virtual images. The latter, automatic imaging, is used when an image is turned on and left on, making visible the camera's stare. This image is anticipating a future; it is an image "waiting for its events to happen." Automatic timing opens up to the event-ness of the virtual-actual circuit. For Dienst, television's imagining of both still time and automatic time allows for the capitalization and consumption of commodified units of image and time, befitting television's economic demand to be on everywhere and all of the time. As Dienst puts it: "Whereas automatic time demands that we keep watching, still time demands that we keep switching; driven by these two pressures, the image on screen extends its claim over other images, near and distant, already past and yet to come."[86]

Turning Deleuze's treatment of cinema into a tele-vision, Dienst, however, leaves off thinking about the relationship of television and Deleuze's proposal that the notion of the past in general has ontological implications — that the notion of the past in general is meant to displace Being or that it is "identical with being in itself," as Deleuze puts it. Dienst is not that interested in Deleuze's treatment of the past in general and nonsubjective memory; he is not that interested in Deleuze's proposal that there is an unconscious or pure recollection of all pasts, that there is an unconscious outside the subject that conditions the possibility of the productivity of the individual subject's unconscious in the actualization or repression of recollection images, or that this unconscious allows for folds in flows of matter and energy, so that a plane of consistency can be unrolled in desiring production. Dienst is uninterested in the way Deleuze draws thought closer to an ontological perspective that brings Being down into finitude or the finite forces of mattering, an ontology that Deleuze himself characterizes by quoting Foucault: "This ontology discloses not so much what gives beings their foundation as what bears them for an instant towards a precarious form."[87]

Dienst does not end his treatment of television with a consideration of the ontological implications of Deleuze's thought; he ends instead with a complaint that television stops thought. He argues that because television has recontained the force of the time-image in its programming, the time-image no longer can surprise as Deleuze imagined it did in avant-garde film. Dienst even argues that if we want to look for the future, "our eyes ought to be trained not on television but on the active and critical powers of thought."[88] But what is this turn against television so suddenly after Dienst himself has shown the way television has drawn the critical and active powers of thought toward it? Is not this last-minute turn against television and its programs made in behalf of Dienst's commitment to Marxist cultural criticism? Is there not a turn against television for what it has done to culture in its doing its part in the transnationalization of capital, that is, in its socialization of free time?

No doubt this is so. Dienst ends his treatment of television by going back to culture and the capitalist organization of production, its relationship to the late twentieth-century globalization of teletechnology and to the transnationalization of capital in postmodernity. Dienst ends his treatment of television with a criticism of late capitalism befitting Marxist cultural criticism; he turns to Deleuze's treatment of "control societies." But Deleuze's treatment of control societies not only pushes Marxist cultural studies beyond the notion of a structured mode of production in its effects; it also returns Marxist cultural studies to the ontological implications of Deleuze's treatment of unconscious memory.

Deleuze's treatment of control societies points to the social situation of societies in postmodernity; it points to the reconfiguration of the arrangement of national and family ideologies, the state and the economy, the public and private spheres presumed in modern western discourse of the ideological apparatuses. Control societies, Deleuze argues, go beyond what Foucault describes as disciplinary societies, where the "governmentalization of the state" allows the state to extend its disciplinary practices through social institutions such as the church, the school, the prison, the family, the union, the party, and the media — what Foucault refers to as the "enclosures" of civil society.[89] This interpenetration of state and civil society, as Michael Hardt points out,[90] still is characterized, albeit weakly, by a politics of representation, by the ideological production of subject identities. But control societies are not. In control societies there is a smoothing out of the arrangement of family and national ideologies, the state and civil society, and private and public spheres beyond that seen in disciplinary societies, thereby making possible the dispersion of control throughout social space, no matter whether and how the arrangement of family and national ideologies, the state and civil society, and the public and private spheres is being reconfigured.

As Hardt suggests, Deleuze's treatment of control societies not only fits the global extension of teletechnology in the late twentieth century, but it also befits a capitalism where human labor is no longer central to production, so that human labor need not be collectivized by the state and socialized through subjection to family and national ideologies. A politics of representation is thereby thrown into crisis and the ideological construction of subject identities made frenetic unto exhaustion. It would seem that it is in control societies that television need not and does not function as a technology of the subject; neither is it primarily or simply a vehicle for national and family ideologies. Television, therefore, calls into question the social structural. Or, as Brian Massumi has argued when commenting on teletechnology and control societies, "If all this adds up to a structure, it is a dissipative structure combining a multiplicity of periodicities in a fluctuating set of highly complex differentiations that are locally implanted following divergent patterns, but resonate globally."[91]

Under these conditions it would seem that resistance to the organization of capital surely must involve thinking of capitalism other than as totality; nor should its history be thought of as becoming universal. Now when capitalism appears to be transnational, indistinguishable from a globalized teletechnology, to think of capitalism as other than totality or a unified identity provides lines of flight. Now when capitalism appears to be transnational, indistinguishable from a globalized teletechnology, its critics need rather to think most about different "capitalizations whose antagonisms are irreducible,"[92] as Derrida puts it, where the social and political situation of each capitalization makes a difference, where "the differentiating process of advanced globalizing capitalism" itself provides what Lisa Lowe and David Lloyd describe as "the potential to rework the conception of politics in the era of transnational capital itself."[93]

But to point to the specific social and political situations of the trans-nationalization of capital as providing lines of flight is not meant only to undermine a definition of capitalism as unified and everywhere the same by highlighting what is not capitalism. This is what J. K. Gibson-Graham do[94] when they focus on nonmarket exchange networks, barter systems, noncommodity production, even family-based relations of commodity production and exploitation and then go on to refuse to treat any of these as simply not-yet transnational capitalism. But to think of capitalism as other than unified also means thinking of some exchange markets as more repugnant than others, and therefore some capitalizations as themselves preferred lines of flight.

All this is to propose thinking of capitalist production in global and local terms or in nonsystemic systemic terms, if not only for the political and cultural differences of localized situations of transnational capital, then for the inextricability of globalized teletechnology and trans-nationalized capital, such that it is impossible to define either as the condition of possibility of the other. Neither, therefore, is a unity or a totality. Each is the other's internal difference, opening each to *différance*, to the finite forces of mattering. Thinking of capitalist production and teletechnology in these terms opens one up to the thought of machinic assemblages, allowing cultural criticism to treat different amalgams or modules of capitalist production and social spaces, that is, as reconfig-ured arrangements of family and national ideologies, the state and civil society, and the private and public spheres.

The thought of machinic assemblages thereby becomes available for treatments of subject identity and unconscious memory in relationship to the glocalization of world cultures. Deleuze's effort to ontologize a past in general gives support to this effort. It meets the teletechnological, giving thought over to an unconscious memory that is neither individ-ual nor merely collective; rather, it is nonsubjective and not necessarily human. As such, the thought of unconscious memory is opened to a rethinking of desire, bodies, and sexuality, responding to the question: How are *différantial* relationships of human and machine, nature and technology, the real and the virtual embodied, and what can these bodies do? It is in this context that the turn in feminist theorizing from a Lacanian treatment of unconscious desire to a treatment of queer bodies in a transnational frame would seem to be a matter of thought reaching for the ontic. It is to feminist theorization of bodies and unconscious memory outside the oedipal narrative that I now want to turn.

True Confession

My nose up close to the screen, I waited for the hand that would pull
back the wooden door, exposing the priest's ear and his other hand,
sweeping up and then down, cutting through the air,
making the sign of the cross.

It was the sign that I was to begin.

At seven, I made my first confession.
I confessed.
I confessed to adultery.

The confessional was no bigger than a closet except there was no ceiling.
Built into one of the church's spires,
the confessional seemed opened to the heavens.
Still, it was close, and while waiting, I felt my knees melt
into the red leather kneeler.
I slid my hand over the wooden frame
around the screen.
Mahogany,
warm to the touch, having been heated up
with the passions left here with secret tellings.

Bless me father for I have sinned.
I have committed adultery.
My body was trembling with shame and horror.
I had disobeyed the sixth commandment,
the no-sex commandment:
Thou shalt not commit adultery.
But my shame and horror would only be turned to confusion

when the priest tried to convince me I had committed no sin —
that I did not know what I was saying.
The word would not make safe passage
across the screen
from profane lips to the sacred ear of the father.
No absolution.
He offered only to fix me with right words, fixing my sin to silence.
No way to move from transgression
to promises of never,
never again
and forgiveness.
Invented in the deaf ear of the father,
I was sentenced to right words.
Here was the domain of the sayable within which I began to speak[1]
the words of passionate attachment to soulful subjection
that becomes a habit.
First one kneels and then one believes.[2]
It is a habit made unconsciously into a whole cloth
for sewing whispered words into pleats and folds of silence.

In the year before they make solemn vows, novices live in silence,
praying, cleaning, and sewing.
Just below the neckline of the long black dress,
the pleats are tacked down,
fourteen in all,
each a quarter of an inch wide.
The leather cincture, or belt, kissed with lips
already moving in silent prayer,
is wrapped around the pleats
holding the black woolen cloth close to my waist.
And the dead crucified Christ body, all but penis naked,
hung from a delicate rope around my neck,
is tucked behind the belt and made to rest just below my heart,
safe and saved,
as if in my body to live forever the unforgiven sin.

I have committed adultery.
I have sinned.
Bless me father, each novice intones as she is handed a small bundle of
black veiling. As the priest blesses and then fixes the veil on my head,
fifteen decades of rosary beads are attached to the leather belt and let to
fall about my right side.
I hear myself singing the hymn so long practiced.
All the joys that the world has to offer,
I now reject with gladness

because I have seen the face of Christ,
my lord and redeemer.
And I have heard the call
in my ear, the words of God
and mine: I promise poverty,
chastity, and obedience.

This is faith in fiction, a pure fiction.
A veiling of the truth,
a dream fugue in the head of a teenage girl,
making costume with bits of Joan of Arc armor
and a virgin's wedding gown,
trimmed with bridal lace and crowned with white gardenias.
Their last sweet bitter odor fills my nostrils and lingers
as my hair is shorn away
and the curls fall around my feet.
I look down.
I look up, as my head is pulled back and wrapped tight in starchy white.
Made to look straight ahead,
I could only see them from the bottom up,
from my cot, which was pulled out each night
and placed at the foot of my parents' bed.
My sister's was placed along the side.
We slept there until she was fifteen
and I was eleven,
in that bedroom
in the small three-room apartment
where my father, my mother, my sister, and I lived.

It was in second grade when I vowed to become a nun. It might have
been because my teacher's name was Sister Patricia. But I think it was
because I won the spelling bee and was given as my prize
the book Sister Patricia had been reading to us —
The Great Women of the Church.
Women who had beheaded kings with the sword of truth.
Women who had honored vows to a husband,
leaving homeland to follow him.
Women who had forsaken children and simple pleasures.
Women who had sinned but promised to live forever repenting.
Women who refused to eat, to rest, to speak, to think
but only one thought of God.

I read the words of the book again and again.
They gave shape to my fevered imaginary,

and form to my impassioned young being,
mixing my futures into the past forever,
giving my will to live over to the repeated
pleasurable painful acts of self-renunciation.
My body took the face of solitude,
a turning inward that goes beyond the self,
with the solitary aim of being only among
the poor and the sinned against,
the hopeless and the depraved,
the ugly and the humbled.
Their downcast eyes between my young wide-open hands.
Sorrow and pity, inequality, and mercy in closed cycle repeated.
Repeated,
repeated until there was a fissure, a breach, and then openings everywhere.
One took me to the outside.

It was sometime after when I first saw it.
I was in a hotel room with a lover when I first saw it.
Perhaps it was a gesture, a shift in the arm,
a lifting of the thigh.
Perhaps it was the sexual position
that made me look at him, that made me watch him.
I did not always watch him.
It was only that last time,
when I felt the distance of the far-off country
from which my lover came,
when I saw that his family and his political commitments
would prevent him from having any other lover
but one so different as me—his exotic other.
I reached my hand beyond his head
and pressed it up against the headboard.
Mahogany.
It was then that I saw it—
my parents' bodies, flashing before my inward eye.

I could only see them from the bottom up
as the sheet was thrown off and in slow-motion waves
landed on my cot.
I could only see them from the bottom up, and then
I could see nothing at all
but bits of metal entwined with twigs and flowers,
braids of fine chain and delicate rope
and bright lights.

My parents' bed was made of mahogany, a rich brownish-red wood.
It was beautifully engraved, swirling into pagodas and flowers —
intimating secret lovers.
Above the bed was a large crucifix.
The wood was twisted to look like tree bark.
The body, cast in burnt bronze, glowed with excruciating pain.
At night,
the wood would let loose the nails
and the body would begin to slip.
The white loin cloth would float up above the thorn-crowned head,
as he fell,
descending into his mother's outstretched arms.
She shrouded him and laid him to rest for a moment,
the pietà, and then
their bodies arose, flashing before my inward eye.
I could only see them from the bottom up,
and then I could see nothing at all
but my hand
writing on white sheets that would not be stilled.
Sentenced to write words,
without forgiveness.

Notes

1. Judith Butler.
2. Judith Butler quoting Louis Althusser quoting Pascal.

CHAPTER THREE

Queer Desire and the Technobodies of Feminist Theory

Unconscious Desires without a Transcendent Phallicity

Over the past three decades, feminist theorists have persistently questioned the "naturalization" of the woman's body; they have argued that it is a masculinist strategy to authorize the privilege given to reason in the modern western discourse of Man. Yet feminist theorists also have been suspicious of postmodern strategies for "denaturalizing" the woman's body; they have claimed that often these strategies are masculinist as well. For example, in her feminist treatment of bodies and technologies, Ann Balsamo[1] echoes Nancy Hartsock's[2] often repeated complaint about postmodern theory—that is, that it put the subject under erasure just as women were attaining a subject status and voicing their subjective identities. Balsamo, however, aims her complaint more specifically at what she refers to as "the postmodern theory of the body" as it is elaborated in the works of Jean Baudrillard, Arthur Kroker, Gilles Deleuze, and Félix Guattari; she questions whether it is "ironic that the body disappears in postmodern theory just as women and feminists have emerged as an intellectual force within the human disciplines?"[3]

Suspected of being informed with a masculine desire to deny the body or to disavow its imperfections and limitations through technological enhancement, the postmodern treatment of bodies as machinic assemblages, technobodies, and cyborgs has appealed only to some few feminist theorists. Only some feminist theorists have wanted to circulate what N. Katherine Hayles refers to as the "metaphoric network" that is borne of the development of teletechnology in the late twentieth century and bears profound implications for bodies as well as relations of space and time.[4] If I am one of these feminist theorists, it is because I want to follow unconscious thought to teletechnology as it crosses over feminist theory, drawing out from it its ontological implications. I want to propose that feminist theory elaborated over the past three decades has ontological implications along the lines Donna Haraway first suggested when, in her 1985 feminist manifesto, she claimed: "The cyborg is our ontology; it gives us our politics."[5]

Although rightly criticized for its masculinist uses in the discourses of science, militarism, and popular culture, the figure of the cyborg, when deployed for feminist ends, can only trouble the presumption of any simple identification of technology with a disavowed unconscious desire for phallicity referred to men or male theorists only. Therefore, the deployment of the cyborg in feminist theory has been joined with the queering of unconscious desire by feminist theorists. Like the feminist deployment of the cyborg, queer theory has emphasized the complexities and difficulties of unconscious identification, what Eve Kosofsky Sedgwick has described as the "intensities of incorporation, diminishment, inflation, threat, loss, reparation and disavowal" with which identifications are "sufficiently fraught."[6]

Like the feminist deployment of the cyborg, the feminist queering of unconscious desire is a deconstruction of the subject, but not one meant merely to dismiss the subject's agency. Drawing on psychoanalysis, queer theory means rather to question the unity of the subject's identity and the simplicity of its unconscious identifications; queer theory deconstructs the subject by drawing it back to the fantasmatic construction of the body. But in doing so, queer theory also has called into question the psychoanalytic configuration of the imaginary, the symbolic, and the real. It is in this sense that queer theory has drawn out the ontological implications of feminist theory for rethinking nature and technology, the body and the machine, the real and the virtual, the living and the inert as *différantial* relationships rather than as oppositional or dialectical ones.

In the next chapter I treat the cultural studies of science in which Haraway's feminist deployment of the cyborg has had its greatest influence. But in the first and second parts of this chapter I want to treat the writings of Judith Butler and Elizabeth Grosz, who in rethinking the body have drawn on male theorists such as Deleuze and Guattari, Jacques Lacan, Michel Foucault, and Jacques Derrida with much less suspicion than many feminist theorists have expressed. In doing so, both Butler and Grosz, I want to propose, have been drawn by the unconscious thought of teletechnology to the ontological implications of feminist theory; their works make more visible the way in which feminist theory has been profoundly linked to the deconstruction of the opposition of nature and culture as it has been deployed in modern western discourse of Man. But neither Butler nor Grosz merely dismisses nature; rather, both rethink nature along the lines suggested by Haraway when she proposes that nature is an "achievement among many actors not all of them human, not all of them organic, not all of them technological"; rather nature is "a construction among humans and nonhumans."[7]

Although neither Butler nor Grosz has explicitly or systematically theorized technology, their treatments of bodies not only deconstruct the opposition of nature and culture in relationship to the human body. But, as Pheng Cheah has suggested, their works also contribute to the deconstruction of the opposition of nature and culture in relationship to matter, making more explicit the dynamism of matter.[8] Butler and Grosz, thereby, bring feminist theory to its ontological implications pertaining to bodies other than the human body, what Grosz refers to as "volatile bodies."

Although Butler and Grosz have borrowed from postmodern male theorists, both are, however, as much indebted to feminist theorists and to the feminist treatment of the sexed body in psychoanalytic terms. Both Butler and Grosz draw on feminist theorists who have elaborated the relationship of fantasy and sexual difference in terms of the oedipal logic of narrativity. They both draw on Lacanian psychoanalysis, but both also struggle to disconnect unconscious desire from the oedipal narrative. They do so by rethinking the constitution of sexed bodies and by treating bodies other than those figured in and prescribed by the oedipal logic of the dominant cultural narrative.

Taken together, Butler and Grosz have opened feminist theory to explore what is proposed by Haraway when she remarks that "the most terrible and perhaps the most promising monsters in cyborg worlds are embodied in non-oedipal narratives with a different logic of repression, which we need to understand for our survival."[9] But both Butler and Grosz interrogate the notion of monstrosity with much more care than even Haraway does. After all, they think to the outside of the oedipal narrative through the abject, the marginal, and the perverse; that is, they rethink bodies by questioning how only some bodies come to matter while others are made monstrous, unintelligible, even "unlivable," as Butler puts it. Whereas Rosi Braidotti has shown that the feminine figure of the mother usually is implicated in treating the body in the machine metaphors of the monstrous,[10] Butler and Grosz show that the figures of the lesbian, the homosexual — the queer — are also implicated.

The deconstruction of the opposition of nature and culture, therefore, poses a certain difficulty for feminist theorists because the opposition of nature and culture already is sedimented with the rhetoric of sexual difference, that is, with the opposition of the feminine and masculine figures of sexual difference. These figures necessarily will be repeated in order to be worked through in the deconstruction of the opposition of nature and culture; no doubt this repetition will likely have unintended consequences for the reconfiguration of masculinity and femininity, heterosexuality and homosexuality. Furthermore, sexual difference is not all that is at issue in the deconstruction of the opposition of nature and culture. Also implicated are the differences of race, class, ethnicity, and nation; these too have been deployed in the opposition of nature and culture in the modern western discourse of Man. It is in terms of these differences that the deconstruction of the opposition of nature and culture becomes linked to issues of racism or sexism as well as to neocolonialism, the globalization of teletechnology, and the particular social and political situations of the transnationalization of capital. That queer theory and postcolonial theory increasingly have been drawn closer together is no doubt a response to the need to think of bodies or bodily matter through the deconstruction of the opposition of nature and culture across the local situations of the globalization of teletechnology and the transnationalization of capital in neocolonialism.[11]

The ontological implications of feminist theory that Butler and Grosz make more explicit, I want to suggest, crisscross with the political effects of thinking of feminist theory in a transnational frame. Butler's and Grosz's efforts to take unconscious desire beyond the oedipal narrative opens feminist theory to what Gayatri Chakravorty Spivak has referred to as "other indigenous regulative fictions of psychobiography"[12] that are linked to the reconfiguration of that arrangement of family and national ideologies, the state and civil society, and the public and the private spheres presumed in the modern western discourse of Man. Although this configuration of social spaces now is being profoundly troubled—deterritorialized and reterritorialized—in the globalization of teletechnology and the transnationalization of capital, it also has long been the object of feminist criticism.

Feminist theorists especially have criticized the ideological separation of the private or domestic sphere from the public sphere; they also have elaborated the implications of the ideology of separate spheres for the engendering of cultural, social, and political economic relations in terms of which not only subjects and bodies are constituted, but also political agency is determined. The criticism of the separation of the private and public spheres in feminist theory, however, has been limited and has had to be opened to rethinking the configuration of state and civil society, family and national ideologies, and the public and private spheres in a transnational frame. In this sense the intersection of postcolonial theory and queer theory not only shows feminist theory reaching for an ontological perspective for rethinking the oppposition of body and machine, nature and technology, and the real and the virtual as *différantial* relationships; it also shows feminist theory drawn to rethinking the configuration of social spaces in terms of which political agency has been constituted in modern western thought, that is, rethinking political agency in terms of the glocalization of cultures in a transnational frame.

Butler's and Grosz's treatments of bodies especially make clear that the reach of feminist theory to rethink both an ontological perspective and a political perspective has engaged feminist theory in the reformulation of materialism or a materialist criticism of nature and culture in the age of teletechnology. Indeed Butler has argued that her treatment of bodies is not meant to be an ontology, but rather aims to politicize ontology, "to recirculate and resignify the ontological operators, if only to produce ontology itself as a contested field."[13] Grosz's treatment of the body, although more explicitly aimed at rethinking an ontological perspective, is nonetheless primarily concerned with politicizing ontology as well. Like Butler, Grosz refuses to put power relations outside bodily matter. In this sense both Butler and Grosz outline a materialist approach to culture and nature befitting the age of teletechnology. Taken together, their work is part of the rethinking of materialism begun with Louis Althusser's treatment of ideology.

Butler traces one thread of her theoretical lineage back to that moment when feminist theorists first followed Althusser in engaging Lacanian psychoanalysis as a way to think of the ideological construction of the subject in a given capitalist mode of production. Butler shares this theoretical lineage with early feminist film theorists on whom she has drawn in her treatment of sexed bodies. She has borrowed from their revisions of Lacanian psychoanalysis and has pointed especially to the importance of Jacqueline Rose's rereading of Lacan, in which Rose underscores Lacan's insistence on the failure of the imposition of the oedipal law in the construction of subject's identity.[14] There also is Butler's reference to Kaja Silverman's treatment of cinema in terms of the oedipal logic of the dominant narrative of western modern capitalism.[15]

Butler has signaled her agreement with Silverman's effort to "pry" the prohibition against incest away from the oedipalized law of the phallus; she has refused, as Silverman also does, to conflate the "lack of being" incurred with the entrance into language or the symbolic order with the lack of the phallus or castration. Butler appreciates Silverman's argument that although the fantasmatic is shaped by the imposition of the oedipal law of the phallus, unconscious fantasy also is informed with the failure of the oedipal law, so that an unconscious resistance to it is to be expected. But Butler has questioned, more than Silverman has, whether the fantasmatic elaboration of unconscious resistance to the law of the phallus does not also reproduce the law; she asks, therefore, whether such resistance is enough to carry out Silverman's proposed political agenda, that is, to pry the incest taboo from the law of the phallus, thereby disconnecting the loss of being in language from an oedipalized castration.

Like Silverman, Butler treats the oedipal narrative, but she treats it in terms of the unconscious or fantasmatic construction of sexed bodies. She therefore takes up the repetition compulsion that she argues is central both to the fantasmatic construction of sexed bodies and the reproduction of the law of the phallus; at the same time, she rethinks the repetition compulsion in terms of Foucault's treatment of power/knowledge and Derrida's treatment of *différance*. Butler thereby collapses the opposition of unconscious fantasy and bodily matter, with ontological implications befitting the age of teletechnology.

To begin, Butler rethinks the oedipal law of the phallus along lines suggested by Foucault; that is, she treats the oedipal law not only as a juridical law, but as a generative one. But where Foucault recognized that the oedipal law constitutes a domain of cultural intelligibility in terms of which bodies are constructed, Butler argues that he does not recognize, as Derrida does, that "principles of intelligibility require and institute a domain of radical unintelligibility."[16] Butler shows that bodies prescribed by the law of the phallus are haunted or encrypted with those bodies that the law excludes from "existence" that is, those bodies for which gender does not follow from sex and the practices of sex do not follow from either sex or gender — that is to say, queer bodies.

Following other feminist theorists who also have engaged Lacanian psychoanalysis, Butler argues that the oedipal law of the phallus imposes sexual identity by prohibiting the incestuous heterosexual object choice — the mother for the boy and the father for the girl. But she also argues that along with this prohibition, let us say even prior to it, there is a prohibition of the incestuous homosexual object choice — the mother for the girl and the father for the boy. The loss of the homosexual incestuous object, unlike the loss of the heterosexual incestuous object, is denied completely, so that what Butler calls "the modality of desire," or what Freud refers to as "the sexual aim," also must be denied. For example, in the case of the boy, not only is the father tabooed as an object choice, but the sexual aim, or the act toward which the sexual drive tends, also is tabooed; in this case the tabooed aim may even be figured as feminine, that is, treated as what a male should not desire to do at all because it is what only a female desires to do or have done to her.[17]

Because the incestuous homosexual object choice and the homosexual aim both are denied, Butler argues that they cannot be grieved, and therefore the loss cannot be internalized and displaced onto others. Rather than grieved, the loss is "melancholically incorporated" and thereby kept alive in and as part of the one who cannot grieve. As Butler puts it, there is an "encrypting of the loss in the body." It is as if "the body is inhabited or possessed by phantasms of various kinds."[18] In the case of the boy, both his father and his desire for the father are kept living by encrypting the deadening loss on the child's body. The child's body thereby becomes a male body: "Incorporation literalizes the loss on or in the body and so appears as the facticity of the body, the means by which the body comes to bear 'sex' as its literal truth."[19] Butler argues that sexual identity is produced on the skin, as if an image or surface of an inner depth or a "true" core of sexuality.

For Butler, therefore, the sexed body is an effect of what she refers to as a "literalizing fantasy." As she puts it, "The belief that it is parts of the body, the 'literal penis,' 'the literal vagina,' which cause pleasure and desire — is precisely the kind of literalizing fantasy characteristic of the syndrome of melancholic heterosexuality."[20] A literalizing fantasy works as a form of forgetfulness; it "forgets the imaginary and with it an imaginable homosexuality." It is in these terms that Butler argues that the oedipalized sexed body is a performance involving the compulsive repetition of unconscious forgetting, which, however, also gives the possibility for difference in the variations of performance, and therefore gives the possibility for change.

Butler's notion of performance, so often understood as an activity of intentional gender role-play or the intended transgression of gender role requirements, is not this at all. It rather follows thought to the indistinguishibility of the body and unconscious fantasy, of matter and the image. Not only does the notion of performance refer to the body as an imaginary matter, a matter of an unconscious repetition compulsion. It also relocates the matter of the unconscious in the interval between repetitions. As Butler puts it: "If every performance repeats itself to institute the effect of identity, then every repetition requires an interval between the acts, as it were, in which risk and excess threaten to disrupt the identity being constituted. The unconscious is this excess that enables and contests every performance, and which never fully appears within the performance itself."[21] Butler's notion of performance suggests that bodily matter is dynamic, more an event or a matter of temporality.

Here, of course, Butler is drawing on Derrida and drawing the unconscious repetition compulsion to *différance* or pure repetition. In this sense Butler argues that the unconscious is to be located "within a signifying chain as the instability of all iterability." The unconscious, therefore, "is not 'in' the body, but in the very signifying process through which that body comes to appear; it is the lapse in repetition as well as its compulsion, precisely what the performance seeks to deny, and that which compels it from the start."[22] In drawing the unconscious back to *différance,* Butler allows for a more general unconscious than the Freudian or Lacanian unconscious. But this rethinking of the unconscious presumes the deconstruction of the psychoanalytic configuration of the imaginary, the symbolic and the real.

It is the opposition of the imaginary and the symbolic that, Butler proposes, disallows an imaginable homosexuality in the construction of an oedipalized melancholic heterosexuality. Therefore, Butler sets out to deconstruct the opposition of the imaginary and the symbolic by elaborating the fiction of the lesbian phallus, making it possible to rethink bodies other than those constituted through an oedipalized heteronormativity, and therefore also to think regulatory psychobiographic fictions other than the oedipal narrative. To elaborate the fiction of the lesbian phallus, Butler traces the way in which the phallus becomes a transcendental signifier of the oedipal law, such that the imposition of the incest taboo becomes an imposition of sexual difference reduced to an opposition of phallic and castrated.

Butler begins with Freud's troubled treatment of narcissism. In "On Narcissism" Butler notices that Freud proposed illness, but also hypochondria, sleeping, and dreaming, as an example of a narcissistic libidinal self-investment. Butler especially emphasizes how Freud saw the same connection between actual pain and erotic self-investment as he did between imaginary pain and erotic self-investment. Butler concludes that, at least at first, Freud proposed that narcissistic erotic self-investment functions as an imaginary construction of any and every body part. Indeed, as Butler sees it, the "body part is delineated and becomes knowable for Freud only on the condition of that investiture."[23] She quotes Freud: "We can decide to regard erotogenicity as a general characteristic of all organs and may then speak of an increase or decrease of it in a particular part of the body."[24]

But Butler also reports that Freud quickly and defensively retreated from his own first thoughts about narcissism, especially because they seemed to elide the difference between the imaginary and the symbolic in relationship to the body. Butler follows Freud to his discussion about a genital organ, seemingly the penis, which he proposed is exemplary of a body part that, although not ill, can be made sensitive to pain through a state of erotic excitation. Not only did Freud make this organ the model or prototype of all erotogenicity; he wound up reducing all the other examples of eroticized body parts, such as those produced in illness and in hypochondria, to the prototypicality of the penis. In other words, the penis became the transcendental phallus of the oedipal law; it became the transcendental signifier of sexual difference, turned into the opposition of phallic and castrated. As Butler puts it: "The Phallus is then set up as that which confers erotogenicity and signification on these body parts, although we have seen through the mytonymic slide of Freud's text the way in which the Phallus is installed as 'origin' to suppress the ambivalence produced in the course of that slide."[25]

Of course, Butler is suggesting that Freud theoretically produced the phallus as a transcendental signifier by libidinally investing the penis; in doing so Freud defensively reproduced the narcissistic process of the imaginary construction of a body part, a process that he himself first described. But in denying that he was doing so, Freud could make the phallus itself the very mark of the opposition between the imaginary and the symbolic, as well as the mark of the opposition between the narcissistic and the social or culturally normative — each elaborated in the figures of phallic masculinity and castrated femininity. If the fiction of the lesbian phallus refuses the opposition of the imaginary and the symbolic, the narcissistic and the social, it is because it allows one figure both to "have" the phallus by which masculinity is marked and to "be" the phallus (for the other still threatened with castration) by which femininity is marked. In collapsing the opposition between having and being the phallus, the lesbian phallus allows erotogenicity to be a property belonging to no particular sexual identity nor to any bodily organ, being defined instead by "its plasticity, transferability, and expropriability."

Having begun the deconstruction of the distinction between the imaginary and the symbolic in Freud's treatment of narcissism, Butler turns next to Lacan's treatment of the body in two of his essays — the one on the mirror stage and the other on the meaning of the phallus. Butler notices that Lacan, like Freud, vacillated over the meaning of the phallus. In his treatment of the mirror stage, Lacan proposed that against the infant's experience of being a body-in-bits-and-pieces, the mirror image offers an idealizing image of unity, what Butler describes as "an idealization or 'fiction' of the body as totality and locus of control."[26] But Lacan not only argued that the mirror image is a psychically invested projection through which the morphology of the body is produced. He also argued that the ego is formed through identification with the image or the imaginary bodily morphe. Therefore, Butler emphasizes that the bodily ego, rather than being "a self-identical substance," is a "sedimented history of relations," locating its center outside in the image.

Lacan also suggested that as an imaginary bodily form the ego is formed in and informs the distinction of the interior and exterior of the subject's identity. That is to say, the image not only gives bodily form to the ego; it also establishes perceptual objects as external objects. Butler quotes Lacan's conclusion: "On the libidinal level, the object is only even apprehended through the grid of the narcissistic relation (of ego to the image)." To this she adds: "This claim offers . . . an irreducible equivocation of narcissism and sociality which becomes the condition of the epistemological generation of and access to objects."[27] The social is a defensive structure against the very narcissism upon which it depends.

In Lacan's discussion of the mirror stage, Butler suggests, the penis enters only as part of the narcissistically invested image, the imaginary center of the body's fiction of totality. That is to say, the penis becomes the phallus as itself an imaginary effect. However, when Butler turns to Lacan's essay on the meaning of the phallus, she finds that he finally refused this thought of the imaginary construction of the phallus. Like Freud, Lacan defensively insisted that the phallus is neither an imaginary effect nor an organ. Instead he proposed that the phallus is a prototype, a transcendental signifier, that distinguishes the symbolic from the imaginary in the first place.

But by this point Butler has made both Lacan's and Freud's treatments of the phallus as a transcendental signifier seem unconvincing; thus the distinction between the imaginary and the symbolic also does not hold, displaced onto an irreducible equivocation of narcissism and sociality. Butler even suggests that the mirror stage, which Lacan treated as preoedipal, seems rather to always already presume the oedipal law and the phallus as its transcendental signifier. After all, the preoedipal body-in-bits-and-pieces is meaningful only against the horizon of the body's fictional totality, which the transcendental phallus signifies.

Although Butler concludes by arguing that the oedipal law of the phallus is a historically and culturally specific regulatory ideal, something more like an ideology, she does not mean to dismiss the unconscious altogether. Her careful and detailed rereading of Freud and Lacan surely suggests this. Instead of dismissing the unconscious, she draws the unconscious outside of its enclosure in an oedipal logic of narrativity. She not only unsettles the heteronormativity of a symbolic order organized by the oedipal law, thereby allowing for queer sexualities. She also seems to propose that there are other regulatory psychobiographic fictions than the oedipal narrative, other regulatory bodily ideals, other symbolic orders.

For Butler, the historical and cultural specificity of any and every symbolic order even raises a question about Lacan's treatment of the real as radically incommensurable with the symbolic.[28] Whereas the symbolic constitutes the cultural norms of intelligibility through which any reality is constituted, the real, as Lacan put it, "resists symbolization absolutely." Butler instead argues that what is unintelligible, nonsymbolizable, or outside the symbolic must also be in the symbolic. Drawing on Derrida, Butler argues that as an outside, the real is the defining limit of the symbolic; it is its constitutive outside. Therefore, the real also is part of the symbolic order; it is part of a culturally and historically specific symbolic order. Butler also argues that Lacan's "resisting real" is, however, a symbolization that institutes a desire for there to be a real referent, a pregiven materiality that transcends historicity and grounds meaning. Indeed, Lacan sometimes referred the real to matter, materialism, even the brute physicality of the human body.[29]

Although Butler recognizes that there always is an outside to the symbolic, she nonetheless proposes that the boundary between the symbolic and its outside is not determined by a universal law. She argues instead that the boundary between the symbolic and the real is culturally and historically variable; there is the possibility, therefore, of rearticulating the boundary. As Butler puts it: "To supply the character and content to a law that secures the borders between the 'inside' and the 'outside' of symbolic intelligibility is to preempt the specific social and historical analysis that is required, to conflate into 'one' law the effect of a convergence of many, and to preclude the very possibility of a future rearticulation of that boundary."[30]

In deconstructing the oedipal law of the phallus, Butler not only proposes that bodily matter be thought of in relationship to symbolic orders other than those organized in terms of the oedipal narrative; she also gives the possibility of different boundaries between the real and the symbolic. What this might be read to propose is that it may now be the case, in the age of teletechnology, that the brute physicality of the body no longer marks the difference between the real and the symbolic. Or, to put it another way, it may no longer be the case that the real is defined only by the brute physicality of the body in every symbolic order. After all, Butler makes it possible to think of bodily matter as an imaginary construction, and therefore to think matter and the image in a *différantial* relationship rather than an oppositional one. She makes it possible, that is, to think of unconscious fantasy and bodies without referring them only to an opposition of sexed subjectivity and brute physicality.

But Butler herself retreats from elaborating these possibilities. Aimed primarily at radicalizing the cultural construction of the human body in terms of an unconscious repetition compulsion, she does not explore the ontological implications of her deconstruction of the psychoanalytic configuration of the imaginary, the symbolic, and the real in relationship to rethinking bodies other than human bodies. Butler even hesitates to fully elaborate the materialism or the materialist criticism of culture that her treatment of bodies seems to imply. This has led a number of Butler's critics to argue that her politics of performance is indifferent to the material conditions of political economic realities. Although her critics, like Butler herself, do not focus on the ontological implications of her treatment of bodily matter, and therefore fail to explore the materialism suggested, they nonetheless draw her work into a discussion of the separation of the private and public spheres, which, after all, has been a presumption and a central concern of the feminist theory of gender, which Butler engages and means to revise.

Nancy Fraser, for example, does not question that Butler's materialist criticism of culture is a treatment of political economy; she rather has questioned whether it offers anything more than a 1970s post-Althusserian feminist Marxism in which it is thought that the construction of subjects functions to support economic production, that is, in which performativity makes production and reproduction seamless. Fraser argues instead for analytically distinguishing what she describes as injustices of economic distribution and injustices of social recognition. These are held separate, Fraser proposes, by the idea of "personal life," "a space of intimate relations, including sexuality, friendship, and love, that can no longer be identified with the family and that is lived as disconnected from the imperatives of production and reproduction."[31]

Fraser's remark about personal life rests on her earlier criticism of the ideology of the separate spheres and her effort to propose models for strengthening the conditions of possibility of democracy beyond the liberal model of the bourgeois public sphere. Fraser rejects the liberal model of the public sphere, in which it is deemed necessary that there be a rigid separation of civil society and the state, the public and the private spheres. Fraser not only argues that this rigid separation of social spaces is not necessary for democracy; she also argues that it has required that some subjects be excluded from social recognition and/or economic opportunity.

Fraser proposes instead that the state needs to be involved in ensuring economic parity and in addressing injustices of social recognition; this is especially important both in post–welfare state politics and where "personal life" seems to have been released from familial ideology or a nonreflexive embeddedness in the institution of the family. Therefore, Fraser also argues that in the liberal model of the public sphere a notion of privacy is presumed, which in fact needs to be adjusted to differences of gender, sexuality, class, ethnicity, and race — recognizing especially how these differences change the very definitions of privacy and publicness. She instead calls for "multiple counter publics" or "subaltern counter publics" that might allow for redressing issues already addressed in the dominant public sphere or permit issues to be raised that have not been addressed at all. Fraser seems to suggest that the very definition of privacy and publicness might be one of the issues, if not the crucial issue, put forward by counter publics.[32]

Fraser's criticism of the liberal model of the bourgeois public sphere is especially concerned to support the political agency of women and to encourage their full inclusion in democratic politics. After all, in the liberal model of the public sphere it has been presumed that private needs will be transformed in terms of public discourse and that the woman will function primarily to link family members both to the public sphere and to the state, promoting familial and national ideologies. Although Fraser criticizes the liberal model especially for the position it assigns women, her treatment of the configuration of state and civil society, family and national ideologies, and the public and private spheres remains within the limits of both the subject-centeredness and the nation-centrism of the modern western discourse of democracy. Postcolonial theorists, however, have argued that this configuration of social spaces presumed in the western discourse of democracy becomes impossible, especially when considering the political agency of women in neocolonial societies. Pheng Cheah, for one, even has suggested a connection between the feminist criticism of the liberal model of the public sphere, the situation of neocolonialism, and the cultural materialism of Butler's treatment of sexed bodies.[33]

Cheah argues that attempts to rethink the liberal model of the public sphere, such as Fraser's, often leave intact the model's idealization of publicness. Focusing especially on Jürgen Habermas's treatment of the public sphere, Cheah points out that an idealization of publicness, elaborated in face of the limitations of actual democracy, is given as a transcendental norm. Furthermore, it is presumed that a critical reason can judge the limitations of actuality against the transcendental norm and thereby support overcoming these actual limitations, which are embedded, as Habermas proposes, in particular capitalist modes of production.

As Cheah sees it, this model of publicness is troublesome especially in neocolonialism under the conditions of the transnationalization of capital and the globalization of teletechnology. Under these conditions, Cheah suggests that often a stable state cannot be presumed for which a public sphere can be thought to offer a resource for criticism of excessive state power. Nor can the private or domestic sphere be presumed to be a protection of the individual, as it often is assumed to be in the liberal model and its feminist critiques. Cheah points out that in many of the situations of neocolonialism, it often falls on women to resist neocolonialism with a reassertion of a nationalist cultural identity, sometimes expressed in a patriarchal or ethnic fundamentalism. And yet some of these same women can also find themselves released from a certain local patriarchalism by their labor force participation in transnational corporations locally situated. Many of these women become open to the culture of a globalized media in which women are sometimes figured quite differently than the local state apparatuses or familial traditions figure them, resulting in a cultural situation that is something like, but more disturbing than what James Clifford describes as a "discrepant cosmopolitanism."[34]

All of this suggests that women are to be found resisting and supporting localizations against globalization as well as resisting and supporting globalization locally situated; indeed, they often use one against the other, draw one into the other. In such situations the derivation of a feminist politics or a democratic politics is not clear, whereas agency, as Cheah suggests, "is not an unproblematic assertion of the co-belonging of freedom and humanity." Instead what is proposed is a rigorous responsibility to what Cheah describes as a "condition of global miredness," that is, where every determination of agency yields an undecidability of effect that frames and reframes further determinations or where forces are "unmotivated but not capricious," as Cheah puts it, borrowing Gayatri Chakravorty Spivak's phrasing.[35]

The criticism of the configuration of social spaces presumed in subject-centered, nation-centric modern western discourse is meant to underscore the way in which a certain idea of democracy is being imposed on neocolonial nations, even made a "privileged point of vantage," as David Scott puts it; as such, it is made "the standard for the assessment of all political institutions and political discourses, not only for those of Europe's own past . . . but for those as well of the non-European worlds whose political presents have been re/constructed in colonialism's wake."[36] This criticism, however, is also meant to point to the contention over and the resistance to compliance with the imposed configuration of social spaces presumed with a certain idea of democracy, thereby producing in neocolonial societies the complex situations out of which political agency is to be determined and to which Cheah refers. But this condition of global miredness refers not only to neocolonial societies, but to postmodern capitalist societies as well, since in both postmodern capitalist societies and neocolonial societies the configuration of social spaces presumed in the modern western discourse of democracy is being deterritorialized and reterritorialized as part of the globalization of teletechnology and the transnationalization of capital.

Therefore, Cheah proposes that the transcendental norm guiding the liberal model of the public sphere would better be drawn back to finitude, back into the forces of a global miredness. With a nod to Derrida, Cheah suggests that these finite forces — unmotivated but not capricious — must be thought of as immanent to matter, that is, referred to the dynamism of matter or mattering. It is in his elaboration of what he proposes to call a "deconstructive materialism" that Cheah critically engages the cultural materialism of Butler's treatment of bodies.

Although Cheah recognizes Butler's contribution to the thought of a dynamic matter, he nonetheless argues that a politics of performance seems to require "a constitutional democracy within passive capitalist relations," and therefore is able to ignore "oppression at the physical level."[37] But Cheah does not pay much attention to Butler's effort to deconstruct the oedipal narrative that functions to position the subject

within the social spaces configured in the discourse of democracy, such
that Butler also initiates the deconstruction of this configuration of social spaces in the direction of psychobiographic fictions other than the
oedipal narrative. In doing so Butler points, in her own way, to the global
miredness of the forces of political agency attending the reconfiguration of social spaces with the realization of various psychobiographic
fictions. If she refers to democracy, it is to what Derrida describes as a
"democracy to come,"[38] when the conditions of its possibility cannot
be preordained. Butler puts it this way: "A social theory committed to
democratic contestation within a postcolonial horizon needs to find a
way to bring into question the foundations it is compelled to lay down."[39]

Although Butler refuses any fixed notion of democracy, she does so
without, however, giving up on the unconscious process of disavowal,
which, in her view, produces subject identities and bodies; from her
position the question would be whether these are not still at issue in
the imposition of the discourse of democracy with its demand for subjects of the nation and bodies for a capitalist economy. Therefore, Butler's focus on subject identity and the unconscious is not a denial of
physical oppression, although she means to make it unnecessary to distinguish physical oppression as such. There is no oppression, she seems
to propose, that is not also a matter of unconscious fantasy, what she
refers to as "the psychic life of power."[40] Butler's deconstruction of the
oedipal logic of narrativity, however, throws the meaning of the psychic life of power into a transnational frame, such that psychic life is to
be understood in terms of various psychobiographic fictions.

This surely is what Fraser misses in her response to Butler's reference
to an Althusserian Marxism. Although Butler first draws on Althusser's
analysis of the subject's ideological interpellation, she finally argues that
it is wrong in its presumption that interpellation works and that unconscious desire can finally be fixed in a subject identity.[41] Butler, after all,
turns to Foucault's treatment of power/knowledge, where the interpen-
tration of civil society and the state in a disciplining society makes the
ideological interpellation of the subject less pressing. In this sense Butler's attempt to move unconscious fantasy beyond an oedipal narrative
also moves it beyond Althusser's treatment of ideology, closer to the
teletechnological, beyond the representation of the subject, nearer to a
network imagination, where the aesthetic of exposure, over- and under-

exposure, operates, befitting a politics of performance. This is also on the way to facing the necessity to rethink "oppression at the physical level" in terms of the becoming indistinguishable of matter and image.

Nonetheless, Cheah is right to suggest that Butler's treatment of bodies remains caught in its reference to the human body, and that therefore it is concerned only with how human bodies become culturally intelligible. In contrast, Cheah points to Grosz's treatment of bodies, endorsing the way in which it moves beyond the human body while shifting thought about the body beyond a concern only for its cultural meaningfulness. But Grosz, like Butler, begins her treatment of the body in psychoanalytic terms even though, like Butler, she means to deconstruct the psychoanalytic configuration of the imaginary, the symbolic, and the real. Grosz, too, is hesitant to leave behind all thought of the unconscious even when finally she turns from a psychoanalytic treatment of unconscious fantasy to engage Deleuze and Guattari's treatment of desire and the body without organs. In doing so Grosz points to something more like what Haraway describes as "an 'unfamiliar' unconscious, a different primal scene, where everything does not stem from the dramas of identity and reproduction."[42] Even more than Butler, Grosz draws from feminist theory its ontological implications while drawing feminist theory closer to the teletechnological.

Unconscious Bodies without Organs

Although Grosz does not end with a psychoanalytic account of the human body, she begins with it. Like Butler, she is drawn to the psychoanalytic treatment of the body in relationship to unconscious fantasy. But in her reference to psychoanalysis Grosz resists dismissing the body's given materiality. Her aim in deconstructing the opposition of nature and culture is to treat culture as the deferral of nature, to draw out the *différantial* relationship of nature and culture. Unlike Butler, Grosz does not risk treating bodies as a meaningless nature that is given meaning in subjection to the forms of cultural intelligibility. Although the body is in no sense "non- or presocial," Grosz proposes, the body also is not "purely a social, cultural, and signifying effect lacking its own weighty materiality."[43]

Without rejecting the thought that there is no "real" material body distinct from the cultural inscriptions that constitute it, and yet holding that nature is neither origin nor causality, Grosz nonetheless refuses to reduce the body to the cultural. For Grosz, the natural and the cultural are "interimplicated, such that their relationship is neither dialectic (in which case there is the possibility of a supersession of the binary terms) nor a relation of identity but is marked by the interval, by pure difference."[44] Whereas Butler points to the dynamism of bodily matter fantasmatically informed with cultural norms, Grosz points to a dynamism of matter, which, as Cheah puts it, is a "nonanthropologistic level of dynamism," without a reduction "to mechanical laws of causality and naturalist teleology."[45]

Therefore, when Grosz turns to giving a psychoanalytic account of the body, she draws on Freud's *Three Essays on a Theory of Sexuality*, emphasizing the way Freud treated unconscious fantasy as an elaboration of the drives, which themselves "lean on" the biological instinct. The often repeated example is the fantasmatic construction of the mother's breast as an object of the sexual drive, leaning on and displacing the nonsexual instinct of hunger. Although feminist film theorists have taken the psychoanalytic treatment of fantasy as a way to deliver psychoanalysis from any biological determinism, Grosz is less interested to do so. It is not that she thinks psychoanalysis is a biological determinism; she is more interested in what the drive can do. She concludes that the drive is to be understood to "mimic" the instinct, even seeming to act like it.

Again, when Grosz turns to criticize psychoanalysis for its oedipalization of unconscious fantasy and, like Butler, aims to queer the law of the phallus, her focus is less on the deconstruction of the subject's sexual identity and more on the sexual drive, what it does in seeking sexual pleasure. Rather than adopting Butler's fiction of the lesbian phallus, Grosz treats the possibilities of "lesbian fetishism."[46] In her effort to reread Freud's treatment of fetishism in order to allow for the possibility of lesbians' sexual practices, Grosz begins to turn thought of the body away from treating it as an imaginary identity, a surface projection of an internalized imaginary, toward thinking of bodies in terms of what they are assembled to do.

Grosz closely follows Freud's treatment of fetishism, a perversion that Freud thought characteristic of men mostly, given the difficulties the male child has in facing the seeming castration of his mother, once imagined to be phallic. Grosz is especially drawn to the capacity of the male fetishist "to have it both ways," that is, to accept the oedipal law, but to repudiate or foreclose its content. The fetishist, after all, really believes in his penis substitutes. He believes they are real. To use Freud's terms, the fetishist has "two attitudes" that "persist side by side . . . without influencing each other."[47] This means that fetishism is not hallucination. Grosz quotes Freud, who argued that the fetishist "did not simply contradict his perceptions and hallucinate a penis where there was none to be seen, he effected no more than a displacement of value—he transferred the importance of the penis to another part of the body," even to an object outside the body.[48]

For Grosz, this displacement of value offers the possibility of "having it both ways," a possibility she would extend to the lesbian in her exploration of lesbian fetishism. To illustrate this, Grosz turns next to Freud's treatment of women's perversions, especially to women whom Freud describes as suffering from "the masculine complex," which he connects to lesbianism. These women, Grosz proposes, are closest to the male fetishist. Not only do such women disbelieve their castration; they also refuse the oedipal demand to shift their libidinal investment from the maternal or feminine object to the paternal or masculine one. They expect to act as men are permitted to do. They libidinally invest in feminine love objects whom they can love as men are permitted to love, as if they were phallic. Grosz concludes that the lesbian fetishist is having it both ways and, as such, the feminist theorist is like her. In accepting and refusing as social reality that which devalues the feminine and oppresses women, the feminist theorist also needs to have it both ways. Both feminist analysis and the psychoanalysis of unconscious fantasy, therefore, are needed.

So Grosz does not elaborate the possibilities of lesbian fetishism simply to give a psychoanalytic account of it or merely to revalue what psychoanalysis treats as perversion. Like Butler's treatment of the lesbian phallus, Grosz's treatment of lesbian fetishism is for political effect. It demonstrates that psychoanalysis, although important for understanding unconscious fantasy, is unable to treat feminine sexuality adequately, "even within the confines of Western capitalism"; and beyond western capitalism, Grosz proposes, "the categories that Freud proposed as universally relevant—the function of the phallus, the Oedipus complex, the ubiquity of the castration threat, and women's status as passive—surely need to be contested."[49]

If Grosz's treatment of lesbian fetishism distances itself from psychoanalysis, it does so to suggest that bodies are formed and reformed in libidinal attachments that are made so that the drive realizes its sexual aim. It also does so to put the drive of sexual desire, or desiring itself, outside the compass of the oedipal law of the phallus. Grosz is on her way to thinking of bodies as matter that is neither organic nor mechanical, but dynamic. She is on her way to refusing the distinction, elaborated in psychoanalysis, that separates what a body is from what a body does.

Grosz even criticizes Butler's revision of psychoanalysis for maintaining this separation in her deployment of a restricted/restricting thought of repetition. As Grosz puts it:

> In separating what a body is from what a body can do, an essence of sorts is produced, a consolidated nucleus of habits and expectations takes over from experiments and innovations: bodies are sedimented into fixed and repetitive relations, and it is only beyond modes of repetition that any subversion is considered possible (this is Butler's position, and its limitation: that subversion is always only a repetition and never in any straightforward way an innovation, a production of the new).[50]

In criticizing Butler, Grosz already shows signs of her engagement with Deleuze's treatment of pure repetition and with Deleuze and Guattari's thought of desiring production; she has already begun to think of bodies and desire at even a greater distance from psychoanalysis than she permitted herself in her treatment of lesbian fetishism. She describes her shift in perspective this way:

> While psychoanalysis relies on a notion of desire as a lack, an absence that strives to be filled through the attainment of an impossible object, desire can instead be seen as what produces, what connects, what makes machinic alliances. Instead of aligning desire with fantasy and opposing it to the real, instead of seeing it as a yearning, desire is an actualization, a series of practices, bringing things together or separating them, making machines, making reality.[51]

Having become more interested in what a body can do, Grosz rethinks the body as "a discontinuous, nontotalizable series of processes, organs, flows, energies, corporeal substances and incorporeal events, speeds and durations."[52] Grosz, therefore, moves from thinking about bodies in terms of fantasy in order to rethink desire. Desire becomes the affective and unconscious movement of thought that assembles bodies. Bodies are what desire assembles in order to do something.

In all this Grosz borrows from Deleuze and Guattari's treatment of "the body without organs" and follows their preference for mapping, for treating assemblages. For Deleuze and Guattari, the body without organs is a plane of consistency "specific to desire (with desire defined as a process of production without reference to any exterior agency, whether it be a lack that hollows it out or a pleasure that fills it)."[53] As Grosz describes it, the body without organs refers to "a field for the production, circulation and intensification of desire, the locus of the immanence of desire."[54] The body without organs is matter open to the flows of intensities; it is a plane of consistency in and for the flows of desire.

It is this getting ready–ness that makes the body without organs explode or refuse organization. It is not organs, therefore, that the body without organs is against, but any organization into organism. Deleuze and Guattari quote Antonin Artaud, from whom they take the notion of the body without organs: "The body is the body. Alone it stands. And in no need of the organs. Organism it never is. Organisms are the enemies of the body."[55] The notion of the body without organs is antioedipal; it is against the oedipalization of the unconscious. It is meant instead to restore the partiality of objects that psychoanalysis aims to turn into whole persons — father, mother, and infant. Deleuze and Guattari rather treat part objects "like the intensities under which a unit of matter always fills space in varying degrees." Part objects are to be understood as "degrees of matter" that are "pure positive multiplicities" and where everything is possible, indifferent to an underlying support, "since this matter that serves them precisely as a support receives no specificity from any structural or personal unity but appears as the body without organs that fills the space each time an intensity fills it."[56]

As Grosz points out, the body without organs is "all the more alive and teeming once it has blown apart the organism and its organization." Or, to put this another way, the body without organs is distinguished by the type of movements it allows, the types of flows to which it is amenable. Deleuze and Guattari argue that although the "full" body without organs is amenable to flows of intensities, the "empty" body without organs is too full to allow for further circulation of intensities. They also propose that exploding organization can be too fast, making the further circulation of intensities impossible. Most likely there will be subject identity, identification, resemblance, and representation that slow down the body without organs. But desire also is always subtracting the body without organs from identity, identification, resemblance, and representation in order to speed the body up, giving it a multiplicity of new directions.

Although Grosz has not fully elaborated the implications of her treatment of bodies for gender, class, ethnicity, sexuality, and race, she at least suggests the possibility of thinking of these in terms of machinic assemblages, bodies without organs, proposing that the given materiality of the human body affects the way it becomes interlaced with the cultural inscriptions of race, class, gender, sexuality, and ethnicity. Therefore, race, ethnicity, sexuality, gender, and class are to be treated politically as elements of a machinic assemblage, matters of a desiring production that does not reduce to an individual's desire, but rather points to the direct links between microintensities and various territories—human bodies, cities, institutions, ideologies, and technologies. In this sense race, class, sexuality, ethnicity, and gender are not simply matters of subject identity and surely not of authentic subject identity. Rather they are rethought in terms of the connections and disconnections on a plane of consistency, the interlacing of given materialities of the human body and cultural inscriptions, given over, however, to the speeds of deterritorialization and reterritorialization, to the vulnerabilities of exposure, under- and overexposure to media event-ness, such that politics involve the when, where, or how of acknowledging, elaborating, resisting, or refusing the visible and invisible markings and effects of desiring production.

In her efforts to take up Deleuze and Guattari's treatment of desire and the body without organs, Grosz, it would seem, labors to give the phenomenological body over to the "technophenomenological," a term Amelia Jones uses to treat bodies as part of the circuitry of the teletechnological flows of sounds, images, and information.[57] To put this in another way, the body without organs is a way of thinking the unconscious thought of teletechnology, where desire is no longer the possession of only the human being, referring only to the human body. Rather, the body without organs makes unconscious thought part of desiring production, part of machinic assemblages. As Deleuze puts it: "The unconscious no longer deals with persons and objects, but with trajectories and becomings; it is no longer an unconscious of commemoration but one of mobilization, an unconscious whose objects take flight rather than remaining buried in the ground."[58]

Thinking of unconscious thought and bodies in this way necessitates redefining the real, even going beyond historicizing the distinction of the real and the symbolic, as Butler proposes be done. Therefore, Grosz gives up on the Lacanian treatment of the real. Following Deleuze, she proposes that the real is productive; it gives virtualities that are already real, although not yet actualized. The real and the virtual are interimplicated; they are in a *différantial* relationship. Grosz argues, too, that to think of virtualities in this way is linked to thinking unconscious thought differently; it is to think of thought as being drawn to the outside, to its virtualities. As Grosz puts it: "Thought, life, is that space outside the actual which is filled with virtualities, movements, forces that need release. It is what a body is capable of doing, without there being any necessity, and without being captured by what it habitually does, a sea of (possible) desires and machines waiting their chance, their moment of actualization."[59]

For Grosz, to think of the real along the lines Deleuze proposes is not only to rethink the body, but also to rethink frames, grounds, figures, and social structures — that is, the constructed, the architectural, and the built. It is in these terms that Grosz's treatment of bodies without organs connects with the deterritorialization and reterritorializations of social spaces in the globalization of teletechnology and the transnationalization of capital. Not surprisingly, then, when Grosz rethinks bodies, she thinks them becoming cities or becoming architecture, whereas she rethinks architecture and cities as productions of desire. She thinks of bodies, cities, and architecture in terms of speeds, allowing and disallowing the actualization of virtualites in the reconfiguration of family and national ideologies, the state and civil society, and the public and private spheres.[60]

Surely Grosz's thinking about bodies without organs fits the ongoing shift in cultural criticism from thinking of culture as a bounded homogenized community to thinking of cultures as contestations of meaning. It also fits the often referenced hybridization of cultural identities and cosmopolitanisms of various kinds as well as the intense rearticulations of nationalism and even the horror of what Arjun Appadurai refers to as "ethnocidal violence," that is, a violence done to the neighbor's body in an effort to produce an intimate enemy body, "a somatic stabilization" or "dead certainty" of ethnic identity that globalization makes both impossible and desired.[61] Here the human being is subject to a violence that Appadurai refers to as "vivisection," making identity out of cut-up pieces of body. Although there is a reassertion of identity, it is not a matter of the interpellation of an ideological narrative that connects the individual to the arrangement of nation and family ideologies, the state and civil society, and the public and private spheres presumed in modern western discourse of democracy. Instead there is a violent suturing of the individual to ethnic group, home, and/or region.

In all these situations the body without organs is the shape of intensities, exposures, and speeds. Its politics is beyond the configuration of state and civil society, national and family ideologies, and the public and private spheres presumed in modern western discourse of democracy; its politics is in being ready to intervene in what takes shape in the reterritorializations of deterritorialized social spaces. Grosz imagines the conditions of the possibility of politics in these terms: "Individuals, subjects, microintensities blend with, connect to, neighborhood, local, regional, social, cultural, aesthetic, and economic relations directly, not through mediation of systems of ideology or representation, not through the central organization of an apparatus like the state or the economic order, but directly, in the formation of desiring machines."[62]

The challenge to the structural that Grosz's thinking poses enjoins Deleuze's treatment of form. Deleuze treats form as a matter of ungrounding, of being without frame or plot or narrative. Deleuze's treatment of form, John Rajchman suggests, fits the reconfiguring of social spaces with the globalization of teletechnology and the transnationalization of capital; Rajchman points especially to the reconfigurations involving increased urbanization worldwide. He also notices how Deleuze's treatment of form "matches not so much the industry and engineering that produced cinema as the new kinds of televisual and digital images that came to displace it."[63]

For Grosz, as for Rajchman, Deleuze delivers form from an overseeing eye or from an overarching organization or plan; form is connected instead to singularities, to iterability, and pure repetition, where there is no origin or end, only virtualities. Form is not negative; it is positive as the form of virtualities to be actualized. Rajchman refers to it as "operative form," because the emphasis is on what form does, and what it does is not planned or expected. It is "a virtual plan." As such, form moves and is moved by sensation, by affects; this "affective space" has mobility and plasticity. For Deleuze, form is the trajectories of bodies prior to, and remaining alongside, relations of subject and object, before and alongside figure, ground, and narrative structure. In this sense Deleuze's notion of form is not merely against narrative; nor is it antifoundational. It is an ungrounding foundation; it provokes formation of an ontological perspective from which matter and form are interimplicated.

The positive sense of form that is deployed by Grosz in her treatment of bodies differs from Butler's deployment of form in her treatment of bodies. Butler treats form as a historical and cultural regulatory norm, an ideal, such as the bodily form that the oedipal law of the phallus imposes. Because the norm is regulatory, it excludes as well as prescribes. It is the excluded bodies that return; their negative force is let loose in the compulsive repetition of the norm, which is, however, the condition of possibility for changing bodily forms. But Grosz engages Deleuze's sense of form in order to go beyond Butler's treatment of bodies as historically and culturally specific forms. For Grosz, therefore, morphology does not give dynamism to bodily matter. Instead bodies are given in their modes of materiality; they are dynamic matter. They are form and matter interlaced.

Both Grosz's and Butler's deployments of form derive from differing readings of poststructuralism: one in the direction of Derridean *différance* and one in the direction of Deleuzian pure repetition. Their different treatments of bodies, however, also arise out of feminist theorizing, which since the 1970s has been itself characterized by contention over the meaning of the woman's body and therefore bodily matter and form more generally. Therefore, feminist theorists either have aimed at deforming and reforming structures of oppression, repression, and domination or they have aimed at informing something else, something new.

If some feminist theorists first asked for equality for women within given structures, it was followed by other feminist theorists' recognizing that reforming such structures is not the same as deforming and ungrounding them; it is not the same as informing something new. If some feminist theorists deconstructed the universalization of thought in what would then be more properly called masculinist thought, gynocriticism quickly produced a universalization of feminine thought. But it, too, was quickly ungrounded by criticisms of essentialism. The deconstruction of essentialized identities, nonetheless, produced a desire for identity and for a politics of identity. The charges that in some feminist theory there were unrecognized exclusions of race, ethnicity, sexuality, class, and nation ironically produced both a celebration of the specificity of cultural identity and the profound uncertainty that any specific identity could withstand deconstructive criticism.

The movement — from reforming to deforming to informing something new — was produced with such speed and intensity that finally feminist theorists found themselves in close contact with the thought of movement, of speed and intensity, in the face of questions about the glocalism of world cultures in the globalization of teletechnology and the transnationalization of capital. It is in this sense that I have proposed that feminist theory has been drawn to the future of thought and has developed an ontological perspective that takes the thought of machinic assemblage seriously, such that neither body nor machine, nature nor technology, the real nor the virtual is ontologically privileged.

Of course, to propose that feminist theory ends up in the thought of machinic assemblages will surely be discomfiting to feminist theorists and return them to where I began, namely, feminist theorists' profound suspicion of the masculinism of the postmodern theory of the body. But the machinic assemblage is not thought to be a resolution of or the end of questions about women that have been taken up by feminist theorists over the last three decades. The thought of machinic assemblage, therefore, is not necessarily beyond sexism, heterosexism, racism, or ethnocentrism, as it sometimes has been imagined to be. Rather, the thought of machinic assemblages goes only beyond thinking of the body as organic or mechanistic matter. It is a postpersonal thought of bodies, evoking an ontological perspective befitting the age of teletechnology.

Postpersonal thought of bodies, however, also has been part of feminist theorizing since the 1970s. It is implied in the linking of the feminine and monstrosity, but also in feminist theorists' revaluation and re-embodiment of monstrosity. That is to say, becoming monstrous has been a feminist strategy to deform the reality that devalues women and refuses them rationality; it also is part of informing a future. However, because such a strategy returns the feminist theorist to the monstrosity that has been projected onto the feminine in modern western discourse, it is not a strategy that can be described as intentional. The feminist theorist surely is unconsciously drawn to it, drawn in a compulsive repetition of monstrous embodiment. It is this link connecting monstrosity, the feminine, and the feminist theorist that has repeatedly brought feminist theory back to psychoanalysis and to the unconscious.

Here, too, there has been contention among feminist theorists over psychoanalysis, whether it should be deployed in a description of what the woman becomes, what becomes of her femininity in her development of a subject identity, or whether it should be deployed in a deconstruction of feminine subject identity in the direction of the drives and singularities — to the unconscious thought of virtualities and then finally to the thought of pure repetition. These differences notwithstanding, one of the most remarkable aspects of feminist theorizing over the last three decades has been the way feminist theorists have rethought psychoanalysis with such close attention to details, the stakes being the deployment of unconscious thought for feminist ends.

No matter how critical they have been of psychoanalysis, feminist theorists have found in the unconscious a marker of repression, oppression, and domination, the marker of what could not be spoken — not ever or not yet — and even of the *différance* between the two. But the unconscious has seemed to them also to be the marker of desire, passion, and affection, the form of virtualities, future possibilities. Feminist theorists have meant to save the unconscious even though they have made every effort to free it from confinement in the oedipal narrative, so that if desire is not to be drawn back to the idea of lack, it surely is to be drawn to a dynamism of matter, to finitude, death, and the machine internal to life. To face these has always been and still is a feminist necessity.

All this is to say that feminist theorists have been so intensely engaged with unconscious thought that they have instigated its migration to the future. There unconscious thought can no longer be understood only in terms of the oedipal narrative, although this does not necessarily mean the end of oedipalized forms of sexism, racism, heterosexism, and ethnocentrism. It only makes necessary also finding ways to understand other bodily matters. Implied is the task of revising science and knowledge practices given an ontological perspective that privileges neither nature nor technology, body nor machine, the virtual nor the real. It means rethinking practices of self-reflection in science and knowledge practices in terms of the teletechnological, that is, learning how "to diffract the rays of technoscience so that we get more promising in- terference patterns on the recording films of our lives and bodies," as Haraway puts it.[64] This is the effort that has characterized the field of cultural studies of science that has developed since the 1970s. It is to this I turn next.

Attic Women

The nuns kept Crazy Mary in the attic,
high above the orphanage playground.
Her warped body, all but disappearing in folds of tattered cloth,
whispered through the large space,
haunting it all through the day, until night,
when she slipped into the corner
where the floor met the steeply pitched roof.

Only Sister Lucia talked to her, when she brought Mary food.
She spoke as if Mary was going to answer.
She never did.
Her eyes only turned blank as if already gone to God.
But Lucia didn't seem to mind.
The blankness only freed her,
making her brighter, even brilliant,
perhaps more than she had ever really been.

Mary's craziness seemed to spread all over her
like a birthmark that didn't know how to stop.
It was molecular:
in the way her nails grew wild and her skin flaked,
in the way her hair twisted from her head out of shape.

If Mary appeared born crazy, Lucia seemed destined to go crazy.
She was still exquisitely beautiful. Her black veil,
like gossamer wings,
floated around her as she walked,
so that her feet seemed barely to touch the ground.
Not that she was naturally angelic.

It was a practiced walk, part of an effort to be holy.
She tried just as hard not to notice
that she was beautiful or brightness itself,
although it mattered that she was.
It kept her engaged in the world
from which she long had tried to lift her spirit,
seeking some sort of freedom.
Her beauty made young novices admire her
and want to be just like her.
Her brightness made her a gift of grace
to old nuns dulled from long years of service.

That summer, Lucia shared all of her thoughts with Mary.
Hunched over a china bowl, Mary clawed at her food.
Lucia spoke.
Words rushed from her feverish mind,
showing itself in the natural rouge
rising up to her cheekbone and brushing red
across her full shaped lips.
She spoke as if Avila to John of the Cross —
theological debate and prayerful wishes for love-pierced hearts —
all running together, humming with an eroticism
in excess of doctrinal limits.
Yet, safe.
Mary was no church censor.

By evening the daily conversation deepened
and darkened what, at breakfast, had passed from Lucia's lips
as sheets of light, crystalline, trembling with eternity
and close to hope that was gone by nightfall,
when the godly words showed signs of human despair,
when Lucia's need for Mary seemed terrifying,
and what they shared even more so.
Lucia was drawn,
like a moth to flame,
to the madness.
It made her comfortable,
a guilty ransom paid for not being born mad.

It was a true exchange.
Mary did give Lucia something back,
a mirroring that Lucia took for understanding.
Mary drew out of her what those who thought Lucia all beauty and
brightness could not bear to see —
that withdrawing gesture

that overwhelming need to retract from love,
although it was often not even recognized to be such,
to extract from life its purest thought of impossibility, impassivity even,
a moment that repeatedly returned
and before Lucia's eyes spread out its pall,
turning everything alive toward death.
A moment whose arrival Mary no longer could even recognize.

She was young when the nuns gave her over to the attic,
snatched from something intolerable for someone of her age.
Something crude and clamoring
and then grunts and whizzing too close to her face.
Something with a pungent odor that did not seem human
always left in her hair clinging.
And this, all after her eyes had been sanded smooth,
blinded by images of something without contour
washing over her, as the waves of her inner horror
met with the waves of the world's terrors
and communicating so completely,
left her senseless,
forever an infant-child but exhausted.

Lucia no longer wondered about Mary's madness, as she once had,
when, for long hours, she had visited that ancient certainty that
she herself would go mad,
when she had tried to meet its point of origin.
But it all had remained incomprehensible.
So she made the certainty of her madness into a calling.
She became a finely tuned ear
for all those in need of the simple sympathy
for which she herself yearned
but seemed incapable of knowing.

What Mary no longer could remember, Lucia no longer cared to.

And so that fateful summer passed, as if time had no passing.
Yet, outside, summer came to an end and in a heat wave.
It made the attic sweltering and then humanly insufferable.
The nuns tried,
but tried in vain, to get Mary down from the attic.
Perhaps, she had that one last choice to make
to let her heart fail
and she did.

It wasn't clear whether Lucia would lose her mind.
She came to the attic every day for weeks. She sat and waited

ready to speak, but no words came.
Her cheeks did not flush,
and her lips remained drawn white-tight
until one night, she began to cry.
A river of tears flooded her face and rushed down
to fill her cupped hands to overflowing. Suddenly, she rose
and slowly walked to the window.
For some time, she looked down at the playground below.
Then, she turned, it seemed, as if from some fate.
She left the attic and never ever returned.

Grace

The first time she saw him, he was standing in the doorway.
He had placed a hand on each side of the door frame,
his emaciated body needing the support.
Yet, just for a moment, his outstretched arms
turned the sleeves of his shirt into a pair of wings.
They blocked the sun,
absorbing the light and shooting it
back, out
from every angle of his body.
He appeared to her as a dying version
of some grand archangel.

"I was thinking about my sins," he said.

She was terribly afraid, but let herself
be drawn to him
through a break in her reason
made by an irresistible painfulness
that she connected to living.

She had meant to bow her head,
let the lids lower over her eyes
and the muscles tighten around her mouth.
But he already had come closer,
and the blue-green of his eyes
faded into the inside purple of his lips,
fixing her gaze
and holding her still.

He already was telling her of the hunger
that made him long
for bits of fruit and nuts
and the words that his mother once had read to him.
He now no longer read or ate.
His stomach and mind had come apart,
all at once, going to pieces
that commingled in the pink and brown stuff he vomited.

"I am starving," he said.

She tried to reach for her pen, but her fingers
could not hold it and it was left to lay
across the forms,
leaving them blank,
not even his name would be written.
She felt his longing and hunger overwhelm her
and fuse with the memories of the fat child
she had been, stuffed
with food she had had no intention of eating.

There had been no one to read to her.
Words had been denied her by those closest to her,
who had sensed her desire for words,
only to experience it, through a mix of envy and rage,
as a vague memory of something they once knew, but no longer.
They had not had much chance for learning
and had turned their deprivation into a mean way of life.
She only would give them fresh reason
to scorn every sort of subtlety
to resist the complexity of meaning.
There was no one to understand
the sweetness and blessedness of the words filling her mind
and dancing in her thoughts.
There had been no one who did not mean to make her despise words
and distrust their pleasures.
Finally she could only hold words in some terrible awe,
while forbidding them to herself.

She practiced at erasing bits
and pieces of letters, breaking up each and every word,
so that they might dance on in her thoughts, disabled,
yet not forever banished from her mind.
She did read in secret,
searching page after page for new words,
words that would not be broken, that would make her whole.

But no matter how much she tried
to save the words just as she read them,
they shattered to pieces,
flying free in her mind,
joining the leftover heaps of fractured letters,
endlessly giving way to word shapes, never repeated
an internal poetry, born of randomness
that frightened and fascinated her,
drawing her to the abstract expressionism of those unsettled minds,
destined for insanity.

She had tried to lift her eyes just above his,
to steady her focus on the braided cloth
that wrapped around his head and
girdled the uneven strands of hair,
holding them close to his face, soft
against the hard of his cheekbone.
She tried not to look at the gaunt face,
the skin stretched so tight that,
although nearly decaying, it
had the transparency of a satiated nursling.

"I had hoped you would not yet leave me," he said.

When he was a young boy, his mother read him rhymes
from a big book, bound in leather and engraved with golden letters.
She read to him each evening,
when the yellow sun set in the brown red earth.
At first, he did not pay much attention
to the meaning of the words.
It was something else he awaited.
It was the feel of her small round breast, lightly touching
his shoulder,
gone and returning
in counterpoint with her breathing,
in and out.
He felt his body go limp,
lost to the mix of rhyme and rhythm of body and words.

But the mother's body never relaxed. He could feel some tension,
some tightness, that would not let her go,
would not let her give herself to him with abandon,
drawn back,
just when he thought himself safe.
Stiffening to attention, he would brush the tears away,
drying his eyes with the words.

Day after day, word after word, he took hold of them,
eyes, mouth, and ears filled with them
until words were all that could make him safe,
in body and soul.
They were all he could remember of her
when she was gone,
one day just gone.

He had been a brilliant student,
fulfilling his teachers' hopes and desires.
He had read poets and philosophers,
studied art and politics.
He had learned music, calculus,
French, and Italian.
He wrote treatises, marked by a careful logic and a steady mind:
thoughts adding up,
conclusions made with force and fervor.

But it was not enough.
He wanted to go beneath the words,
to what held them up to their momentary meanings,
to the stuff below, which moved words slow and steady,
light and fast, lifting them through
layer after layer to the surface.
He craved the rhyme and rhythm of words,
searching text after text,
until he wanted to do nothing but read.
The drugs were a help at first. And then,
they gave him everything he had ever desired.
Everything he had read and written
turned into the exquisite abstraction of sounds and sights
of moving intensities,
ending in a moment that was nothing less than eternity
in which he could feel his lips curve
around the erect nipple of a lover's breast.
He could feel the touch of her long fingers
drawing his face close,
gently caressing his cheek,
softening the hardness of the bone.
He could hear love sonnets, all at once,
whispered so tenderly, so sweetly.
He felt awash with grace, forever saved,
as his lover retraced the curve of his brow with her thumb.
Undulation and quick quivering.
A bittersweet fragrance never smelled before,

only now,
heightened the pleasures vibrating along the invisible tendrils
that wrapped around his mind
until eternity ended. All its pleasure suddenly escaped him,
become some garish architecture of death and horror
coming at him from the outside.
His hunger and longing his, but no longer his,
coming at him from the outside.
Terrifying and unforgiving,
coming at him from the outside.

She felt the ecstasy go to anguish and pain.
She no longer knew if it were his or hers.
She only knew she wanted to put her lips
so gently to his and to kiss him forever.
She thought of washing him, part by part,
and swathing him in white gauzelike strips of cloth,
cradling him in her arms.
She had never had a lover.
The desire for one, long ago folded up
once, twice
and placed deep inside her.
Forgotten, now unfolding.
He had moved her.

She might have been a brilliant student.
But her teachers had quickly grown impatient.
They gave no encouragement, finding her writing confused,
if any meaning there,
not worth the difficulty of piecing it together.
They scratched lines through phrase after phrase,
red circled word after word.
She felt ashamed, but she did not want to give up.
She could still feel the pressure of the memory of words,
sweet and blessed.
But somehow without knowing it, she did give up.
The memory faded into a blankness.
It made her face plain and her hair dull.
It turned her skin gray and made her lips lose their moistness.
She had become bent from devotion
to a practical profession,
thinking herself that way safe from the envy,
except now and again,
when she would see it flash in someone's eye,
when she said something, and with its unusual turn of phrase,

made a world, never imagined, unfold before them.
A sudden and surprising fresh creation of thought
that made someone jealous,
then angry and dismissive,
leaving her, although for a moment more alive,
in the end, more frightened and withdrawn.
He was her last chance, her only chance.

"I am dying," he said.

She meant to hold him in her gaze
and to speak with what eloquence was left her,
poetry so simple,
a word sculpture made of the commonest things of life.
To start again at the beginning.
To go step by step from death to living.
But words failed her; they broke apart before she could speak them.
She never promised to share her life with him,
to give him back his mind
and by that, gain the hope of recovering hers.
Words had failed her, and then he was gone,
just gone.
"I will help you," she said.

CHAPTER FOUR

The Ontological Perspective of Knowledge Objects

The Writing Criticism of Technoscience

The early criticism of ethnographic writing drew on literary theory in order to analyze the way in which the ethnographic text constructed the scientific authority of western anthropology. The intended political aim, however, was to adjust anthropology's ethnographic form to decolonization when the "west," in James Clifford's often repeated words, "no longer [could] present itself as the unique purveyor of anthropological knowledge about others."[1] Although the criticism of ethnographic writing has led to a recognition of the interrelationship of colonialization and the production of western anthropology's scientific authority, a more general cultural studies of science, of which the early criticism of ethnographic writing is a part, has more often been met with a disapproving response; specifically, it has been accused of being politically quieting.

The Alan Sokal hoax involving *Social Text*, a journal of Marxist cultural studies, might be taken as one example, if not the most outrageous example of such a response. In an essay appearing in a special issue of *Social Text* on science studies, Sokal drew connections between the field of modern physics and Derridean deconstruction, feminist theory, and Marxist cultural studies;[2] later Sokal claimed that in the essay published in *Social Text* he had purposely offered insupportable arguments and had drawn illogical conclusions, which nonetheless had gone unrecognized as such by the editors of the journal. He proposed that this had occurred because of the editors' unquestioned presumption of the political correctness of the cultural studies of science. Sokal claimed that his aim in perpetrating the hoax was to teach the "leftists" involved in science studies that they do not know science or politics, not if they mean to turn the latter against the former and break with what he described as "the two century old identification" of the left with science in order to lay bare "the mystifications promoted by the powerful."[3]

In the wake of the Sokal hoax, a number of leftist critics expressed concern that the only result of science studies might be an epistemological relativism that makes science impotent in the cause of leftist politics. They called for a return to a materialist analysis of political economy, without, however, taking account of Sokal's own deployment of a cultural politic aimed at manipulating the vulnerabilities of exposure to media event-ness, a politic characteristic of the age of teletechnology. Nor did they pay much attention to the shift in ontological perspective connected to teletechnology, which has informed the materialism elaborated in the cultural studies of science. Indeed, restricting the treatment of science studies to its epistemological implications not only relegates science studies to a cultural politics of representation; it thereby makes it more difficult to realize the ontological implications of science studies for rethinking the relationships of nature and technology, body and machine, the virtual and the real as *différantial* relationships, given out of the dynamism of matter.

Of course much of the response to the early criticism of ethnographic writing also has focused on its epistemological implications, which has made it more difficult to recognize it as part of the cultural criticism of science, having ontological implications linked to the teletechnological and enabling the thought of the *différantial* relationship of nature and technology, body and machine, the real and the virtual, and the living and the inert. Such response has relegated the early criticism of ethnography, like that of the cultural studies of science, to a cultural politics of representation, where writing and textuality are all too narrowly conceived and treated as "merely cultural."[4]

True, the early criticism of ethnographic writing focused on the historical development of western anthropology in relationship to colonialism by tracing the connection between ethnographic writing and European literary forms of the eighteenth and nineteenth centuries—travel writing, letters, diaries and the realist novel. But this focus on textuality and writing also links the early criticism of ethnographic writing to teletechnology and to "a world of generalized ethnography" where, as Clifford first noted, "expanded communication and intercultural influence" are allowing people to "interpret others, and themselves, in a bewildering diversity of idioms—a global condition of what Mikahail Bakhtin called 'heteroglossia.' "[5]

Clifford's remarks suggest that a globalized teletechnology has drawn the early criticism of ethnographic writing to it, such that its treatment of writing and textuality reaches to the larger sense of these terms, that is, to the global miredness of finite forces, a global condition of heteroglossia, out of which political agency arises. His remarks suggest thinking about the early criticism of ethnographic writing as well as the experimentations in writing under the condition of a generalized ethnography—that is, the postmodern ethnography, the autoethnography, the literarization of a politics of identity and a politics of location—all as part of the globalization of teletechnology and the transnationalization of capital in neocolonialism, where the configuration of family and national ideologies, the state and civil society, and the public and the private spheres presumed in modern western discourse is being imposed, but often refused and reconfigured.

But all this also means that we must think about the early criticism of ethnographic writing as well as the writing experiments of a generalized ethnography as part of social situations in which the agency of the knowing subject is not only a matter of the individual's embeddedness in intersubjective relations of knowledge production, connected to face-to-face communities as well as to large organizations such as the research university, health care institutions, corporations, or government bureaucracies. Rather, the agency of the knowing subject also refers to an embeddedness in environments of "knowledge objects," where agency and reflexivity refer as much to an "interobjectivity" or "the sociality of objects" as it does to intersubjectivity.[6]

This is a sociality under the conditions of technoculture and technonature, where technoscience has become a primary agency of power/knowledge—a sociality that often has been connected to what sociologists refer to as postmodern societies or "knowledge societies," that is, societies each of which, as Karin Knorr-Cetina describes it, "is not simply a society of more experts, of technological infra- and information structures, and of specialist rather than participant interpretations. It also means that knowledge cultures have spilled and woven their tissue into society, the whole set of processes, experiences and relationships that wait on knowledge and unfold with its articulation."[7]

It is under these conditions, I want to suggest, that modern western sociology and modern western anthropology have been opened to each other and both to worldly matters at the intersection of the criticism of ethnographic writing and the elaboration of the cultural studies of science. And no doubt there has been an intersection. The most crucial input to the cultural studies of science, after all, has come from what is referred to as "the new sociology of science," which first traced the lines of sociality characteristic of knowledge societies. It also is the case that the early criticism of ethnographic writing transmigrated from western anthropology quickly to western sociology, where it not only influenced the new sociology of science, but also provoked experiments in ethnographic writing that are thought by some sociologists, myself included, to be linked to teletechnology. That is to say, the experiments in writing are thought to be experimentation with the technical substrate of unconscious memory given with teletechnology.[8]

In the first and second parts of this chapter I want to explore the links between the cultural criticism of science and the criticism of ethnographic writing. I want to propose that the former has overseen the becoming of technoculture and technonature, as well as the becoming of technoscience as a primary agency of power/knowledge, whereas the latter has led to the experimental writings of a generalized ethnography that is the condition of possibility of any self-criticism in the practices of knowledge production in the age of teletechnology. The writing experiments of a generalized ethnography are not, therefore, merely about textuality or writing in the narrow sense of these terms. Nor are they only vehicles for authorizing the voices and experiences of subjects long excluded from the authority of modern western discourse.

Rather, the emphasis in experimental ethnography on the poetic, on experimental writing forms or textual devices, as well as on performances of self-exposure, suggests that the self-criticism being elaborated for practices of knowledge production is an effort to reach beyond the hyper-self-reflexivity of the teletechnological in order to critically engage it through a more direct intervention in, or a more direct interference with, the human and nonhuman agencies of the machinic assemblages of technoscience, technoculture, and technonature. The question of how to critically engage machinic assemblages while also attending to subjects' desire for cultural identity, although now an even more pressing question, has from the start drawn the criticism of ethnographic writing and the cultural studies of science together to the unconscious thought of teletechnology.

Reflexivity and the New Sociology of Science

It was in the 1970s that sociologists, among other scholars, began to rethink the sociology of science established by Robert K. Merton in the mid-1940s. Merton had proposed to study science in terms of the social relationships between knowledge practitioners, the effects of science on society, and the institutional development of science, including the political dynamics of funding. The next generation of sociologists doing science studies, however, shifted the focus of science studies to the content of science, that is, to the social production of scientific knowledge itself. This made it possible, if not necessary, to rethink the reflexivity of knowledge and the self-reflection of the scientist, including the self-reflection of the sociologist doing science studies. A displacement would take place; there would be a shift in focus in the domain of reflexivity and self-reflection, turning attention from the scientist's reflection to the scientist's deployment of inscription devices or knowledge objects embedded in the relationship of writing, textuality, and technology.

Michael Lynch and Steve Woolgar have suggested that although poststructuralism is to be counted as an influence on the new sociology of science, perhaps even more important at first were philosophers and historians of science, such as Thomas Kuhn, Ludwig Fleck, Michael Polanyi, Imre Lakatos, and Paul Feyerabend. They shifted the focus of science studies to "the importance of 'agreement' in communities of scientists on matters of fact and procedure, and of efforts to enlist such agreement through persuasive appeals."[9] These historians and philosophers suggested that scientific knowledge is socially produced. With the development of the Edinburgh "strong programme" of science studies,[10] science would no longer be studied in terms of its truth or its mimetic relationship to "reality" or "nature"; scientific knowledge would be treated instead in terms of the local processes of its production. The contents of scientific knowledge would be treated as an accomplishment, as a doing. Researchers were to question how scientists actually produce models or do experiments, how they represent and make use of technologies in the production of scientific knowledge, how they represent the authority of scientific knowledge.

Given this focus on the way science is socially produced at local sites, ethnography became the preferred method for science studies. Researchers, among them Michael Lynch, Harry Collins, Trevor Pinch, Karin Knorr-Cetina, John Law, Bruno Latour, and Steve Woolgar, turned the ethnographic method to the study of lab work. Ethnography perhaps is most creatively deployed in the observation of what Latour and Woolgar first called "inscriptions."[11] Inscriptions or inscription devices, such as graphs, machines, or narrative forms, function to "draw things together," as Latour puts it;[12] they make possible a transference of elements across different surfaces — the literary, the cultural, the political economic, and the biological — in such a way that the connection between matter and image, discourse and institution, and text and world is without break.

In Latour's discipline-defining study of Louis Pasteur's lab work, which Latour describes as an "ethnography of inscription," he shows how inscription is to be treated in order to uncover what and how it produces. Latour suggests that inscription allowed Pasteur to translate across spaces that would seem impossible to connect, such as "a dirty, smelling, noisy, disorganized nineteenth century animal farm and the obsessively clean Pasteurian laboratory.[13] Once these were connected, it was possible to control the farm environment from the lab by isolating and controlling the disease-causing microorganism within the lab. Pasteur was thereby able to make the lab a "theatre of proves," staging a spectacle of the dead unvaccinated animals and the living vaccinated ones. This finally authorized the movement of the vaccine from the lab back to the farm and then to the entire society. Through its dissemination of the vaccine, the lab controlled the social life of farmers and all of France. All of France, Latour argues, was drawn into a "network much like a commercial circuit."[14] For Latour, all this is to say that Pasteur's lab work defined a society, one dependent on a certain kind of science.

Beyond showing the inseparability of the contents of science from inscription devices, Latour's study of "the pasturization of France" also allows him to draw the stunning conclusion that one of the effects of science is that in its lab work science and society are reconfigured. For example, when what is at the farm is moved to the lab and what is in the lab is moved back to the farm, the distinction between the inside and the outside of science is reconfigured, and with it a society is defined. Against Merton's much-defended view that political and economic forces influence science only from outside of science, Latour suggests that there is no outside of science that science itself does not construct; what appears as an outside of any science is only the displacement of prior extensions of former sciences.

Through its lab work, then, science plays an important part in the territorialization, the deterritorialization, and the reterritorialization of social spaces. Science is a primary force in the configuration and the reconfiguration of the arrangement of family and national ideologies, the state and civil society, the private and public spheres presumed in modern western discourse. Thus, lab work is itself both politically and economically productive; as Latour puts it, "it is in laboratories that most new sources of power are generated."[15] And, in lab work, it is inscription that is central to the generation of power.

Inscription is central to the generation of power because inscription is "flat" or two-dimensional, but it nevertheless manipulates three-dimensional objects. Its ability is to modify the scale without changing the internal proportions of three-dimensional objects; therefore, inscriptions can transform and transfer three-dimensional objects from one surface to another until matter and image, discourse and institution, text and world are inextricable. Latour describes this capacity of inscription as its "merge with geometry."[16] But although inscription devices are crucial to the generation of power, Latour argues that they are so only as "the fine edge" of the mobilization or "machination of forces" that draws disciplinary knowledge, the scientist's actions, technical objects, and interest groups together without distinction, on the same plane of consistency, as Deleuze and Guattari might put it. Each time a society is defined, Latour argues, all that is thereby named social will have passed through a powerful machination of forces.[17]

The power of the machination of forces is to be understood in terms of speed, driven as it is by "immutable mobiles" that are meant to increase the movability and the immutability of traces of arguments, representations, and technical devices. Latour offers, for example, the many identical copies produced by the printing press, which make possible an acceleration of the transfer of information, but also incite a desire for debate over the compatibility of any copy with reality in the various contexts of the different times and places of a copy's migration.

These debates only increase the circulation, and therefore the value, of traces of arguments, representations, and technical devices. It is precisely the way Latour treats inscription in the machination of forces, or what he also simply calls "machines," that gives a sense of the relationship of science studies to the becoming of technoscience; moreover, there is a sense of technoscience as an agency of power/knowledge in that it can materialize or surface various elements from various series, on the same plane, such that Latour argues: "Going from science to technology is not going from a paper world to a messy, greasy concrete world. It is going from paperwork to still more paperwork, from one centre of calculation to another which gathers and handles more calculations of still more heterogeneous origins."[18]

Inscription, of course, also is deployed in the authorization of the scientist and his or her representation of findings. Inscription devices produce scientific authority through the production of what Latour describes as the "double text" of scientific representation. On one hand, there is the argument presented, and on the other hand, there are diagrams, tables, images, and rhetorics — all to make the reader more easily grasp the argument presented as scientific facts about nature or reality. Although Latour does not account for it, there seems to be a blindness induced in the reader by the representation, so that the reader does not become conscious of the functioning of the double text. In other words, Latour's treatment of the double text indicates, but does not treat, the fictionality of the scientist's realist account, its deployment of a narrative logic to engage the reader's unconscious or blind(ing) identifications. Latour's treatment of the double text points to, but does not critically engage, the links between the scientist's account and cultural forms, like the realist novel with its powerful literary mechanism, both to maintain the opposition between "what is made up" and "what is really out there," and by that means authorizes whoever is figured as the narrative's subject.[19]

But it is these links between science and cultural forms that will become more troublesome when questions arise about the use of inscription devices in the field of science studies itself. For one, how is the ethnography of lab work to frame the loops of reflexivity between what the ethnographer represents and the productivity of his or her own science studies or lab study? Will the framing of the reflexive link between the science that studies and the science that is studied turn into the ethnographer's endless self-reflection, threatening to undermine the scientificity of science studies itself? Such questions, although avoided by Latour and many of those doing science studies, finally are brought into science studies by those who would engage the early criticism of ethnographic writing and further explore the links that connect the early criticism of ethnographic writing with various cultural theories, such as feminist theory, postcolonial theory, queer theory, and critical race theory.

It is not surprising, then, that along with a number of other feminist theorists, Donna Haraway takes Latour to task for his inattention to his own writing form; she would argue that he and others in doing science studies often fail to examine their own inscription devices. Especially significant is their failure "to draw from the understandings of semiotic, visual culture, and narrative practice, coming specifically from feminist, postcolonial and multicultural opposition theory."[20] Haraway insists that it is important to engage these critical theories in approaching "the potent incarnated fictions of science," especially those "that run riot through technoscience." This is because technoscience smooths out the arrangement of social spaces presumed in modern western discourse, such that the opposition between science and politics, science and society, and science and culture are collapsed, and no one of the opposed terms can any longer "be reduced to the status of context for the other."[21]

In not attending to these critical theories, Haraway argues that ethnographers doing lab studies often reproduce in their ethnographies the very figures of scientific authority that science studies means to deconstruct. Latour, for example, deploys an ethnographic realism that constructs scientific authority and objectivity by means of the narrative fiction of the ethnographer who enters the field and becomes submerged in what is other, foreign, or mysterious, then struggles to free himself in order to finally return from the field with the scientific facts about nature or reality — his struggle legitimating the authority of his knowledge. Haraway notices that it is in the terms of a "virile heroics" that Latour describes his own work as an ethnographer; she refers to his description of lab work: "Surprisingly few people have penetrated from the outside the inner workings of science and technology, and then got out of it to explain to the outside how it all works."[22]

Haraway cannot help but notice that those who do science studies often figure as feminine whatever in their own lab studies seems to demand a struggle, what in their own studies seems foreign or mysterious. This enables the deployment of a narrative logic by which the lab researcher figures himself and his acts of research in terms of virile heroics; the lab researcher thereby appears as a warrior, "testing the strength of foes and forging bonds among allies, human and nonhuman, just as the scientist hero does."[23] Indeed, various of the figurative devices used in the production of scientific authority are also deployed in lab studies; Haraway points to " 'the self-birthing of man,' 'war as his reproductive organ,' and 'the optics of self-origination.' "[24] All these figures, of course, fit what feminist theorists, especially feminist film theorists, have referred to as the oedipal logic of narrativity, suggesting that this narrative has been a dominant cultural narrative because it not only functions to produce an authorized subject for mass media, such as film; it also functions to produce the subject of scientific authority.

Haraway's criticism notwithstanding, there has been some experimentation in the writing of science studies. Steve Woolgar's attempts are notable. Woolgar writes himself into the text of his ethnographic studies in order to correct for the presentation of science studies, as if there is not a perspective from which observations are made and a location from which the ethnographer writes. Woolgar instead autographs or personalizes the text with dialogue and commentary about the writing, ruffling the smooth surface of the text that is usually offered as scientific facts about nature or reality. But Woolgar also expresses concern that his efforts at experimenting with writing might be no more than "a self-conscious and clever device" of introspection that, in its more "benign" form, has always been tolerated in science as a way of improving research.[25]

And this is not all that is problematic. Woolgar's autographic and personalized gestures are also textual devices, which can be further deconstructed. Although they allow him to critically engage the formulas taken for granted both in producing the authority of an ethnographic realism and in distributing the relevancy of various realities, Woolgar's deconstructive strategies do not put an end to the production of authority in writing, once and for all. Even when, after the deconstruction of the authority of ethnographic realism, the ethnographer produces a more self-conscious text, the text produced is a text. It still refers; it still incites and fulfills the desire for reference to a reality outside the text. As such, it is still open to further deconstruction.

It is for this reason that Woolgar's experiments are not exactly what Haraway had in mind when she criticized Latour and others for not paying enough attention to their own writing form. She is not fully supportive of what she describes as Woolgar's "relentless insistence on reflexivity, which seems not to be able to get beyond self-vision as the cure for self-invisibility."[26] Haraway argues that reflexivity and self-reflection are "bad tropes" for the cultural studies of science because they only sustain "the search for the authentic and the really real."[27] It is in this context that Haraway offers the notion of "diffraction," which is meant to go beyond both hyperreflexivity and self-reflection in order to meet the implosion of the distinction of matter and image, discourse and institution, and text and world that is characteristic of the technoscientific production of material-semiotic objects. Although still retaining its link to vision, diffraction (as in light passing through a prism) is more about recording the movement of forces (such as recording light passing through the various slits of a prism); diffraction is about recording histories of the movement of the forces of material-semiotic objects with the aim of intervening in or directing and redirecting the movement.

Like her deployment of the cyborg, Haraway's deployment of diffraction is meant to recognize human and nonhuman agencies, to recognize the *différantial* relationships of body and machine, nature and technology, the virtual and the real, the living and the inert. The notion of diffraction is meant to allow for a critical engagement or a direct interference with the speeds of deterritorialization and reterritorialization of social spaces and with the vulnerabilities to exposure to media eventness. It is meant to make it possible to distinguish the terrors of certain embodiments of technoscience while generally valorizing "techno-organic kinships."

In offering the notion of diffraction as a correction to those efforts, like Woolgar's, to personalize the ethnographic text, Haraway means to short-circuit the accusations that a self-conscious science studies only produces an epistemological relativism and a political paralysis, although she does recognize that diffraction also involves "the constructed and never finished credibility of those who do it, all of whom are mortal, fallible and fraught with the consequences of unconscious and disowned desires and fears."[28] Nonetheless, Haraway wants to respond to the becoming of technoscience, technoculture, and technonature. So she offers the notion of diffraction as a way to do cultural criticism of science, given that technoscience has become a primary agency of power/knowledge and when, therefore, technoculture and technonature need to be the focus of cultural criticism. But of course, when Woolgar first experimented with the ethnographic writing of science studies, it was when the early criticism of ethnographic writing in anthropology was first being offered and when it had just begun to provoke various experimentations, among them autographing and personalizing the ethnographic text. Although experimentation with autographing the ethnographic text would have a limited career among those doing lab studies, it would have a richer and longer career in anthropology, sociology, and cultural criticism generally.

Writing Criticism of Ethnographic Writing

Ironically, just as ethnography was becoming the preferred method for the new sociology of science, a criticism of ethnographic writing in anthropology was being elaborated. In his introduction to the now famous collection of essays entitled *Writing Culture: The Poetic and Politics of Ethnography,* James Clifford makes use of footnotes to draw a connection between his criticism of ethnographic writing and science studies. He especially refers to Latour and Woolgar's treatment of inscription devices.[29] Clifford's criticism of ethnographic writing might itself be considered a treatment of ethnography as a potent inscription device for producing the scientific authority of western anthropology.

But Clifford is much more concerned with the fictional or imaginary aspects of science's authorizing narrative, especially the way it figures the observed "native/other." In his 1983 essay "On Ethnographic Authority," Clifford treats the narrative form that ethnography shares with the realist novel and that anthropologists developed for ethnography beginning in the 1920s with Malinowski's *Arogonauts of the Western Pacific.*[30] Clifford argues that for the generation of ethnographers following Malinowski, the realist narrative provided a form with which to present a seemingly objective picture of "the captured reality of the other."

The realist narrative allows a picture of the other's experiences to be projected without the ethnographer's being visibly present in the picture, although his or her absence is to be felt as a present absence. This felt absence is made even more palpable when supplemented with a story that frames the picture of the other; the story tells of the anthropologist's heroic activity of entering the field, suffering its difficulties, and struggling to leave it. With the deployment of this framing device, Clifford suggests, "the predominant mode of modern field work authority is signaled: 'You are there, because I was there.' "[31]

The authority of anthropology as an empirical science is grounded, therefore, in its ethnographic narrative of observation, rendering the ethnographer's hero-making struggle a matter of being able to see, to see deeply and widely, without becoming lost in the seeing or in the scene of observation, that is, without the observer's being observed in his or her observations. But the realist narrative also makes it possible to see more quickly; that is, it allows one to make, with much fewer observations, a structural generalization about any culture as a whole way of life.[32] Clifford proposes that the deployment of a realist narrative in the composition of the ethnographic text is inextricable from the notion of culture as a closed structure of reality. In sum, the realism of the ethnographic narrative underwrites the method of participant observation, giving it what appears to be its power to pierce through the surface of behaviors, and therefore to see quickly the larger structure of which these behaviors are only an effect. So, along with structuring the reality of a culture as a whole, the realist narrative makes the control of speed and the adjustment to exposure to media event-ness scientific values of modern western anthropology.

Ethnography, therefore, did not borrow only from the realist narrative of the novel developed in the eighteenth and nineteenth centuries. It would seem that it also borrowed from realist narrative cinema, which was developed in the first half of the twentieth century. The cinema provided a resource for a certain understanding of vision that is deployed in the textuality of ethnographic realism. In his introduction to *Writing Culture,* Clifford again makes use of footnotes to draw a connection between ethnography and realist narrative cinema. He points to the way that feminist film theorists already had deepened "the critique of visually based modes of surveillance and portrayal, linking them to domination and masculine desire."[33] He gives as an example Laura Mulvey's treatment of the voyeuristic aspects of Hollywood realist narrative cinema. Also mentioned is Teresa De Lauretis's treatment of cinema's oedipal logic of realist narrativity and its deployment of a rhetoric of sexual difference. After all, De Lauretis already had argued that the oedipal logic of realist narrativity is "predicated on the single figure of the hero who crosses the boundary and penetrates the other space. In so doing the hero, the mythical subject, is constructed as human being and as male; he is the active principle of culture, the establisher of distinction, the creator of differences."[34] The oedipal logic of realist narrativity that is operative in classical Hollywood cinema is easily recognized as being operative in ethnographic realism as well, there to stage the erotic fantasy of the ethnographer's penetration of the field and the final traumatic freeing of himself from immersion with the other in order to return home, a man and a hero, with a vision of the truth of reality.

Clifford's reference to feminist film theory raises questions that, however, he did not pursue, about the relationship of ethnographic writing to the various technical substrates of unconscious memory given with various technologies and changing the relationship of narrative and images, time and space, and being and historicity. With only a few exceptions, such as Christian Hansen, Catherine Needham, and Bill Nichols's treatment of ethnography and the pornographic aspects of realist narrative cinema,[35] there has not been much criticism of ethnographic writing oriented to an analysis of it beyond its links to the realist narrative of the novel, that is, an analysis of ethnographic writing in relationship to film or teletechnology. Yet Lisa Cartwright[36] has shown that a visual culture usually is shared by science, professional expertise, and the popular mass media. Her focus, however, is on the visual culture shared by cinema and medical sciences. Although Cartwright does not address what is shared by the popular media and the ethnographic method of social science, her treatment of visual culture, nonetheless, suggests the possibility of extending the early criticism of ethnographic writing to the technical substrate of unconscious memory given with teletechnology.

Cartwright illustrates her argument about the visual culture shared by medical science and cinema by treating film technology beginning with the film motion studies of the late nineteenth century and ending with radiography, ultrasound, and mammography, especially as these are applied in the field of women's health in the late twentieth century. Not only does Cartwright suggest that the technologies of the late twentieth century now make it necessary to see the human body as part of a machine technology in that "one can no longer speak of bodies or objects of knowledge without acknowledging the in-built technologies through which their health and life are regulated and disciplined."[37] She also suggests that it is only with the attachment of narrative to film technology in the early twentieth century that interest in the complexities of a movie's content begins to override the pleasure of film's spectacularization of motion itself, a pleasure that was apparent to the viewers of the film motion studies. It would seem that adding narrative to film technology subordinates film's technical substrate of unconscious memory to a narrative logic, the oedipal logic of realist narrativity.

Cartwright's argument, therefore, also suggests that narrative makes it more difficult to grasp the various technical substrates of unconscious memory given with various technologies. Narrative even may be deployed to make it more difficult to realize that unconscious memory has become attached to a new technical substrate. That is, narrative may be deployed in slowing down the realization of the changed relationships of temporality and spatiality brought on by the introduction and development of a new technology. The oedipal logic of realist narrativity works this way. At least it has worked this way in its transmigration from the novel to cinema and then even to teletechnology. Yet it is the electronic time-image of teletechnology that makes it possible again to think of an analytic separation of the technical substrates of unconscious memory from narrativity, the oedipal logic of realist narrativity in particular. It is this possibility, I want to suggest, that has overseen the early criticism of ethnographic writing and has drawn it to the future.

After all, both the time-image of teletechnology and the early criticism of ethnographic writing force a deconstruction of the oedipal logic of realist narrativity, making it possible to distinguish the oedipal narrative from unconscious memory and its various technical substrates. Both teletechnology and the early criticism of ethnographic writing point to a different configuration of bodies, space and time, and being and historicity than those that the oedipal logic of realist narrativity inform; questions even are raised as to whether narrative form still is the dominant form of authorizing scientific knowledge or whether vision still is central to establishing this authority.

All this suggests that the early criticism of ethnographic writing was not only about writing or literary form in any narrow sense. Rather, the early criticism of ethnographic writing seems to point to a larger issue, that is, the teletechnological condition of possibility of the authority of

empirical scientific knowledge. In this sense it might be argued that although the early criticism of ethnographic writing made ethnography
an object of literary criticism, it also made it necessary to further problematize the literary in relationship to writing technologies other than
that of the novel, such as those connected with cinema and teletechnology. Surely the early criticism of ethnographic writing would have
to be extended beyond a literary criticism narrowly conceived in order
to be rearticulated in terms of the glocalization of world cultures as
part of the globalization of teletechnology and the transnationalization
of capital. As Trinh T. Minh-ha argues, when the issues raised by feminist theorists and postcolonial theorists about neocolonialism are made
into literary concerns, it becomes "quite easy for anthropologists to by-
pass, if not dismiss, the issues raised by confining them to the realm of
literature."[38] But the issues raised, Trinh suggests, are not resolvable only
by fixing the literary style of ethnography.

What Trinh finds problematic, however, is the reduction of relations
of power/knowledge to writing style when writing or textuality are narrowly conceived. Trinh instead wants to deconstruct the apparatus that
administers power/knowledge and critically engage the machinery for
interlacing the differences of nation, race, class, ethnicity, sexuality, and
gender with the globally distributed relations of power/knowledge. In
other words, after the deconstruction of western anthropology's authority,
the question of ethnographic writing remains a question about power/
knowledge in relationship to the globalization of teletechnology and
the transnationalization of capital in neocolonialism.

Although concerned not to reduce questions of power/knowledge to
writing style, Trinh does not, however, refuse to acknowledge experimental writing altogether. Her experimentation usually involves the mixing
of technologies to create a space of movement in the in-between, a space
to play with the eye/I of the various visually oriented apparatuses and
the various subject positions they elaborate in the production of knowledge of the self and the other. Trinh is concerned to bring different
technologies or genres into the same space in order to put into question
any "framing of consciousness" and thereby lead theory to "dangerous
places."[39] This mixing of technologies and genres, however, is not only
a matter of Trinh's being a filmmaker. It is also about her desire to perform the complexities of the "native woman other," as she comes to

write for herself and others in the wake of colonialism, the deconstruction of the authority of western anthropology and neocolonialism in the age of teletechnology.

Trinh's experiments, therefore, are exemplary of a tension between the deconstruction of the ethnographic authority of western anthropology and giving writing over to the multiplicity of voices and sights of those whose very existence and everyday life practices have been the object of western anthropology. The complexity of her work is an elaboration of the lessons of the deconstruction of the ethnographic authority of western anthropology while being drawn, nonetheless, by the desire to give the sights and sounds of the self and others — if not to "speak about," then to "speak nearby," as Trinh puts it. Trinh thereby engages in what Gayatri Chakravorty Spivak, among others, has referred to as "strategic essentialism." For Spivak, strategic essentialism is a critical practice that seemingly allows the assertion of the cultural critic's authorized subject identity, but is something more like a performative framing, functioning "not as descriptions of the way things are, but as something that one must adopt to produce a critique of anything."[40]

Spivak also warns, therefore, that in ethnographic encounters, even when these are encounters between "radicals and the oppressed in times of crisis," what must be recognized is what Derrida describes as an ethic in "the experience of the impossible," or what he also refers to as "the secret." The experience of the impossibility of full subject identity, as well as the impossibility of fully disclosing encounters between the ethnographer and others, is something like an experience of the unconscious, an indefinitely deferred nonknowing, which Spivak suggests is to inform the writing of ethnography.[41] Trinh also wants cultural criticism to be guided by this ethic, and therefore she wants ethnographic writing to be ghosted by the nonknowable.

In Trinh's work there is a deferral of the authorizing identity of the ethnographic writer, the endless displacement of self-same identity. As she puts it: "In writing close to the other of the other, I can only choose to maintain a self-reflexively critical relationship toward the material, a relationship that defines both the subject written and the writing subject, undoing the I while asking 'what do I want *wanting* to know you or me?'"[42] Trinh's displacement of identity is even a refusal to be the representative of "the third world woman" or "the woman of color":

"There is no real me to return to, no whole self that synthesizes the woman, the woman of color and the writer; they are instead, diverse recognitions of self through difference, and unfinished, contingent, arbitrary closures that make possible both politics and identity."[43] Again Trinh's remarks seem to resonate with Spivak's advice that when confronted with the subaltern, what matters is "not to represent (*verstreten*) them, but to learn how to represent (*darstellen*) ourselves."[44]

But Trinh also wants to tell stories. To do so she produces something like documentaries — on Vietnam, Africa, and China — but not without critically deconstructing the presumed identities of the documentary form. Trinh's film *Surname Viet Given Name Nam* is a well-known example. Not about the war between Vietnam and the United States as might be expected by those socialized on Hollywood Vietnam War movies, the film is rather a more complicated exploration into the identities of Vietnamese people, especially Vietnamese women. It is organized around a set of interviews with women, which, however, were filmed by someone other than Trinh and then translated. In the film the interviews actually are performed by Vietnamese women who are in the United States, but who give the appearance that they are speaking from Vietnam, speaking about and for themselves. In a later part of the film, even the women performers are interviewed about their performances as well as their lives in the United States.

Like all of Trinh's films, *Surname Viet Given Name Nam* foregrounds not only the performative aspects of identity, but also the hybridity of a cultural location, including Trinh's location as professional filmmaker. *Surname Viet Given Name Nam* is exemplary of Trinh's deployment in ethnographic film of film techniques that have been exhausted in avantgarde film and that already have been transferred to television. This gives Trinh's films a look of mixed media across multiple planes of time, space, and place, a critically complex practice of knowledge production that allows for both the deconstruction of identity and the strategic use of identity and that shows the dependency of such on the mixed deployment of technologies.

There is in Trinh's films, therefore, a multiplicity of voices, a heteroglossia, that evokes a sense of identities; there also is the elaboration of an ethic of impossibility, which is expressed in the refusal of identities. For this reason Herman Rapaport argues that in Trinh's work deconstruction "has given way to multiplicity, coalition, hegemony, collabo-

ration and hybridization." The result is "that deconstruction is prohib-
ited from becoming one with itself as an objectifiable ensemble or totality
that can be associated with highly developed institutional practices."[45]
For Rapaport, Trinh's work gives a different destiny to deconstruction
than Derrida imagined. But perhaps this is the only destiny deconstruction could have, one that is not predetermined and that is realized, therefore, in unexpected ways in the wake of colonialism and in
the face of neocolonialism.

Indeed, Trinh's works seem to be a proposal for a cultural criticism
that is deeply resistant to conceptions of culture as universal or whole.
She rather prefers to recognize cultures as fluid and mobile; therefore,
the critic is committed to engagement with generalities and particular-
ities of the local and the global, drawn into various configurations both
by the worldly intensities — the political, the economic, and the social —
and by desire moving through the critic, making her a critic. These configurations cannot be predetermined, and in this sense there must be an
experimentation in form, a formal differing with and deferring of the
given. These experiments in cultural criticism arise out of the *différance*
between, not the opposition of, the deconstruction of the ethnographic
authority of western anthropology and the resulting experimental writing of a generalized ethnography.

This *différance*, although recognized in Trinh's works, often has gone
unrecognized in some experimental writings, such that the early criticism of ethnographic writing became enmeshed in, if not lost to, the
debates over standpoint epistemologies and a politics of location, identity politics, and its antiessentialist reversals. Although these debates
have only complicated the relationship of the early criticism of ethnographic writing and the cultural studies of science, they also have been
the condition of possibility of the elaboration of autoethnography and
the return, after deconstruction, of an intense desire for voice and identity that is itself symptomatic of the becoming of the teletechnological.

The Autoethnographic Turn in Cultural Criticism

Whether expressed in the work of postmodern sociologists, postcolonial anthropologists, or cultural critics long excluded from the authority of western modern discourse on the basis of race, class, gender, sexuality, ethnicity, or nation, the autoethnographic turn of a generalized
ethnography has not been an altogether surprising departure from post-

structuralism. Indeed, those writing autoethnographically often refer to poststructuralism as a resource. They take for granted the deconstruction of the subject, and they write in the displacement of the authority of the modern western discourse of Man.

But the autoethnographic turn also has been grounded in standpoint epistemologies and the discourses of identity politics and a politics of location, such that autoethnography differs from poststructuralism. Often autoethnographic writing even reproduces the oedipal logic of realist narrativity, albeit for different purposes than those of traditional ethnography; that is, an autoethnographic realism often has been deployed in a description of a writer's experiences of oppression, exploitation, and domination that is also meant to authorize the writer. Not only does autoethnographic writing give voice to the writer's experience; it also makes the validity of the presentation of experience the very ground of the writer's authority. The autoethnographic turn of a generalized ethnography refers, therefore, to writing experiments that often are claimed to be empirical science and, although not always strictly autobiographical, usually are personal, poetic, or evocative expressions of cultural identity and experience. The texts produced are given over to emotional matters, often focusing on the tragedies and insights borne of the experience of oppression, exploitation, and domination, as well as the everyday traumas of life and death.

Both in drawing on poststructuralism and in differing with it, autoethnographic writing, I want to propose, is drawn to the teletechnological; autoethnographic writing draws cultural criticism to unconscious thought in the age of teletechnology. Surely the personal, emotional, and experiential focus of autoethnographic writing is also the signature of teletechnology, television especially. Just as television has been linked to, if not often blamed for, displacing history with nostalgia, reason with melodrama, deep meaning with "flickering signifiers,"[46] autoethnographic writing has been criticized for devaluing rational discourse, thereby sharpening the contentious debates that, over the last three decades of the twentieth century, have put to question the privilege given both empirical science and reason in academic and intellectual discourses. In doing so, however, autoethnographic writing shifts the focus of cultural criticism of science to engagement with the speeds of deterritorialization and reterritorialization of the arrangement of the private and public spheres, the state and civil society, and family and national ideologies, as well as with the vulnerabilities of exposure to media event-ness.

That is to say, autoethnographic realism makes empirical scientific research go even faster than when realist narrativity first was deployed in the ethnographic text of western anthropology. Although autoethnographic realism often reproduces the oedipal logic of narrativity, which in traditional ethnography served to produce the present absence of the ethnographer in the text, in autoethnographic realism the oedipal logic of narrativity functions to produce a full exposure of the ethnographer, his or her personal experiences, in the text. Autoethnographic writing thereby collapses the temporal and spatial distance between the presentation of observed experience and the reflexive self-criticism of the observer. It is in this sense that autoethnography shifts the focus of cultural criticism of science to speeds and exposures.

As a result, autoethnographic writing strains the limits of the oedipal logic of realist narrativity that it itself deploys. In doing so, the link of autoethnography to writing and textuality in the larger sense of these terms is realized, returning autoethnographic writing to questions about its relationship to the teletechnological and the becoming of technoscience, technoculture, and technonature. Autoethnographic writing, that is, is made to engage the demand for a more direct intervention in the globalization of teletechnology and the transnationalization of capital. Autoethnography meets up with a politics of location, made to travel with the speeds of the globalized network of glocalized cultures in diaspora.

The autoethnographic turn in postcolonial anthropology, postmodern sociology, and cultural criticism therefore is a response to the early criticism of ethnographic writing, especially for the way it establishes scientific authority by absenting the ethnographer's presence from the textualization of his or her observations. Autoethnography proposes to return the ethnographer's I/eye to the writing surface. This is, of course, a provocative proposal. After all, not only has empirical science been defined in opposition to fictional forms, such as the autobiography; its objectivity has been grounded in the scientist's capacity to eliminate the effects of his or her personal life on research. Absenting the presence of the ethnographer from the ethnography, therefore, is meant to demonstrate the ethnographer's compliance with the tenets of empirical science; the scientific objectivity of the ethnography is thereby established.

However, if science is opposed to autobiography in the institution of scientific authority, science and autobiography are also interimplicated. Their opposition not only defines both. It also allows both to disavow what each shares with the other. For example, there is the logic of realist narrativity that both science and autobiography share with each other as well as with other fictional genres, such as the novel. Indeed the development of realist narrativity often has been connected to the rise of European bourgeois ideology; realist narrativity is thought to have been a vehicle for images of the subject as a self-possessed, self-identified individual and the only candidate for political, social, and cultural agency. Circulating these images from the autobiography to the novel, to history and scientific discourse, realist narrativity framed the figure of the authorized subject of knowledge in terms that privileged European empire, whiteness, masculinity, heterosexuality, and property wealth.

As for its past, therefore, the realist narrativity of the autobiographical form is neither innocent nor transparent. Nor is the autobiographical form presently free from the complications that have arisen with its deployment in mass media other than the novel, such as those connected to globalized teletechnology and transnational capital in the glocalization of world cultures. Although the autobiographical form is meant to be a critical device in the writing of ethnography, it nonetheless is haunted by its history and its present complicities connected to a mass-mediated circulation of its formula for a self-possessed, self-identified subject whose identity is authorized in the full knowledge of his or her experience elaborated with a realist narrativity.

But these complicities often go unnoticed as autoethnographic writers find support in various standpoint epistemologies, referring to them as a critical resource much more than to poststructuralism. Autoethnographic writers especially draw on feminist theorists who have claimed authority for the subordinated knowledge of women, arguing that women's experiences give them the possibility of a more adequate and more accurate understanding of the structures of dominance, exploitation, and oppression, especially as these affect everyday life. The works of feminist theorists such as Dorothy Smith or Nancy Hartsock have not only been a resource for autoethnographic writing; their works also have been central to a feminist criticism of science, especially since Sandra Harding first treated Smith's and Hartsock's works as standpoint epistemologies when Harding first compiled early feminist criticisms of the practice of empirical science.[47] Although Smith and Hartsock both focus more on fixing the relationship between Marxist political economy and feminist theory, both also have criticized the approaches to empirical research in each of their respective disciplines. Smith offers a reformulation of sociology,[48] while Hartsock rethinks the presumptions of political science.[49] In doing so, both elaborate what Harding would refer to as the "strong objectivity of a feminist standpoint epistemology."[50]

In the early 1970s Smith turned feminist theory to rethink women's work and ways of knowing. She drew on Marx's assertion that laborers have the potential for a fuller understanding of capitalism due to their relationship to the mode of production; she argued that, similarly, it is women's position in the mode of production that gives women's way of knowing a strong objectivity. As Smith saw it then, this is because women's work in both crucial and devalued; that is, although women's work as housewives, mothers, and office administrators is devalued, nonetheless, it is this work which makes it possible for men to produce abstract knowledge. Smith put it this way: "To a very large extent the direct work of liberating men into abstraction . . . has been and is the work of women. The place of women, then, in relation to this mode of action is where the work is done to facilitate men's occupation of the conceptual mode of action."[51]

By maintaining the local and particular existence of actors on behalf of the abstract conceptual mode, women maintain what Smith called "the relations of ruling." But she also argued that in coming to realize their oppression and in refusing their long-held position in exploitative and dominating relations, women can transform the mode of ruling; they can transform the conditions of knowledge production. According to Hartsock, this would involve taking the women's psychic inclination to nurturing and mothering as the model for the production of knowledge, thereby offering an antidote for the abstractness of the disciplines that are organized from the perspective of the powerful, that is, men. A feminist standpoint, Hartsock would argue, provides a "vision" that is characterized not only by a "valuation of concrete, every-

day life," but also by "a sense of a variety of connectednesses and conti-
nuities both with other persons and with the natural world."[52]

In these early feminist treatments of knowledge production, it is perhaps not so surprising but surely ironic that there was a presumption
of the same configuration of family and national ideologies, state and
civil society, the private and public spheres that is also presumed in the
modern western discourse of Man. It was only the place of women in
this configuration that was to be rethought and revalued; it was only
the ideological separation of the public and private or domestic spheres
that was to be reconsidered for feminist ends. It was, after all, the presumption of women's seclusion in the home, given with the separation
of the public and private spheres, that allowed both Smith and Hartsock to elaborate a women's standpoint and to revalue women's work
as a counter to the capitalist organization of production.

It was not long, however, before feminists of color, of different classes,
and of marginalized sexual orientations and different ethnicities questioned whether each and every woman has the same experience of oppression, exploitation, or domination. Some of these feminist theorists
also argued that many women have never been secluded in the private
or domestic sphere; others, especially those in situations of neocolonialism, criticized the way that the configuration of social spaces presumed in modern western discourse and imposed with the transnationalization of capital and the globalization of teletechnology also structures
the standpoint epistemologies of the early feminist criticism of science.
These criticisms of women's standpoint epistemologies, when taken together, suggest that the differences among women, as well as the different configurations of social spaces that shape these differences, need be
allowed to inform the feminist criticism of science, indeed the practice
of empirical science itself.

In the early works of Smith and Hartsock there was a failure to account for the differences among women, as well as a resistance to seeing the relevancy of these differences in feminist criticism of science,
and therefore neither Smith nor Hartsock was prepared to treat the reconfiguration of family and national ideologies, the state and civil society, and the private and public spheres in the glocalization of cultures
along with the transnationalization of capital and the globalization of
teletechnology. Nor were they prepared to treat the transformation of

technology in the late twentieth century and its effects on bodies, being, thought, historicity, and the relations of time and space. Ironically,
although both Hartsock and Smith had treated women's standpoint or
women's way of knowing in relationship to women's place in the mode of
production, neither had taken into account the change in political economic analysis that their own analyses of the mode of production as a
mode of knowledge production implied.

The elaboration of women's standpoint epistemologies in the early
feminist criticism of science does imply, however, that knowledge has
become central to the mode of production, such that women's place in
the production of knowledge can be thought to make it possible for
women to more adequately understand the structure of capitalist production as a whole. Yet in treating women's work Smith and Hartsock
draw on Marx's labor theory of value without noticing that Marxist theorists have begun to recognize that human labor is no longer central to
production and that technoscience or abstract knowledge has instead
become central to production, such that labor has become stratified along
lines of technical knowledge.

Although the exploitation of workers around the world, women workers especially, has continued to be recognized as a political issue, Marxist theorists no longer have been taking the labor theory of value as the
method for treating the capitalists' adjustment of exchange relationships against the falling rate of profit, not at least without focusing more
on what Jonathan Beller refers to as "the productive value of attention"
connected to mass media and teletechnology, with implications for
"biosocial (cybernetic) modification at all levels of social interaction."[53]
In the early 1990s, Stanley Aronowitz would argue that in the 1970s and
1980s, Marxist theorists of the new left became increasingly involved
with cultural studies, including the cultural studies of science, as an effect of a growing awareness that abstract knowledge had become central
to production and that technoscience had become the primary agency
of power/knowledge in what would be referred to as postmodern capitalism; Aronowitz even proposes that the new left's resistance to the
Viet Nam War already had engaged Marxist theorists in a cultural criticism of technoscience.[54]

But neither Hartsock nor Smith recognized these same conditions as
the conditions of possibility of their early feminist criticism of science;
focused on the seclusion of women in the domestic sphere as well as on

the ideological separation of the private and public spheres, neither noticed that the configuration of social spaces that they presumed and wanted to critically engage was already being smoothed out, deterritorialized and reterritorialized in the globalization of teletechnology and the transnationalization of capital. Still, it was their focus on the ideological separation of the private and public spheres, and on the mode of caring characteristic of both women's work and women's knowing, that made the early feminist criticism of science a resource for autoethnographic writing, especially among feminist researchers in sociology, anthropology, and cultural studies.

It is not, however, only for their treatment of women's work and women's way of knowing that Smith's and Hartsock's works are referenced in autoethnographic writing; it is also because of their proposal that women understand better than men the experience of oppression, exploitation, and domination because these are uniquely women's experiences. Thus, women's standpoint means to put those studied on the same plane as those who study; as Harding put it: "The woman inquirer interpreting, explaining, critically examining women's condition is simultaneously explaining her own condition."[55] Such a proposal could be generalized and at the same time made more specific. That is, standpoint epistemologies might be generalized to groups suffering various marginalizations, making their specific experiences the ground for an authorized knowledge of oppression, domination, and exploitation generally.

In her feminist criticism of science, Harding argued that standpoint epistemologies allow for strong objectivity by making the social location of the knower the starting point of inquiry; she thereby generalized standpoint epistemology to those who are marginalized on the basis of class, sexuality, ethnicity, race, and nation. In doing so Harding was responding to criticisms of Smith and Hartsock for essentializing women's identity, for ignoring differences, even antagonisms, among women. She was responding to theorists who were elaborating an identity politics in the terms of standpoint epistemologies, such as Patricia Hill Collins in her treatment of a black feminist standpoint and Gloria Anzaldúa in her treatment of mestiza consciousness.[56] Harding also considered standpoint epistemologies in their intersection with postcolonial theory and multiculturalism, still arguing that a "robust reflexivity" is possible in empirical science.[57]

But in extending standpoint epistemologies to include those who experience a range of oppressions and marginalizations, Harding gives a different treatment of standpoint epistemology than Smith and Hartsock do; they ground the value of women's way of knowing in women's location in the mode of production, in the context of the separation of the public and private spheres. Harding's treatment of standpoint epistemologies presumes instead that experience puts members of an oppressed group in essentially the same position to know and that each member of the group is able to be an authorized subject, having a unified, self-same identity both given with and given to a more adequate grasp of oppression. Whereas Norma Alarcón argued that "to be oppressed is to be disenabled not only from grasping an identity, but also from reclaiming it," and therefore "the theory of the subject of consciousness as a unitary and synthesizing agent of knowledge is always already a posture of domination,"[58] in her treatment of standpoint epistemologies Harding proposed otherwise.

Arguing for the strong objectivity of knowing when it is located in the experiences of a knowing subject, Harding made it possible for standpoint epistemologies to find expression in the autoethnographic writing of science. She made it probable that standpoint epistemologies will be deployed in experimental writing where and when the autobiographic is allowed to be the only or the primary site of the self-reflection of the scientist. In sociology, for example, there has been a number of efforts at experimental writing of autoethnography. Although Susan Krieger's *The Mirror Dance*,[59] an ethnographic study of a lesbian community, first led her to treat the "self in social science" by means of the artful writing of personalized essays,[60] feminist autoethnography is more often connected to sociologists Carolyn Ellis and Laurel Richardson.

Ellis's writing is presented as an "honest and open" expression of the sociologist's feelings about events in her own life, such as death, illness, and personal trauma, events that usually have been connected, at least for white middle-class women, to the private or domestic sphere, and therefore have been devalued as being part of women's work of nurturing, what Ellis refers to as "emotional work." For example, in Ellis's *Final Negotiations*[61] she produces the story of her relationship to her husband, Gene Weinstein, focusing on the nine years of his illness that ended in his death. Here Ellis proposes to render her own emotional experience of being engaged over a long period of time in the care of a chronically ill and dying loved one. But Ellis does not only want to tell of her experiences; she also hopes to coax the reader "to be open to your feelings as you take this narrative journey."[62]

Autoethnographies, like Ellis's, challenge the tradition of empirical sociology. They mean to turn the eye of the sociological imagination back on the ethnographer; they mean as well "to personalize and humanize sociology," as Ellis puts it. Although they are hesitant about treating other subjects as objects of observation, autoethnographers nonetheless expose themselves and those closest to them as part of a self-conscious reflection on doing ethnography. If these autoethnographies aim to revalue the work of women, they also aim to revalue the work of women sociologists. This has even led sociologists, Laurel Richardson for example,[63] to produce autoethnographies of local academic settings, to expose the politics of university departments, and to reevaluate women's participation in or exclusion from lines of academic and disciplinary authority, all in personal terms.

These autoethnographers do not, however, explicitly address the configuration of family and national ideologies, the state and civil society, and the private and public spheres, which, nonetheless, allows them to presume that nurturing or emotional work is the work of women only or that women's work and only women's work is hidden and devalued. These autoethnographers do not address the relationship of their experimental writing to the globalization of teletechnology and the transnationalization of capital. They do not treat changes in technologies or the various technical substrates of unconscious memory and the effects of these on autoethnographic writing. The autoethnographic form remains unexamined as to its relationship to changed relations of time and space or the shift in aesthetic concerns from narrativity to the speeds of territorialization and reterritorialization, along with the adjustment to the vulnerabilities of exposure to media event-ness. Yet all this seems to be at issue in these self-exposing writings, which turn sociological theory almost exclusively to theorizing autoethnography in terms of the criticism of traditional ethnographic methods of research and writing as depersonalizing or dehumanizing.

It would seem possible, therefore, to trace the focus in autoethnographic writing on emotions and personal experience to the emotional realism or melodramatics of television. Television, after all, has been described by its critics in terms of an intensified emotionalism characteristic of melodrama.[64] They have argued that television's overexposure of the private sphere and personal experience provokes an anxiety that can be soothed only by a further intensification of television's emotional appeal such that tragedy, catastrophe, and death become the only touchstones of the really real of a hyperreality.[65] It is therefore also possible to connect autoethnographic writing to what sociological theorists have described as social situations, where there is a progressive freeing of agency from embeddedness in traditional social relationships as well as in larger social structures and a further embeddedness in global and local cultural networks of information and communication.[66]

Anthony Giddens, for one, argues that this retraction of the social in the face of the expansion of the teletechnological in postmodern societies refers reflexivity and self-reflection to what he calls the "pure relationship," which is entered into for itself and which, with the help of the therapeutic practice of expert systems, no doubt including talk television, takes as its horizon open and honest communication, or what Giddens refers to as an "emotional democracy."[67] A symptomatic expression of the retraction of the social, autoethnography suggests that the social structural approaches of modern sociology have been displaced by the teletechnological; as a response, autoethnography turns sociology into a therapeutic expertise tuned into the personal or the interpersonal. Autoethnography aims sociology in the direction of the different sociality of postmodernity, where sociality is inextricable from knowledge objects, rhetorical mechanisms, writing technologies, therapeutic expert systems, even machine agencies other than human agencies, and where the personal is the touchstone of hyperreality.

However, if autoethnography seems much more a reflection of the retraction of the social than a critical intervention in postmodern societies, it does at least suggest that a more critical response necessarily would have to engage the speeds of the teletechnological and the vulnerabilites to exposure to media event-ness. That is to say, autoethnography urges a recognition of the reconfiguration of the conditions of possibility of sociology as an empirical science in the age of teletechnology. Therefore, autoethnography is linked to the new sociology of science, along with which it is rethinking the methods of self-reflection and reflexivity befitting a sociology of postmodernity. In this sense autoethnography is both an effect of the new sociology of science and its excess. It turns the new sociology of science on sociology itself; autoethnography is an emotional exposure of the sociologist as well as the local institutional sites of the production of sociological knowledge.

Sociology, however, is not the only discipline in which autoethnography has been developed. In anthropology, too, there has been experimental ethnographic writing, including autoethnography. Feminist anthropologists have engaged in experimental writing and often refer to the early criticism of ethnographic writing, especially Clifford's, as one of the inducements. After all, in his introduction to *Writing Culture,* Clifford tried and failed to explain the absence of any essay in the volume that treated feminist ethnography; he argued that "feminist ethnography has focused either on setting the record straight about women or on revising anthropological categories (for example, the nature/culture opposition). It has not produced either unconventional forms of writing or a developed reflection on ethnographic textuality as such."[68] The anger Clifford drew from feminist anthropologists, nonetheless, led them more explicitly to rethink ethnographic writing.

In the introduction to a collection of essays titled *Women Writing Culture,* Ruth Behar suggests that Clifford's remarks did call women anthropologists, disciplined in a certain way of writing, to experiment beyond both the limits imposed in their training and those self-imposed in the need for acceptance in the empirical scientific discourse of anthropology. As Behar puts it, "In truth, the *Writing Culture* project was a sullen liberation."[69] But a more joyful, although perhaps more disturbing call to experimental writing, Behar reports, was the 1983 publication of *This Bridge Called My Back.*[70]

Of course *This Bridge Called My Back* was to have a profound effect on feminist theorists generally, calling them to rethink the relationship of gender to race, class, sexuality, ethnicity, nation, and diaspora. Part of its power to affect, however, was the forms of writing it displayed and made available as models for academic and intellectual writing. This book of essays, poems, letters, and diary entries by women of color not only forcefully taught western anthropologists about issues affecting women of color "at home" as well as abroad. But in making use of personal voices and personal experiences, it also made clear that there is a necessary connection between exploration in writing and articulating identities and experiences that have been excluded from disciplinary discourses. In contrast to traditional ethnography, Behar argues, *This Bridge Called My Back* made western anthropologists especially aware that "first world women had unself-consciously created a cultural other in their images of 'Third World' or 'minority' women."[71]

Writing Culture and *This Bridge Called My Back* together inform the essays in *Women Writing Culture*; there are essays in the first person — autoethnographic treatments of experiences of marginality and diaspora offered by anthropologists who have themselves had these experiences. There are rereadings of the canon of works produced by female anthropologists — Elsie Clews Parsons, Ruth Landes, Ruth Benedict, and Margaret Mead — and fictional and poetic writings by Ella Cara Deloria, Mourning Dove, Zora Neale Hurston, and Alice Walker, among others. The collection is a mix of genres, fact and fiction, conventional anthropology, and personal testimony by anthropologists, much of which is meant to unsettle the distinction between the researcher and her subjects. The collection refuses to privilege the anthropologist, even the feminist anthropologist, as the only authorized subject of anthropological knowledge.

But beyond this, there is no further discussion about what is to be made of these experimental writings in relationship to a cultural criticism of science. As autoethnographic expressions of standpoint epistemologies, the essays do not problematize writing enough to connect writing to technology and to the glocalization of cultures in the transnationalization of capital and the globalization of teletechnology in neocolonialism. Yet Behar's coeditor, Deborah Gordon, suggests in a conclusion to the collection that the distinction of conventional and experimental ethnographic writing increasingly is not easily defined. Or what is defined as experimental writing is increasingly subject to the market in the "fast paced intellectual exchange" of late twentieth-century academic and intellectual discourse.[72] But how the fast pace of intellectual exchange affects the slower-paced anthropological method of ethnography is not addressed. Rather, the collection seems to insist on slowness, taking a long look back to the history of women in the discipline and introducing speed only inadvertently in allowing the faster autoethnographic insights to stand in for the observations of the slower traditional ethnography of western anthropology.

Other responses to the early criticism of ethnographic writing, however, are more apparently linked to the teletechnological; they seem more engaged with the speeds of deterritorialization and reterritorialization of social spaces and with the vulnerabilities to exposure to media event-ness. These responses have been more engaged with postcolonial theory, which is itself confronting the globalization of teletechnology and the transnationalization of capital in neocolonialism. These experimental writings emphasize the movement of persons and things in the transnational flows of information and power/knowledge; they focus on a world situation that is characterized by what Inderpal Grewal and Caren Kaplan refer to as "scattered hegemonies."[73] These responses turn the early criticism of ethnographic writing to the thought of what Clifford refers to as "traveling cultures," where a politics of location is given over to the speeds and exposures of the teletechnological.[74]

The notion of a politics of location, first articulated by Adrienne Rich[75] in an effort to open feminist theory to rethinking the relationship of gender, race, sexuality, ethnicity, class, and nation, has enabled cultural critics not only to recognize, but to render in their writing, the ways they are situated or located in the production of knowledge. Although the politics of location can be expressed in autoethnographic writing informed by a standpoint epistemology, it also has led to a more subtle thinking about location itself in relationship to place, to the possibilities and impossibilities of staying, leaving, and moving, as these are connected to various displacements, especially those of the late twentieth century, such as exile, emigration, immigration, border crossing, homelessness, bondage, forced labor, and tourism. In this sense a politics of location has cast the subject as a multiplicity of movements against a series of displacements that evoke speeds and exposures, those of the information and communication networks of the teletechnological. Paul Gilroy, for example, connects teletechnology to the transformation of a unidirectional model of diaspora to a more "chaotic model" that allows the links between separated black populations around the world to be thought of in terms of "excentric communicative circuitry."[76]

A politics of location that is engaged with speeds and exposures provides a way to rethink the local or locality not as a privileged identity of place, person, or group, but as a mapping of what Chandra Mohanty describes as "multiple locations."[77] As such, a politics of location draws on the localism of every location while not forgetting the globalization of teletechnology and the worldwide transmission of technoscience, technoculture, and technonature. Haraway makes the point this way: "Remembering that located does not necessarily mean local, even while it must mean partial and situated, and that global means not general or universal but distributed and layered, seems the fundamental point to me for binding together the co-constitutive insights of cultural studies, antiracist feminist studies, and science studies."[78] What is required is what Bruce Robbins describes as an imagination of "different modalities of situatedness-in-displacement."[79] What is to be encouraged is that critical discourses crisscross each other, looping through each other, as a matter of the timing and intensities of form more than as a mere interdisciplinarity.

Ethnography is thereby given over to the challenge of informing a cultural criticism that is locally and globally critical all at once. However, this involves not only mapping the global and the local in terms of the speeds of deterritorialization; it also calls for attentiveness to the reterritorializations of social spaces, the reformulations of national and family ideologies, and the reconfiguration of civil society and the state, at the same time remembering that the speeds and exposures of the teletechnological are no longer solely about travel, but about the intensities of seeing, hearing, perceiving, and conceiving. Therefore, ethnography is sent into an endless circuiting from poetry, autobiography, sociology, anthropology, and cultural studies of science, judging the risks and promises of technoscience, technonature, and technoculture.

It is here that Haraway's notion of diffraction returns, when self-criticism in knowledge practices must go beyond reflexivity and self-reflection, when it is no longer only a matter of human agencies caught up in the state apparatuses and a national hegemony, but when it also is about nonhuman agencies that inhere in finite forces, speeds, and exposures that are productive of technonature and technoculture. Diffraction, it would seem, would allow for a recognition of the global miredness out of which political agency arises, but that cannot itself be predetermined. Diffraction would summon an imagination of a critical practice capable of a more direct engagement with semiotic-material objects, locally and globally situated all at once.

Diffraction would make cultural critics ready to think the thought of machinic assemblages, to become engaged with the speeds of deterritorialization and reterritorialization of social spaces, and to meet the vulnerabilites of the exposures to media event-ness — all as a way to critically engage the "time-space regime of technobiopower."[80] Therefore, diffraction seems part of a rhizomatic writing, a composing and recomposing that cuts into and cuts away from genres, technologies, images, and scenes so that the movement is never simply narrative or life story. If it is for subjects, it is not subject centered. Nor are the cuts to or cuts away merely for literary effect in the narrow sense of writing and textuality. They are rather the traces of the movement of desire in the construction of machinic assemblages in order to do something.

Diffraction would enable thought where and when identity politics and standpoint epistemologies are crossed through with the forces of nonhuman agencies, where affectivity is a matter of techno-organic kinships and a more general unconscious. In ecofeminism, for example, there is this mix of movements of identity with a politics of technonature, such that feminist activism aimed at ecosystems is folded into contentions over identity, desire, bodies, class, gender, ethnicity, race, sexuality, and nation.[81] Arturo Escobar[82] offers another example in his discussion of the black and indigenous mobilizations in Colombia, where the politics of identity is also a politics of nature focused on the biodiversity of the rainforest and the surrounding region.

Escobar argues that these movements are not only an effect of and a response to the government's opening up to world markets in the effort to integrate Colombia into the Pacific Basin economies; they also are about the indigenous groups' becoming involved in local capitalizations in an effort to benefit from, while trying to control, the interest in conservation and sustainable development, especially the patenting of "nature," with which the rainforest region is rich. Not only are local ways of knowing defended by indigenous groups, but indigenous groups also make alliances with the northern advocates of the technological conservation of biodiversity.

Escobar argues that it is unclear what will be the effects of these movements of black and indigenous groups, who already have engaged with their own practices of knowing, and also have become engaged with expert systems and the practices of "development" of the northern nations. But whatever the effects may be, this is a social movement where a politics of identity is linked to a politics of nature. The relationship between identity and territory are thereby made stronger, although just as identity is conceived, in nonessentialist terms, as a matter of preserving cultural difference, territory is conceived as something other than homeland. It is an economic resource for sure, but also something more, something like the ground for writing futures. Escobar refers to Deleuze and Guattari's treatment of territorialization and deterritorialization, suggesting that presently the rainforest is a machinic assemblage, a desiring production, a plane of consistency for politics, economics, and cultural transformations, where multiple agencies are assembled to dream a future to come.[83] These examples point to an ongoing rethinking of politics in the age of teletechnology. They point to the effort to make thought practical, an effort that has drawn the cultural criticism of science and the critique of ethnography together to the future.

Imagine the cultural criticism of science joined with the critique of ethnography writing, along with three decades of cultural criticism, philosophical and highly abstract, all to make thought practical, to put thought and practice on the same plane of desire. How can it be? How can it be that it would take so much theoretical effort to produce this practical effect? No doubt it has been difficult to deconstruct the modern western discourse of Man — the rules of empirical science that it gives and the aesthetic methods of mastery that it supports. It has been difficult to think against the insistence on the opposition of nature and culture, body and machine, nature and technology, the real and the virtual, the living and the inert. It has been difficult to follow unconscious thought to its future in the age of teletechnology, to affect an ontological shift and then to let thought begin anew. But it is this future of thought to which poststructuralism has been drawn and that has had elaboration in the cultural criticisms engaged with poststructuralism over the last three decades of the twentieth century.

And if autoaffection would return in the yearning for self-same identity and in the desire to speak with one's own voice, it is not to be unexpected or simply to be scorned. Autoaffection is to be deconstructed, as Derrida puts it, only "to the extent that its power of repetition *idealizes* itself" ... and "appears as my spontaneity." Then repetition must be urged beyond itself to its *différance,* to a different thought of repetition. In such a way was the thought of pure repetition given: to think repetition without origin and without end, letting loose the unconscious from its embodiment in the subject of the modern western discourse of Man and thereby to reach for a cultural criticism that goes beyond, while going through, the unconscious thought given in the joining of Marxism and psychoanalysis.

It is not clear what future will come beyond the future given with the unconscious thought of teletechnology. It is being written; no doubt somewhere it is just starting itself up. In fits and starts, across mystic writing-pads, it is producing bodies, assemblages of scenes, screens, and moving machines. And surely some

> small space is saved for the subject,
> some autoaffectionate bits,
> made out of desire for one more word,
> that there be just one more word between us
> even though there already have been so many.
> I did not want to feel the flow of words ebb
> and go so near to the end,
> so near to death.
> There still is the hope of a voice to be born,
> more than a voice.
> This is the hope of any analysis,
> of psychoanalysis for one.
> The one that befriended me and more
> gave me the transferential love of a lifetime.
> Still, it ended
> at the infinite point that an I needs to be, with
> mommies and daddies and sons and daughters, too,
> but more than two,
> many more visions in blind sights.
> And in the end, the end was not prepared for,
> no matter the ends along the way,
> it was sudden.
> Suddenly she was gone, no longer sitting nearby me,
> my hair brushing against her knees, her knees pressing
> thoughts to my lips
> and the whispers ascending and taking flight over my head
> to her ear,
> the other to hear the other in my ear.
> Together to keep the speaking going without its turning back
> to muteness,
> until the sound of my silence is different
> than when I first tried to speak it to her.
> Until in the end of every end, there also is beginning.
> I began in the end.

Notes

Introduction

1. Derrida's remark appears in a discussion with Christie McDonald, who also is translator. The discussion is entitled "Choreographies" and first appeared in *Diacritics* 12 (1982): 62–76. I am drawing from a reprint in *The Ear of the Other* (Lincoln: University of Nebraska Press, 1988), 172.

2. Rosi Braidotti, *Nomadic Subjects* (New York: Columbia University Press, 1994), 101.

3. I am referring here to Gilles Deleuze, "Postscript on Societies of Control," *October* 59 (1991): 3–7. The notion of "control societies" is discussed more fully in later chapters. Let me say, however, that when I refer to the structural configuration of social spaces presumed in the modern western discourse, I mean to refer to discussions from F. W. Hegel on, where what has been at issue has been the spacing of the public sphere, civil society, private life, capitalism, and the individual and the family in relationship to the state and where it is presumed that there is an ideal separation of these social spaces that allows for the individual's rights, freedoms, and obligations in relationship to a national collectivity. Deleuze's treatment of control societies suggests that the arrangement of social spaces is being smoothed out and thereby prepared for profound reconfiguration. Although Deleuze often gives a negative cast to his notion of control societies, he also recognizes it to be a situation for various virtualities to be actualized in various futures. For further discussion of "smooth" as contrasted with "striated," see Deleuze with Félix Guattari in *A Thousand Plateaus: Capitalism and Schizophrenia,* trans. Brian Massumi (Minneapolis: University of Minnesota Press, 1987), 474–500. For Sassen's discussion of unbundling, see *Globalization and Its Discontents* (New York: New Press, 1998), 81.

4. In the situation described in the previous note, it is possible but also necessary to treat the speeds of the territorialization and deterritorialization of social spaces. As for the notion of media event-ness, for now let me say that I mean to refer not only to the changes connected to television and its logic of exposure, such as the effect on the distinction of private and public spheres, national and family

ideologies, and therefore the effect on identities and bodies of subjects and collectivities. I also mean to put forward the notion of event as a displacement of the objectness of object; that is, borrowing from Deleuze and Guattari, I mean to refer to event as an object that comes into being only with a theoretical apparatus or device; *event* is a better term than *object* when temporality matters, when the ontology of the object in volatile, unfolding. More generally, then, event-ness refers to a shift in the temporal and spatial relationships of Being effected with the teletechnological. I am proposing that Being-ness be thought of in terms of speed and exposure — media event-ness — bringing teletechnology and Being-ness closer together.

5. C. W. Mills, *The Sociological Imagination* (New York: Oxford University Press, 1959). For an interesting treatment of Mills's sociological imagination in postmodernity, which is relevant to my sense that Mills's treatment of the sociological imagination has been displaced, see Charles Lemert, *Sociology after the Crisis* (Boulder, Colo.: Westview Press, 1995).

6. Pheng Cheah's treatment of "mattering" appears in his "Mattering," *Diacritics* 26 (1996): 108–39, where he treats works of Judith Butler and Elizabeth Grosz. His essay is an extremely provocative one that much encouraged my treatment of the ontological implications of poststructuralism. But I am even more indebted, as is Cheah, to Gayatri Chakravorty Spivak's treatment of poststructuralism in relationship to the finite forces of matter. See her *Outside in the Teaching Machine* (New York: Routledge, 1993), 25–51. I discuss Cheah's work, along with Butler's and Grosz's, in chapter 3.

7. See Manuel de Landa, "Immanence and Transcendence in the Genesis of Form," in *The Deleuzian Century*, ed. Ian Buchanan (Durham: Duke University Press, 1997), 499–523.

8. Manuel de Landa, "Nonorganic Life," in *Incorporations*, eds. Jonathan Crary and Sanford Kwinter (New York: Zone Books, 1992), 134.

9. Donna J. Haraway, *Modest Witness@Second Millennium: FemaleMan© Meets OncoMouse™* (New York: Routledge, 1997), 129.

10. Spivak, *Outside in the Teaching Machine*, 30. Here Spivak is referring only to Derrida's and Foucault's works in relationship to the ontic, but her remark, I believe, is appropriate for discussing Deleuze and Guattari's work as well.

11. Derrida's treatments of the "gift" or the "given" appear in a number of his writings, such as *Glas*, trans. John P. Leavey and Richard Rand (Lincoln: University of Nebraska Press, 1986); *Spurs*, trans. Barbara Harlow (Chicago: University of Chicago Press, 1979); and *The Post Card*, trans. Alan Bass (Chicago: University of Chicago Press, 1987). I am drawing most specifically from Jacques Derrida, *Given Time: 1. Counterfeit Money*, trans. Peggy Kamuf (Chicago: University of Chicago Press, 1992). Although difficult, the notion of the given is one that allows ontological implications or an ontological perspective to be drawn without stipulating an originary-ness. It therefore can refer rather to the finitude of beings and to what Derrida describes, and I discuss in later chapters, as the "contamination" of Being with finitude or technicity.

12. Derrida, *Given Time: 1. Counterfeit Money*, 162.

13. I am referring to Derrida's discussion of hauntology in *Specters of Marx: The State of the Debt, the Work of Mourning, and the New International*, trans. Peggy Kamuf (New York: Routledge, 1994), 51.

14. See Avital Ronell, *The Telephone Book: Technology-Schizophrenia-Electronic Speech* (Lincoln: University of Nebraska Press, 1989); Gregory Ulmer, *Teletheory: Grammatology in the Age of Video* (New York: Routledge, 1989); Mark Poster, *The Mode of Information* (Chicago: University of Chicago Press, 1990); Manuel de Landa, *War in the Age of Intelligent Machines* (New York: Zone Books, 1991); Samuel Weber, *Mass Mediauras: Form, Technics, Media* (Stanford: Stanford University Press, 1995); Sadie Plant *Zeros + Ones: Digital Women + the New Technoculture* (New York: Doubleday, 1997); Charles J. Stivale, *The Two-Fold Thought of Deleuze and Guattari* (New York: Guilford Press, 1998).

15. Gilles Deleuze and Félix Guattari, *What Is Philosophy?*, trans. Hugh Tomlinson and Graham Burchell (New York: Columbia University Press, 1994), 18.

16. Ibid., 18.

17. Following Deleuze and Guattari, I am not carefully defining *plane of consistency* or *machinic assemblage,* but letting these terms accumulate, condense, and disperse various meanings along the way. For Deleuze and Guattari's discussions of these concepts, at least see *Anti-Oedipus: Capitalism and Schizophrenia,* trans. Robert Hurley, Mark Seem, and Helen Lane (New York: Viking Press, 1972); Deleuze and Guattari, *Thousand Plateaus: Capitalism and Schizophrenia.*

18. Deleuze and Guattari, *What Is Philosophy?*, 82.

19. Jacques Derrida, *Archive Fever: A Freudian Impression,* trans. Eric Prenowitz (Chicago: University of Chicago Press, 1995), 25.

20. I am indebted to Pheng Cheah's discussion of global miredness in relationship to Jürgen Habermas's treatment of the public sphere both here and in later chapters, especially chapter 3. See Cheah's "Violent Light: The Idea of Publicness in Modern Philosophy and in Global Neocolonialism," *Social Text* 43 (1995): 163–90; see also "Given Culture: Rethinking Cosmopolitical Freedom in Transnationalism," in *Cosmopolitics, Thinking and Feeling beyond the Nation,* ed. Pheng Cheah and Bruce Robbins (Minnesota: University of Minnesota Press, 1998), 290–328.

21. See for example, Teresa Ebert, "Ludic Feminism, the Body, Performance, and Labor: Bringing Materialism Back into Feminist Cultural Studies," *Cultural Critique* 23 (1992): 5–50.

22. Spivak wrote of the "international frame" of feminism in "French Feminism in an International Frame," *Yale French Studies* 62 (1981), 154–84. She has returned to this earlier essay in *Outside in the Teaching Machine,* 141–71. In both of these essays Spivak's discussions of Luce Irigaray, Julia Kristeva, and Hélène Cixous provide links to the works of Butler and Grosz, who also draw on these French feminist theorists. See also Spivak's treatment of Marxism in an international frame in "Scattered Speculations on the Question of Value," *In Other Worlds* (New York: Methuen, 1987), 154–75. Spivak's treatments of poststructuralism, feminism, Marxism, and neocolonialism are important influences on what follows; her work, more than anyone's, has encouraged my ongoing reading of poststructuralist critics.

23. Bruno Latour, *Science in Action* (Cambridge: Harvard University Press, 1987).

24. I am especially indebted to Karen Knorr-Cetina's discussion of "knowledge societies"; see her "Sociality with Objects: Social Relations in Postsocial Knowledge Societies," in *Theory, Culture, and Society* 14 (1997), 1–30. Daniel Bell's *The Coming of Post-Industrial Society: A Venture in Social Forecasting* (New York: Basic Books, 1973) is the reference for the term *knowledge societies.* But other works are also relevant. See Ulrich Beck, *Risk Society: Towards a New Modernity* (London: Sage, 1992);

Anthony Giddens, *The Consequences of Modernity* (Stanford: Stanford University Press, 1990); Jean-François Lyotard, *The Postmodern Condition* (Manchester, England: Manchester University Press, 1984); Krishan Kumar, *From Post-Industrial to Post-Modern Society* (Cambridge, England: Blackwell, 1995); Steven Seidman, *Contested Knowledge: Social Theory in the Postmodern Era* (Cambridge: Blackwell, 1994); Jeffrey C. Alexander, *Fin de Siècle Social Theory: Relativism, Reduction, and the Problem of Reason* (New York: Verso, 1995); Manuel Castells, *The Information Age: Economy, Society, and Culture,* vol. 3 (Oxford: Blackwell, 1998).

25. These are terms suggested by Knorr Cetina in "Sociality with Objects: Social Relations in Postsocial Knowledge Societies," 1–30.

26. Haraway, *Modest Witness@Second Millennium: FemaleMan© Meets OncoMouse™,* 16.

27. This term first arrived in intellectual discussion with Adrienne Rich's "Notes towards a Politics of Location," *Blood, Bread, and Poetry: Selected Prose, 1966–1978* (New York: Norton, 1986).

28. The notion of autoaffection as "hearing oneself speak" in a closed circuit between mouth and ear appears in Jacques Derrida, *Speech and Phenomena and Other Essays on Husserl's Theory of Signs,* trans. David B. Allison (Evanston, Ill.: Northwestern University Press, 1973), 78. The notion of autoaffection as "the giving-oneself-a-presence or a pleasure" appears in Jacques Derrida, *Of Grammatology,* trans. Gayatri Chakravorty Spivak (Baltimore: Johns Hopkins University Press, 1976), 165. There is also Derrida's eroticized version— "auto-fellatio" —discussed in his *Margins of Philosophy,* trans. Alan Bass (Brighton, England: Harvester Press, 1982), 289.

29. Sigmund Freud, *Three Essays on the Theory of Sexuality,* trans. James Strachey (New York: Basic Books, 1962), 92, n. 1.

30. See especially Armand Mattelart, *Mapping World Communication: War, Progress, Culture,* trans. Susan Emanuel and James A. Cohen (Minneapolis: University of Minnesota Press, 1994). Mattelart draws a connection between war and the historical development of world communication systems. See also Manuel de Landa, *War in the Age of Intelligent Machines* and Paul Virilio, *War and Cinema, The Logistics of Perception,* trans. Patrick Camiller (London: Verso, 1989); Paul Virilio, *Speed and Politics,* trans. Mark Polizzotti (New York: Semiotext(e), 1986); Paul Virilio and Sylvère Lotringer, *Pure War,* trans. Mark Polizzotti (New York: Semiotext(e), 1983).

31. Derrida has provided an interesting treatment of his own resistance to psychoanalysis and its relationship to deconstruction. See Jacques Derrida, "Resistances," in *Resistances of Psychoanalysis,* trans. Peggy Kamuf, Pascale-Anne Brault, Michael Naas (Stanford: Stanford University Press, 1998), 1–38. The essay also treats the resistances of psychoanalysis, that is, the resistances psychoanalysis treats, especially in terms of the repetition compulsion of the death drive.

1. The Technical Substrates of Unconscious Memory

1. Jacques Derrida, *Archive Fever: A Freudian Impression,* trans. Eric Prenowitz (Chicago: University of Chicago Press, 1996).

2. Jacques Derrida, "Freud and the Scene of Writing," in *Writing and Difference,* trans. Alan Bass (Chicago: University of Chicago Press, 1978). Freud's essay is "A Note upon the 'Mystic Writing-Pad,'" in *The Standard Edition of the Complete*

Psychological Works of Sigmund Freud, vol. 19, trans. James Strachey (London: Hogarth, 1925), 227–32.

3. Derrida, "Freud and the Scene of Writing," 228.

4. Richard Beardsworth, *Derrida and the Political* (New York: Routledge, 1996), 145–57.

5. I am indebted to Cheah's treatment of mattering, here especially for the way in which he connects it to Derrida's treatment of *différance.* Cheah, however, does not treat Derrida's indebtedness to Freud, and so he draws no connection between the unconscious and technology. See Pheng Cheah, "Mattering," *Diacritics* 26 (1996): 108–39.

6. Elizabeth A. Wilson, *Neural Geographies, Feminism, and the Microstructure of Cognition* (New York: Routledge, 1998).

7. Ibid., 162.

8. Ibid., 201.

9. Derrida, "Freud and the Scene of Writing," 222.

10. Sigmund Freud, "Project for a Scientific Psychology," in *The Standard Edition,* vol. 1, 295–397.

11. Derrida, "Freud and the Scene of Writing," 200.

12. Ibid., 201.

13. Ibid., 201.

14. Ibid., 201.

15. Ibid., 203.

16. Gilles Deleuze, *Difference and Repetition,* trans. Paul Patton (New York: Columbia University Press, 1994).

17. Jacques Derrida, *"Différance,"* in *Margins of Philosophy,* trans. Alan Bass (Chicago: University of Chicago Press, 1982), 13.

18. Derrida, "Freud and the Scene of Writing," 198.

19. Michel Foucault, *The History of Sexuality,* vol. 1, trans. Robert Hurley (New York: Vintage Books, 1980), 93. This connection between Derrida and Foucault is noticed by Gayatri Chakravorty Spivak in *Outside in the Teaching Machine* (New York: Routledge, 1993), 25–51. Drawing on Spivak, Cheah also points to the connection.

20. Freud's remarks are quoted in Derrida, "Freud and the Scene of Writing," 218.

21. Ibid., 211.

22. Ibid., 224.

23. Jacques Derrida, *Of Grammatology,* trans. Gayatri Chakravorty Spivak (Baltimore: Johns Hopkins University Press, 1976), 158.

24. Derrida, "Freud and the Scene of Writing," 222.

25. Jacques Derrida, "Living On: Border Lines," in *A Derrida Reader: Between the Blinds,* ed. Peggy Kamuf (New York: Columbia, 1991), 257.

26. Jacques Derrida, "Structure, Sign, and Play in the Discourse of the Human Sciences," in *Writing and Difference,* 279–80.

27. Jacques Derrida, "Signature, Event, Context," in *Limited Inc,* trans. Samuel Weber (Evanston: Northwestern University Press, 1972), 8.

28. Jacques Derrida, "Freud's Legacy," in *The Post Card,* trans. Alan Bass (Chicago: University of Chicago Press, 1987), 292–337.

29. Derrida, "Freud and the Scene of Writing," 228.

30. Derrida, *"Différance,"* 17.

31. Derrida, "Freud and the Scene of Writing," 227.

32. Ibid., 228.

33. Jacques Derrida, "The Theater of Cruelty and the Closure of Representation," in *Writing and Difference*, 232–50. This essay about Antonin Artaud follows "Freud and the Scene of Writing" in *Writing and Difference* and is an important intertext in Derrida's reading of Freud's treatment of scenes.

34. Jacques Derrida, *Specters of Marx: The State of the Debt, the Work of Mourning, and the New International*, trans. Peggy Kamuf (New York: Routledge, 1994), 51.

35. Derrida, "Freud and the Scene of Writing," 199.

36. Richard Beardsworth's treatment of Derrida's notion of aporia and time has been very helpful to me. See Richard Beardsworth, *Derrida and the Political* (New York: Routledge, 1996).

37. Derrida, *Specters of Marx*, 74–75.

38. Jacques Lacan, *The Seminars of Jacques Lacan*, book 2: *The Ego in Freud's Theory and in the Technique of Psychoanalysis*, 1954–1955, trans. Sylvana Tomaselli (Cambridge: Cambridge University Press, 1988), 170.

39. Jacques Derrida, " 'To Do Justice to Freud': The History of Madness in the Age of Psychoanalysis," *Critical Inquiry* 2 (1994): 265–66. Derrida is referring here to Foucault's often-quoted remark from *The History of Sexuality*, vol. 1: *An Introduction*, trans. Robert Hurley (New York: Random Books, 1980), 159: "We need to consider the possibility that one day, perhaps, in a different economy of bodies and pleasures, people will no longer quite understand how the ruses of sexuality and the power that sustains its organization, were able to subject us to that austere monarchy of sex, so that we became dedicated to the endless task of forcing its secret, of exacting the truest of confessions from a shadow."

40. Ibid., 266.

41. Jacques Derrida, "La Facteur de la Verité," in *The Post Card: From Socrates to Freud and Beyond*, trans. Alan Bass (Chicago: University of Chicago Press, 1987), 413–96.

42. Patricia Mellencamp, "TV Time and Catastrophe, or *Beyond the Pleasure Principle* of Television," in *Logics of Television: Essays in Cultural Criticism*, ed. Patricia Mellencamp (Bloomington: Indiana University Press, 1991), 243.

43. Kaja Silverman, *Male Subjectivity at the Margins* (New York: Routledge, 1992), 15.

44. Ibid., 14.

45. Ibid., 41.

46. Kaja Silverman, "What Is a Camera?, or, History in the Field of Vision," *Discourse* 15 (1993), 3–56. In this essay Silverman returns to her essay, "Fassbinder and Lacan: A Reconsideration of Gaze, Look, and Image," *Camera Obscura* 19 (1989): 55–84.

47. Laura Mulvey, "Visual Pleasure and Narrative Cinema," *Screen* 16 (1975): 6–18.

48. Jacques Lacan, *Ecrits*, trans. Alan Sheridan (New York: Norton, 1977), 2.

49. Ibid., 2–3.35.

50. Christian Metz, "The Imaginary Signifier," *Screen* 2 (1975), 51. Teresa De Lauretis, *Alice Doesn't: Feminism, Semiotics, Cinema* (Bloomington: Indiana University Press, 1984), 144.

52. Ibid., 121.

53. Jane Gallop, *Reading Lacan* (Ithaca: Cornell University Press, 1985), 74–92.

54. Teresa De Lauretis, *Technologies of Gender: Essays on Theory, Film, and Fiction* (Bloomington: Indiana University Press, 1987), 2.

55. Jacqueline Rose, *Sexuality in the Field of Vision* (London: Verso, 1986), 90–91.

56. Kaja Silverman, *The Acoustic Mirror: The Female Voice in Psychoanalysis and Cinema* (Bloomington: Indiana University Press, 1988), 216.

57. Silverman, *Male Subjectivity at the Margins*, 41.

58. In her recent treatment of "black novels" deploying a psychoanalytic perspective, Claudia Tate points to "the misgivings that many African Americans have about the relevance of psychoanalysis to black liberation, thus the general absence of psychoanalytic models in black intellectual discourse" (5). In a footnote she adds, "[M]any would contend that the imposition of psychoanalytic theory on African American literature advances Western hegemony over the cultural production of black Americans, indeed over black subjectivity" (192, n. 6.) Although I agree with Tate, who argues that to have no analysis of the unconscious in treating race seems a terrible loss, I also question the relevancy of the oedipal narrative as a universal narrative. Questions of racial difference, like those of sexual difference, are part of my uneasiness about the oedipal narrative, but that is not my only uneasiness. There also is the historical and cultural specificity of the oedipal narrative in relationship to technology. Of course there are scholars who have treated race and psychoanalysis. Besides Claudia Tate, whose work I have been quoting — *Psychoanalysis and Black Novels: Desire and the Protocols of Race* (Oxford: Oxford University Press, 1998) — there are scholars who have written about race and psychoanalysis in treating the works of Franz Fanon, such as Stuart Hall, bell hooks, Kobena Mercer, and Homi Bhabha; see *The Fact of Blackness: Franz Fanon and Visual Representation*, ed. Alan Read (Seattle: Bay Press, 1996). Among Black feminist theorists, Hortense Spillers has been a provocative commentator on psychoanalysis and race; see especially "Mama's Baby, Papa's Maybe: An American Grammar Book," *Diacritics*, 17 (1987), 65–81. For my own discussion of Black feminist thought, see *Feminist Thought, Desire, Power, and Academic Discourse* (Cambridge, Mass.: Blackwell, 1994). Among early feminist film theorists, Jane Gaines brought attention to the exclusion of questions of race in feminist film theory in "White Privilege and Looking Relations: Race and Gender in Feminist Film Theory," *Cultural Critique* 4 (1986), 59–79. There also is the statement about feminist film criticism by a black feminist theorist; see bell hooks, *Black Looks: Race and Representation* (Boston: South End Press, 1992). There were responses among early feminist film theorists; along with Silverman, see Mary Ann Doane, "Dark Continents: Epistemologies of Racial and Sexual Difference in Psychoanalysis and the Cinema," in *Femmes Fatales, Feminism, Film Theory, Psychoanalysis* (New York: Routledge, 1991), 209–48. See also the more recent book by Sharon Willis, *High Contrast: Race and Gender in Contemporary Hollywood Film* (Durham: Duke University Press, 1997). In the journal *Screen*, where early feminist film theory often was treated, the question of race and film theory was discussed in relationship to colonial discourse in an early essay by Homi Bhabha, "The Other Question: The Stereotype and Colonial Discourse," *Screen* 24 (1983), 18–36. Special issues of this journal on race continued this discussion, the last of which was *Screen* 29 (1988), edited by Isaac Julien and Kobena Mercer. There have been more recent works by early feminist theorists dealing with colonial and neocolonial discourses, film, gender, and race; see E. Ann Kaplan, *Look-*

ing for the Other: Feminism, Film, and the Imperial Gaze (New York: Routledge, 1997). The questions of race and gender are, of course, not the only questions of difference brought to bear on feminist film theory. There also has been the treatment of differences of ethnicity, sexual orientation, and nation in relationship to film, which have drawn on much of the debate indicated above; for example, see Diane Carson, Linda Dittmar, and Janice Welsch, eds., *Multiple Voices in Feminist Film Criticism* (Minneapolis: University of Minnesota Press, 1994)

59. When I say that the question of technology was not engaged in the above debates about psychoanalysis and early feminist film theory in relationship to exclusions of various differences, I mean that these debates were not seen as part of the becoming of the teletechnological. But for an exception see the discussion of questions of difference and changes in technologies of representation in Ella Shohat and Robert Stam, *Unthinking Eurocentrism: Multiculturalism and the Media* (New York: Routledge, 1994).

60. Silverman, "Fassbinder and Lacan: A Reconsideration of Gaze, Look and Image." I am drawing from a reprint of this essay in *Male Subjectivity at the Margins*, 125–56.

61. Jacques Lacan's diagrams appear in his two essays "The Line and Light" and "What Is a Picture?" in *The Fundamental Concepts of Psycho-Analysis*, trans. Alan Sheridan (New York: Norton, 1978), 91–119.

62. Silverman, "What Is a Camera?, or, History in the Field of Vision," 12.

63. Lacan, *The Fundamental Concepts of Psycho-Analysis*, 89.

64. Silverman, *Male Subjectivity at the Margins*, 150.

65. Silverman, "What Is a Camera?, or, History in the Field of Vision," 11.

66. Silverman is referring to Jonathan Crary, *Techniques of the Observer: On Vision and Modernity in the Nineteenth Century* (Cambridge: MIT Press, 1990).

67. Ibid., 175.

68. Silverman, "What Is a Camera?," 12.

69. Ibid., 14.

70. Derrida reports that before Freud took the writing machine as metaphor for the unconscious, he made use of an "optical machine," that is, a camera. But when Derrida emphasizes that the writing machine will better suit the unconscious, it is hard not to think that television is a writing machine with a picture, see "Freud and the Scene of Writing," 215–20 and 330, n. 18.

71. Richard Dienst, *Still Life in Real Time* (Durham: Duke University Press, 1994), 146.

72. Gilles Deleuze, *Difference and Repetition*, trans. Paul Patton (New York: Columbia: 1994), 8.

2. The Generalized Unconscious of Desiring Production

1. Jacques Lacan, *Television: A Challenge to the Psychoanalytic Establishment*, trans. Denis Hollier, Rosalind Krauss, and Annette Michelson (New York: Norton, 1990).

2. Richard Dienst, *Still Life in Real Time* (Durham: Duke University Press, 1994). Dienst reports that Lacan's television talk was given in 1974, but in *Television* it is reported that Lacan appeared on television in 1973.

3. This translation of Lacan's remarks is offered by Dienst in *Still Life in Real Time*, ix. See also Lacan, *Television*, 3.

4. Fredric Jameson, "Surrealism without the Unconscious," in *Postmodernism; or, The Cultural Logic of Late Capitalism* (Durham: Duke University Press, 1991), 67–96.

5. Jacques Derrida, *Specters of Marx*, trans. Peggy Kamuf (New York: Routledge, 1994), 54.

6. Stuart Hall, "Coding and Decoding in the Media Discourse," stenciled paper 7, Birmingham, Center for Contemporary Cultural Studies, 1973. Reprinted in *Culture, Media, Language*, ed. Stuart Hall, Dorothy Hobson, Andrew Lowe, and Paul Willis (London: Hutchinson, 1980), 128–38.

7. Raymond Williams, *Television: Technology and Cultural Form* (New York: Schocken, 1974), 92.

8. Tony Bennett and Janet Woollacott, *Bond and Beyond: The Political Career of a Popular Hero* (London: Macmillan, 1988).

9. Ibid., 264.

10. Jacques Derrida, "Signature, Event, Context," *Limited Inc*, trans. Samuel Weber and Jeffrey Mehlman (Evanston, Ill.: Northwestern University Press, 1988), 12.

11. Ibid., 12.

12. Ibid., 20–21.

13. John Fiske, "Television and Popular Culture: Reflections on British and Australian Critical Practice," *Critical Studies in Mass Communication* 3 (1986), 213. See also John Fiske, "British Cultural Studies and Television," in *Channels of Discourse: Television and Contemporary Criticism*, ed. Robert C. Allen (Chapel Hill: University of North Carolina Press, 1987), 254–89. See also these earlier works: John Fiske, *Television Culture* (New York: Methuen, 1987); and John Fiske and John Hartley, *Reading Television* (London: Methuen, 1978).

14. Meaghan Morris, "Banality in Cultural Studies," ed. Patricia Mellencamp *Logics of Television* (Bloomington: Indiana University Press, 1990), 23.

15. Richard Johnson, "What Is Cultural Studies Anyway?" *Social Text* 16 (1986–1987), 68.

16. Stephen Heath, "Representing Television," in *Logics of Television: Essays in Cultural Criticism* (Bloomington: Indiana University Press, 1991), ed. Patricia Mellencamp, 290.

17. David Harvey, *The Condition of Postmodernity* (Cambridge, Mass.: Blackwell, 1989).

18. Stuart Hall, "The Meaning of New Times," in *Stuart Hall: Critical Dialogues in Cultural Studies*, ed. David Morley and Kuan-Hsing Chen (1989; reprint, New York: Routledge, 1996), 225–37.

19. Fredric Jameson, "Surrealism without the Unconscious," 68.

20. Ibid., 67.

21. Fredric Jameson, *The Political Unconscious: Narrative as a Socially Symbolic Act* (Ithaca, N.Y.: Cornell University Press, 1981).

22. Ibid., 24–25. These remarks are not, of course, only Althusser's, but I follow Jameson in referring them to him. See Louis Althusser and Etienne Balibar, *Reading Capital*, trans. Ben Brewster (London: Verso, 1979).

23. Jameson, *The Political Unconscious*, 35.

24. Ibid., 19.

25. Ibid., 56.
26. Ibid., 76.
27. Ibid., 83.
28. Ibid., 95.
29. Ibid., 95.
30. Ibid., 95.
31. Ibid., 66.
32. Ibid., 68.
33. Ibid., 70.
34. Ibid., 70.
35. Ibid., 74.
36. Fredric Jameson, "The Cultural Logic of Late Capitalism," *Postmodernism; or, The Cultural Logic of Late Capitalism,* 15.
37. Ibid., 18.
38. Ibid., 35.
39. Ibid., 18.
40. Ibid., 35.
41. I am drawing my discussion on technology from Martin Heidegger, *The Question of Technology and Other Essays,* trans. William Lovitt (New York: Garland, 1977).
42. Jameson quotes from Martin Heidegger, "The Origin of the Work of Art," in *Philosophies of Art and Beauty,* ed. Albert Hofstadter and Richard Kuhns (New York: Harper and Row, 1964), 663.
43. Jameson, "The Cultural Logic of Late Capitalism," 9.
44. Ibid., 9.
45. Ibid., 34.
46. Ibid., 8.
47. Ibid., 37–38.
48. Neyer Schapiro, "The Still Life as a Personal Object," in *The Reach of Mind: Essays in Memory of Kurt Goldstein* (New York: Springer), quoted in Jacques Derrida, *The Truth in Painting,* trans. Geoff Bennington and Ian McLeod (Chicago: University of Chicago Press, 1987), 276.
49. Ibid., 288.
50. Ibid., 299.
51. Jacques Derrida, "Remarks on Deconstruction and Pragmatism," in *Deconstruction and Pragmatism,* ed. Chantal Mouffe (London: Routledge, 1986), 81–82.
52. Jameson, "Surrealism without the Unconscious," 91.
53. Ibid., 88.
54. Ibid., 87.
55. Ibid., 94.
56. Ibid., 76.
57. Ibid., 76.
58. Ibid., 76.
59. Paul Virilio, "The Third Interval: A Critical Transition," in *Rethinking Technologies,* ed. Verena Andermatt Conley (Minneapolis: University of Minnesota, 1993), 7.
60. Ibid., 5.
61. Ibid., 5.

62. Jameson, "Surrealism and the Unconscious," 95.

63. Rosalind Kraus, "The Master's Bedroom," *Representations* 28 (1989), 63.

64. Ibid., 63.

65. Dienst, *Still Life in Real Time,* 60.

66. Ibid., 59.

67. Ibid., 179, n. 70.

68. Ibid., 179, n. 70.

69. Jonathan Beller, "Capital/Cinema," in *Deleuze and Guattari: New Mappings in Politics, Philosophy, and Culture,* eds. Eleanor Kaufman and Kevin Jon Heller (Minneapolis: University of Minnesota Press, 1998), 91.

70. See Stanley Aronowitz and William DiFazio, *The Jobless Future: Sci-Tech and the Dogma of Work* (Minneapolis: University of Minnesota Press, 1994).

71. I am thinking here of the works of Antonio Negri, especially "The Physiology of Counter-Power: When Socialism Is Impossible and Communism So Near," in *Body Politics: Disease, Desire, and the Family,* eds. Michael Ryan and Avery Gordon (Boulder, Colo.: Westeview Press, 1994), 225–50. See also Toni Negri's *Marx beyond Marx,* trans. Harry Cleaver, Michael Ryan, and Maurizio Viano (South Hadley, Mass.: Bergin and Garvey Publishers, 1984).

72. Jonathan Cutler and Stanley Aronowitz, "Quitting Time: An Introduction," in *Post-Work,* eds. Stanley Aronowitz and Jonathan Cutler (New York: Routledge, 1998).

73. Beller, "Capital/Cinema," 93.

74. Heath, "Representing Television," 294.

75. Ibid., 294.

76. Dienst, *Still Life in Real Time,* 151.

77. Ibid., 148.

78. I am referring to Gilles Deleuze, *Cinema 2: The Time-Image,* trans. Hugh Tomlinson and Robert Galeta (Minneapolis: University of Minnesota Press, 1989).

79. Ibid., 41.

80. Ibid, 110–111.

81. Ibid., 81–82.

82. Ibid., 91.

83. For a discussion of time and the past in general, see Gilles Deleuze, *Bergsonism,* trans. Hugh Tomlinson and Barbara Habberjam (New York: Zone Books, 1991).

84. Dienst, *Still Life in Real Time,* 162.

85. Ibid., 159–66.

86. Ibid., 169.

87. Deleuze is quoting from Michel Foucault, *The Order of Things,* trans. Alan Sheridan (New York: Pantheon, 1970), 278.

88. Dienst, *Still Life in Real Time,* 169.

89. See Gilles Deleuze, "Postscript on the Societies of Control," *October* 59 (1991): 3–7.

90. Michael Hardt, "The Withering of Civil Society," *Social Text* 45 (1995), 27–44.

91. Brian Massumi, "Requiem for Our Prospective Dead," in *Deleuze and Guattari: New Mappings in Politics, Philosophy, and Culture,* 40–64. Although I find Massumi's reading of late capitalism, like Hardt's and Dienst's, most provocative, I am not fully assuming their criticisms of it.

92. Jacques Derrida, *Specters of Marx,* 59.

93. Lisa Lowe and David Lloyd, "Introduction," in *The Politics of Culture in the Shadow of Capital*, ed. Lisa Lowe and David Lloyd (Durham: Duke University Press, 1997), 25.

94. J. K. Gibson-Graham, *The End of Capitalism (As We Knew It)* (Cambridge, England: Blackwell, 1996).

3. Queer Desire and the Technobodies of Feminist Theory

1. Anne Balsamo, *Technologies of the Gendered Body* (Durham: Duke University Press, 1996). See also Susan Bordo's suspicious treatment of poststructuralism in her treatment of bodies in *Unbearable Weight: Feminism, Western Culture, and the Body* (Berkeley: University of California Press, 1993).

2. Nancy Hartsock, "Rethinking Modernism," in *Cultural Critique* 7 (1987): 187–206.

3. Balsamo, *Technologies of the Gendered Body*, 31.

4. N. Katherine Hayles, "Text out of Context: Situating Postmodernism within an Information Society," *Discourse* 9 (1987): 24–36.

5. Donna Haraway, "Manifesto for Cyborgs: Science, Technology, and Socialist Feminism in the 1980s," *Socialist Review* 80 (1985): 65–108.

6. Eve Kosofsky Sedgwick, *Epistemology of the Closet* (Berkeley: University of California Press, 1990), 61.

7. Donna Haraway, "The Promises of Monsters: A Regenerative Politics for Inappropriate/d Others," in *Cultural Studies*, eds. Lawrence Grossberg, Cary Nelson, and Paul Treichler (New York: Routledge, 1991), 297.

8. I am indebted to Pheng Cheah for his treatment of both Butler's and Grosz's treatments of bodies in "Mattering," *Diacritics* 26 (1996): 108–39. But I am taking his arguments in a direction that I think is quite different than what his own politic would seem to allow and what my interest in the unconscious demands. See also a discussion among Elizabeth Grosz, Pheng Cheah, Judith Butler, and Drucilla Cornel in "Interview," *Diacritics* 28 (1998), 19–42.

9. Haraway, "Manifesto for Cyborgs," 166.

10. Rosi Braidotti, "Mothers, Monsters, and Machines," in *Nomadic Subjects* (New York: Columbia, 1994), 75–94. Braidotti is also a feminist theorist who has treated technology, the woman's body, and poststructuralism. Butler and Grosz, however, have worked through psychoanalysis more systematically with more interesting results for ontology.

11. Take, for example, a spring 1998 conference hosted by the Center for Lesbian and Gay Studies at the Graduate School and the University Center of the City University of New York entitled "Queer Globalization/Local Homosexualities: Citizenship, Sexuality, and the Afterlife of Colonialism." It also would be interesting to articulate the differences between two texts. The first, an early treatment of homosexualities and the rhetoric of nationalisms, is *Nationalisms and Sexualities*, ed. Andrew Parker, Mary Russo, Doris Sommer, and Patricia Yaeger (New York: Routledge, 1992). The second is *Dangerous Liaisons: Gender, Nation, and Postcolonial Perspectives*, ed. Anne McClintock, Aamir Mufti, and Ella Shohat (Minneapolis: University of Minnesota Press, 1997). The latter, at least, is much more aware of transnational capitalism and diaspora, and therefore treats postcolonial theory as itself problematic.

12. Gayatri Chakravorty Spivak, *The Post-Colonial Critic: Interviews, Strategies, Dialogues*, ed. Sarah Harasym (New York: Routledge, 1990), 71.

13. These are remarks that Butler made in an interview with Irene Costera Meijer and Baukje Prins in "How Bodies Come to Matter: An Interview with Judith Butler," *Signs* 23 (1998), 279.

14. Butler has remarked on the importance of Rose's remark, which I quoted in chapter 1. Butler suggests rightly that Rose showed feminist theorists the way to use psychoanalysis in order to think about unconscious desire as a potential for resistance to undesirable social reality. See Judith Butler, *The Psychic Life of Power: Theories in Subjection* (Stanford: Stanford University Press), 97.

15. Butler's comments about Kaja Silverman appear in a footnote to Butler's *Bodies That Matter* (New York: Routledge), 268–69, n. 10.

16. Ibid., 35.

17. Butler's treatment of encrypting is in her *Gender Trouble* (New York: Routledge, 1990), chapter 2.

18. Ibid., 68.

19. Ibid., 71.

20. Ibid., 70.

21. Judith Butler, "Imitation and Gender Insubordination," in *Inside/Out: Lesbian Theories, Gay Theories*, ed. Diana Fuss (New York: Routledge, 1991), 28.

22. Ibid., 28.

23. Judith Butler, "The Lesbian Phallus and the Morphological Imaginary," *Differences* 6 (1992), 124.

24. Ibid., 137.

25. Ibid., 136.

26. Ibid., 143.

27. Ibid., 150.

28. I am drawing from Butler's discussion of Slavoj Žižek's reading of Lacan in her *Bodies That Matter*, 187–222.

29. See especially Lacan's treatment of the real in *The Seminars*, book 2, *The Ego in Freud's Theory and in the Technique of Psychoanalysis, 1954–1955*, trans. Sylvana Tomaselli (New York: Norton, 1988).

30. Butler, *Bodies That Matter*, 206–7.

31. Nancy Fraser, "Heterosexism, Misrecognition, and Capitalism," *Social Text* 52–53 (1997), 284.

32. I am drawing primarily on Nancy Fraser's *Justice Interruptus: Critical Reflections on the "Postsocialist" Condition* (New York: Routledge, 1997), but also on her *Unruly Practices: Power, Discourse, and Gender in Contemporary Social Theory* (Minneapolis: University of Minnesota Press, 1989).

33. I am drawing here on both Cheah, "Mattering," 108–39; and Cheah, "Violent Light: The Idea of Publicness in Modern Philosophy and in Global Neocolonialism," *Social Text* 43 (1995).

34. James Clifford, "Traveling Cultures," in *Cultural Studies*, eds. Lawrence Grossberg et al., 108. See also Rey Chow, "Violence in the Other Country: China as Crisis, Spectacle, and Woman," in *Third World Women and the Politics of Feminism*, eds. Chandra Talpade Mohanty, Ann Russo, and Lourdes Torres (Bloomington: Indiana University Press, 1991); and Aihwa Ong's discussion of multiple modernities in "Anthropology, China, and Modernities: The Geopolitics of Cultural Knowledge,"

in *The Future of Anthropological Knowledge,* ed. Henrietta Moore (New York: Routledge, 1996), 60–92.

35. Cheah, "Violent Light," 79. This formulation appears in many of Spivak's writings, but here the reference is Spivak's treatment of the subindividual forces in Foucault's relations of the forces of power and Derrida's *différance.* See Gayatri Chakravorty Spivak, "More on Power/Knowledge," in *Outside in the Teaching Machine* (New York: Routledge, 1993), 25–51.

36. David Scott, "The Aftermaths of Sovereignty: Postcolonial Criticism and the Claims of Political Modernity, *Social Text* 48 (1996), 17–18.

37. Cheah, "Mattering," 121.

38. Jacques Derrida, "Remarks on Deconstruction and Pragmatism," in *Deconstruction and Pragmatism,* ed. Chantal Mouffe (New York: Routledge, 1996), 77–88.

39. Judith Butler, "Contingent Foundations: Feminism and the Question of 'Postmodernism,'" in *Feminist Contentions,* ed. Linda Nicholson (New York: Routledge, 1995), 41.

40. Butler, *The Psychic Life of Power,* 1–30.

41. Butler has returned to Althusserian Marxism in her *The Psychic Life of Power,* 106–31, where her treatment of Althusser is more provocative than her remarks in the essay "Merely Cultural," *Social Text* 52–53 (1997), 265–77, to which Fraser is responding.

42. Donna Haraway, *Modest Witness@Second Millennium: FemaleMan© Meets OncoMouse™* (New York: Routledge, 1997), 265.

43. Elizabeth Grosz, *Volatile Bodies: Towards a Corporeal Feminism* (Bloomington: Indiana University Press, 1994), 21.

44. Ibid., 21.

45. Cheah, "Mattering," 120.

46. Elizabeth Grosz, "Lesbianism Fetishism?" *Differences* 3 (1991): 39–54. I draw from a reprint of this essay in Grosz, *Space, Time, and Perversion* (New York: Routledge, 1995), 141–54.

47. Ibid., 148.

48. Ibid., 149.

49. Ibid., 154.

50. Elizabeth Grosz, "Experimental Desire: Rethinking Queer Subjectivity," in *Supposing the Subject,* ed. Joan Copjec, (New York: Verso, 1994), 152.

51. Grosz, *Volatile Bodies,* 165.

52. Ibid., 164.

53. Gilles Deleuze and Félix Guattari, *Thousand Plateaus,* trans. Brian Massumi (Minneapolis: University of Minnesota Press, 1987), 164.

54. Grosz, *Volatile Bodies,* 171.

55. Gilles Deleuze and Félix Guattari, *Thousand Plateaus,* 158.

56. Gilles Deleuze and Félix Guattari, *Anti-Oedipus* (New York: Viking, 1977), 309.

57. In her treatment of body art Amelia Jones draws on Judith Butler, but adds to Butler's treatment of the body a certain reading of Merleau Ponty. When it comes to treating technology, her conclusion is: "The body/self is technophenomenological: fully mediated through the vicissitudes of bio- and communications technologies, and fully engaged with the social (what Merleau Ponty would call 'enworlded.') The body/self is hymenal, reversible simultaneously both subject and object." I like this summation except that it seems to refer to the human body, showing the effect

of a return to phenomenology. See Amelia Jones, *Body Art: Performing the Subject* (Minneapolis: University of Minnesota Press, 1998), 235.

58. Gilles Deleuze, "What Children Say," in *Essays Critical and Clinical*, trans. Daniel W. Smith and Michael Greco (Minneapolis: University of Minnesota Press, 1997), 63.

59. Grosz, *Volatile Bodies*, 165.

60. Grosz, "Architecture from the Outside," in *Space, Time, and Perversion*, 125–37.

61. Arjun Appadurai, "Dead Certainty: Ethnic Violence in the Era of Globalization," *Public Culture* 25 (1998), 225–47.

62. Grosz, *Volatile Bodies*, 180.

63. John Rajchman, *Constructions* (Cambridge: MIT Press, 1998).

64. Donna Haraway, *Modest Witness*, 16.

4. The Ontological Perspective of Knowledge Objects

1. James Clifford's remark appears in "On Ethnographic Authority," *Representations* 1 (1983): 118–46. I am drawing on the reprint of this essay in Clifford's *The Predicament of Culture: Twentieth-Century Ethnography, Literature, and Art* (Cambridge: Harvard University Press, 1988), 22.

2. Alan Sokal, "Transgressing the Boundaries: Toward a Transformative Hermeneutics of Quantum Gravity, *Social Text* 46–47 (1996): 217–52.

3. Alan Sokal, "A Physicist Experiments with Cultural Studies," *Lingua Franca* May–June (1996), 64.

4. I take this phrase from Judith Butler's essay entitled "Merely Cultural," *Social Text* 52–53 (1997): 265–77. Butler responds to Marxist critics of her work and, by the way, offers a wonderful criticism of the Sokal hoax, showing how Sokal's parody of cultural critics is ambivalent because of his desire to copy them so closely.

5. James Clifford, *The Predicament of Culture*, 22–23.

6. Karen Knorr-Cetina, "Sociality with Objects: Social Relations in Postsocial Knowledge Societies," *Theory, Culture, and Society* 14 (1997): 1–30. Particularly interesting is Knorr-Cetina's treatment of knowledge objects in terms of Jacques Lacan's notion of "lack." She argues that because knowledge objects are always incompletely given to their users, because of the "temporal volatility and unfolding ontology of these objects," they produce a series of lacks for the user. She then proposes that it is not necessary, however, to draw these lacks back to Lacan's treatment of the narcissistic wounding of the infant-child. I would read this remark as an effort to get beyond the oedipal narrative in order to get to the desire of knowledge objects themselves, to get to their desire to become fully themselves for another.

7. Ibid., 8.

8. My own earlier effort to connect the early criticism of ethnographic writing with other writing technologies connected to film and teletechnology appears in *The End(s) of Ethnography: From Realism to Social Criticism* (Newbury Park, Calif.: Sage, 1992), reprinted with a new preface as *The End(s) of Ethnography* (New York: Peter Lang, 1998). For a careful review of experimentation in ethnographic writing, especially in sociology, which is sensitive to changes in technology, see Norman K.

Denzin, *Interpretive Ethnography: Ethnographic Practices for the Twenty-First Century* (Newbury Park, Calif.: Sage, 1997).

9. Michael Lynch and Steve Woolgar, "Introduction: Sociological Orientations to Representational Practice in Science," in *Representation in Scientific Practice*, ed. Michael Lynch and Steve Woolgar (Cambridge: MIT Press, 1990), 3.

10. The "strong programme" is attributed to the works of Barry Barnes, Steven Bloor, and David Shapin. See Barry Barnes and Steven Shapin, eds., *Natural Order: Historical Studies of Scientific Culture* (London: Sage, 1979); David Bloor, *Knowledge and Social Imagery* (London: Routledge and Kegan Paul, 1976).

11. Bruno Latour and Steve Woolgar, *Laboratory Life: The Construction of Scientific Facts* (London: Sage, 1976).

12. Bruno Latour, "Drawing Things Together," in *Representation in Scientific Practice*, 19–68.

13. Bruno Latour, "Give Me a Laboratory and I Will Raise the World," in *Science Observed*, ed. Karen Knorr-Cetina and Michael Mulkay (London: Sage, 1983), 145.

14. Ibid., 152.

15. Ibid., 160.

16. Latour, "Drawing Things Together," 46.

17. Bruno Latour, *Science in Action* (Cambridge: Harvard University Press, 1987), 128–44; 215–57.

18. Ibid., 253.

19. During the 1980s there were a number of important literary criticisms of realist narrativity and its oedipal logic. These served well in the criticism of what was then referred to as ethnographic realism. See my *The End(s) of Ethnography*, chapter 1; there I especially draw on Mark Seltzer's essays, for example, "Reading Foucault: Cells, Corridors, Novels," *Diacritics* (spring 1984): 78–89.

20. Donna Haraway, *Modest Witness@Second Millennium: FemaleMan© Meets OncoMouse™* (New York: Routledge), 35.

21. Ibid., 62.

22. Bruno Latour, *Science in Action*, 15. For Haraway discussion, see *Modest Witness*, 35.

23. Haraway, *Modest Witness*, 34–35.

24. Ibid., 35.

25. Steven Woolgar, *Knowledge and Reflexivity: New Frontiers in the Sociology of Knowledge* (London: Sage, 1988). See particularly the introduction with Malcolm Ashmore, "The Next Step: An Introduction to the Reflexive Project," 1–11. See also Woolgar, *Science: The Very Idea!* (London: Tavistock). Although Woolgar is perhaps the best known among the new sociologists of science for exploring experimental writing practices, my favorite self-reflexive work is Malcolm Ashmore's *The Reflexive Thesis: Wrighting Sociology of Scientific Knowledge* (Chicago: University of Chicago Press, 1989).

26. Haraway, *Modest Witness*, 33. Woolgar, along with Keith Grint, has responded to Haraway's criticism by arguing that politicized positions like Haraway's show "a lack of nerve" in refusing a full reflexivity. See "On Some Failures of Nerve in Constructivist and Feminist Analyses of Technology," *Science, Technology, and Human Values* 3 (1995): 286–310. Two different kinds of politics are implied here, one like Haraway's feminist, Marxist, and antiracist politics and another politics about refusing authority in writing. Although surely these are not opposed, they are not,

however, reducible one to the other. I argue later that it is impossible to successfully refuse all authority in writing, but it also must be recognized that to take positions such as Haraway's is to foreclose at least temporarily any further deconstruction of one's own position of authority. Of course this is the argument
Haraway herself makes against Latour, and it is in hopes of getting beyond this
aporia that Haraway offers the notion of diffraction. I return to this *aporia* in my
discussion of Trinh T. Minh-ha's experimental works later in this chapter.

27. Haraway, *Modest Witness*, 16.

28. Ibid., 267.

29. James Clifford, "Introduction: Partial Truths," in *Writing Culture: The Poetic
and Politics of Ethnography* ed. James Clifford and George Marcus (Berkeley: University of California Press, 1986), 4, n. 3.

30. Clifford, "On Ethnographic Authority," 21–54.

31. Ibid., 22.

32. Ibid., 33–34.

33. Clifford, "Introduction: Partial Truths," 19–20, n. 9.

34. Teresa De Lauretis, *Alice Doesn't: Feminism, Semiotics, Cinema* (Bloomington: Indiana University Press, 1984), 119.

35. Christian Hanson, Catherine Needham, and Bill Nichols, "Skin Flicks: Por-
nography and Ethnography and the Discourses of Power," *Discourse* 11.2 (1989):
54–79.

36. Lisa Cartwright, *Screening the Body: Tracing Medicine's Visual Culture* (Minnesota: University of Minnesota Press, 1995).

37. Ibid., 28.

38. Trinh T. Minh-ha, *Woman, Native Other: Writing Postcoloniality and Feminism* (Bloomington: Indiana University Press, 1989), 157, n. 64.

39. Trinh T. Minh-ha, *Framer Framed* (New York: Routledge, 1992), 123.

40. Gayatri Chakravorty Spivak, *The Post-Colonial Critic: Interviews, Strategies,
Dialogues,* ed. Sarah Harasym (New York: Routledge, 1990), 51,

41. Gayatri Chakravorty Spivak, "Translator's Preface and Afterword to Ma-
hasweta Devi, *Imaginary Maps,*" in *The Spivak Reader,* ed. Donna Landry and Ger-
ald Maclean (New York: Routledge, 1996), 270.

42. Trinh, *Woman, Native Other,* 76.

43. Trinh, *Framer Framed,* 157.

44. Gayatri Chakravorty Spivak, "Can the Subaltern Speak?" in *Marxism and the
Interpretation of Culture,* ed. Cary Nelson and Larry Grossbery (Urbana: University
of Illinois Press, 1988), 288–89.

45. Herman Rapaport, "Deconstruction's Other: Trinh T. Minh-ha and Jacques
Derrida," *Diacritics* 25 (1995), 102.

46. I take the phrase "flickering signifiers" from N. Katherine Hayles's essay
"Virtual Bodies and Flickering Signifiers," *October* (fall 1993): 69–91. Hayles links
teletechnology to flickering signifiers, marking a shift from the "floating signifiers"
to which Lacan refers and that already indicate the destabilization of meaning. The
flickering signifiers of teletechnology further destabilize meaning, as Hayles sees it.

47. Sandra Harding, *The Science Question in Feminism* (Ithaca, N.Y.: Cornell
University Press, 1986), 136–62.

48. Dorothy Smith, *The Everyday World as Problematic: A Feminist Sociology*
(Boston: Northeastern University Press, 1987).

49. Nancy Hartsock, *Money, Sex, and Power: Toward a Feminist Historical Materialism* (Boston: Northeastern University Press, 1985).

50. Sandra Harding, *Whose Science? Whose Knowledge? Thinking from Women's Lives* (Ithaca, N.Y.: Cornell University Press, 1992).

51. Smith, *The Everyday World as Problematic*, 83.

52. Hartsock, *Money, Sex, and Power: Toward a Feminist Historical Materialism*, 242.

53. Jonathan Beller, "Capital/Cinema," in *Deleuze and Guattari: New Mappings in Politics, Philosophy, Culture,* eds. Eleanor Kaufman and Keven Jon Heller (Minneapolis: University of Minnesota Press, 1998), 91.

54. Stanley Aronowitz, "The Politics of the Science Wars," *Social Text* 46–47 (1996): 177–97. See also Stanley Aronowitz, *Science as Power: Discourse and Ideology in Modern Society* (Minneapolis: University of Minnesota Press, 1988).

55. Harding, *The Science Question in Feminism*, 157.

56. Patricia Collins, *Black Feminist Thought, Knowledge Consciousness, and the Politics of Empowerment* (New York: Routledge, 1990); Gloria Anzaldúa, *Borderlands/La Frontera* (San Francisco: Aunt Lute Books, 1987).

57. Sandra Harding, *Is Science Multicultural? Postcolonialisms, Feminisms, and Epistemologies* (Bloomington: Indiana University Press, 1998), 188–94.

58. Norma Alarcon, "The Theoretical Subject(s) of *This Bridge Called My Back and Anglo-American Feminism,*" in *Making Face/Making Soul/Haciendo Caras: Creative and Critical Perspectives by Women of Color,* ed. Gloria Anzaldúa (San Francisco: Aunt Lute Books, 1990), 364.

59. Susan Krieger, *The Mirror Dance: Identity in a Women's Community* (Philadelphia: Temple University Press, 1983).

60. See Kreiger's *Social Science and the Self: Personal Essays on an Art Form* (New Brunswick, New Jersey: Rutgers University Press, 1991); see also *The Family's Silver: Essays on Relationships among Women* (Berkeley: University of California Press, 1996).

61. Carolyn Ellis, *Final Negotiations* (Philadelphia: Temple University Press, 1995).

62. Ibid., 4.

63. Laurel Richardson, *Fields of Play: Constructing an Academic Life* (New Brunswick, N.J.: Rutgers University Press, 1997).

64. It was especially in early criticism of television that its emotionalism was often a topic of discussion. For example, see Ien Ang, *Watching Dallas* (New York: Hill and Wang, 1985); Lawrence Grossberg, "The In-Difference of Television," *Screen* 28 (1987): 28–45; Lynne Joyrich, "All That Television Allows: TV Melodrama, Postmodernism, and Consumer Culture," *Camera Obscura* 16 (1988): 129–53.

65. I am drawing on Mary Ann Doane's extremely provocative essay on television, "Information, Crisis, and Catastrophe," in *Logics of Television,* ed. Patricia Mellencamp (Bloomington: Indiana University Press, 1990).

66. I am especially drawing on the essays by Ulrich Beck, Anthony Giddens, and Scott Lash collected by them in *Reflexive Modernization: Politics, Tradition, and Aesthetics in the Modern Social Order* (Stanford: Stanford University Press, 1994).

67. Anthony Giddens, *The Transformation of Intimacy* (Stanford: Stanford University Press, 1992). For a more detailed treatment of Ellis's *Final Negotiations* in relationship to recent sociological theory about reflexivity in "high modernity," see

my "Autotelecommunication and Autoethnography: A Reading of Carolyn Ellis's *Final Negotiations*," *Sociological Quarterly* 38 (1996): 95–110.

68. Clifford, "Introduction: Partial Truths," 21.

69. Ruth Behar, "Introduction: Out of Exile," in *Women Writing Culture*, ed. Ruth Behar and Deborah Gordon (Berkeley: University of California Press), 5.

70. Cherrie Moraga and Gloria Anzaldúa, ed., *This Bridge Called My Back* (New York: Kitchen Table, 1983).

71. Behar, "Introduction: Out of Exile," 4.

72. Deborah Gordon, "Conclusion: Culture Writing Women: Inscribing Feminist Anthropology," in *Women Writing Culture*, 329–441.

73. Inderpal Grewal and Caren Kaplan, "Introduction: Transnational Feminist Practices," in *Scattered Hegemonies*, ed. Inderpal Grewal and Caren Kaplan (Minnesota: University of Minnesota Press, 1994), 1–33. The term "scattered hegemonies" was first proposed by Inderpal Grewal and then taken as the title of this collection of essays.

74. James Clifford, "Traveling Cultures," in *Cultural Studies*, ed. Lawrence Grossberg, Cary Nelson, and Paula Treichler (New York: Routledge, 1992), 96–112. For a detailed treatment of cultural theory, diaspora, travel, and feminism in postmodernity, see Caren Kaplan, *Questions of Travel: Postmodern Discourses of Displacement* (Durham: Duke University Press, 1996).

75. Adrienne Rich, *Blood, Bread, and Poetry: Selected Prose, 1979–1985* (New York: Norton, 1986).

76. Paul Gilroy, "Routework: The Black Atlantic and the Politics of Exile," in *The Post-Colonial Question: Common Skies, Divided Horizons*, ed. Iain Chambers and Lidia Curti (New York: Routledge, 1996), 22.

77. Chandra Talpade Mohanty, "Women Workers and Capitalist Scripts: Ideologies of Domination, Common Interests, and the Politics of Solidarity," in *Feminist Genealogies, Colonial Legacies, Democratic Futures*, ed. Jacqui Alexander and Chandra Talpade Mohanty (New York: Routledge, 1997), 3–29.

78. Haraway, *Modest Witness*, 121.

79. Bruce Robbins, "Comparative Cosmopolitanisms," in *Cosmopolitics: Thinking and Feeling beyond the Nation*, ed. Pheng Cheah and Bruce Robbins (Minneapolis: University of Minnesota Press, 1998), 246–64.

80. Haraway, *Modest Witness*, 190–91.

81. I am drawing on Noel Sturgeon's review of ecofeminism in her *Ecofeminist Natures: Race, Gender, Feminist Theory, and Political Action* (New York: Routledge, 1997).

82. Arturo Escobar, "Cultural Politics and Biological Diversity: State, Capital, and Social Movements in the Pacific Coast of Colombia," *The Politics of Culture in the Shadow of Capital*, ed. Lisa Lowe and David Lloyd (Durham: Duke University Press, 1997), 201–26.

83. Ibid., 217–18.

Index

actual-real circuit, 101
age of teletechnology, 2–5, 116–17, 186
Alarcón, Norma, 178
Alexander, Jeffrey, 191–92n24
Althusser, Louis, 9, 45, 70–72, 77–79, 112, 117–18, 129
anthropology, 152–54, 163; and the criticism of ethnographic writing, 163–70, 181–87
Anzaldúa, Gloria, 16, 177
aporia of time, 29, 41, 91
Appadurai, Arjun, 136–37
Aronowitz, Stanley, 6, 176
Artaud, Antonin, 40, 134; and the theater of cruelty, 40
Ashmore, Malcolm, 204n25
attention theory of value, 98, 176
autoaffection, 17–20, 34, 43, 70, 186–87, 192n28
autoethnographic realism, 171–78
autoethnographic turn, 170–83
autoethnography, 16–17, 170–78; and anthropology, 181–82; and the oedipal narrative, 166, 172–73; and poststructuralism, 171; and standpoint epistemologies, 173–78; and television, 179–80
automatic imagining, 103

Bakhtin, Mikhail, 154
Balsamo, Ann, 113

Baudrillard, Jean, 91–93, 113
Beardsworth, Richard, 30
Beck, Ulrich, 191–92n24
Behar, Ruth, 181–82
Bell, Daniel, 191–92n24
Beller, Jonathan, 98–99, 176
Bennett, Tony, 73
Bergson, Henri, 101–3
Birmingham Cultural Studies. See Marxist cultural studies
Black feminist criticism: and feminist film theory, 195–96n58; and psychoanalysis, 195–96n58
Black feminist standpoint epistemology, 177
bodies without organs, 12, 134–38
Braidotti, Rosi, 3, 116
Butler, Judith, 6, 11, 12, 112, 114–30, 133, 138

Cartwright, Lisa, 165–66
Castells, Manuel, 191–92n24
centers of calculation, 14, 159
Cheah, Pheng, 4, 6, 9–10, 13, 115, 126–30, 131, 191n20, 193n5, n19, 202n35
Chow, Rey, 201–2n34
chronoscopical, 92–93
Clifford, James, 6, 127, 152, 154, 163–65, 181, 183
Collins, Patricia Hill, 16, 156, 177
concepts, 7

Ronald M. Morritt, MA, MBA, DBA

Segmentation Strategies
for Hospitality Managers
Target Marketing
for Competitive Advantage

Segmentation Strategies for Hospitality Managers
Target Marketing
for Competitive Advantage

THE HAWORTH PRESS
Haworth Series in Segmented, Targeted, and Customized Marketing: Conceptual and Empirical Development
Art Weinstein
Editor

Handbook of Market Segmentation: Strategic Targeting for Business and Technology Firms, Third Edition by Art Weinstein

Handbook of Niche Marketing: Principles and Practice edited by Tevfik Dalgic

Lifestyle Market Segmentation by Dennis J. Cahill

Segmentation Strategies for Hospitality Managers: Target Marketing for Competitive Advantage by Ron Morritt

Segmentation Strategies for Hospitality Managers
Target Marketing for Competitive Advantage

Ronald M. Morritt, MA, MBA, DBA

The Haworth Press®
New York • London • Oxford

For more information on this book or to order, visit
http://www.haworthpress.com/store/product.asp?sku=5716

or call 1-800-HAWORTH (800-429-6784) in the United States and Canada
or (607) 722-5857 outside the United States and Canada

or contact orders@HaworthPress.com

Published by

The Haworth Press, Inc., 10 Alice Street, Binghamton, NY 13904–1580.

PUBLISHER'S NOTE
The development, preparation, and publication of this work has been undertaken with great care. However, the Publisher, employees, editors, and agents of The Haworth Press are not responsible for any errors contained herein or for consequences that may ensue from use of materials or information contained in this work. The Haworth Press is committed to the dissemination of ideas and information according to the highest standards of intellectual freedom and the free exchange of ideas. Statements made and opinions expressed in this publication do not necessarily reflect the views of the Publisher, Directors, management, or staff of The Haworth Press, Inc., or an endorsement by them.

Cover design by Kerry E. Mack.

Library of Congress Cataloging-in-Publication Data

Morritt, Ronald M.
    Segmentation strategies for hospitality managers: target marketing for competitive advantage / Ronald M. Morritt.
        p. cm.
    Includes bibliographical references and index.
    ISBN-13: 978-0-7890-2216-5 (hard : alk. paper)
    ISBN-10: 0-7890-2216-8 (hard : alk. paper)
    ISBN-13: 978-0-7890-2217-2 (soft : alk. paper)
    ISBN-10: 0-7890-2217-6 (soft : alk. paper)
    1. Hospitality industry—Marketing. 2. Market segmentation. I. Title.

TX911.3.M3M685 2007
647.94068'8—dc22

                                                                    2006031753

CONTENTS

ABOUT THE AUTHOR

Dr. Ron Morritt, is Associate Professor and Chair of the Business Department of Touro College South located in Miami Beach, Florida, where he teaches marketing, management, and ethics. Dr. Morritt has advanced degrees from Cornell University, Nova Southeastern University, and the University of Phoenix. He has also served as doctoral faculty for Walden University and the School of Advanced Studies (SAS) at the University of Phoenix for the past decade. Dr. Morritt taught courses at Walden University in marketing management, international marketing, and services marketing and personally designed all the elective marketing courses for the Walden University online MBA program. He designed the doctoral course in marketing for SAS as well as an online marketing course for the prestigious McKinsey & Co. business consultants. Doctor Morritt also taught courses in marketing, strategy, and ethics for the MBA program of University of Phoenix. Dr. Morritt has consulted in the real estate and hospitality sectors for several years. He is a former CEO of a New York City real estate development firm and has also served as CFO and Chairman of the Board of a condo-hotel in Jamaica. He is the author of numerous scholarly articles in the area of hospitality and marketing strategy. Dr. Morritt now resides in South Florida with his wife and three children.

Foreword

Ron Morritt is one of the most authoritative voices in hospitality marketing. I met him when I was Director of Tourism for Jamaica, and he was vacationing in Montego Bay. Our professional paths crossed, and our marketing interests immediately collided. He told me of his occupational pursuits in the area of hospitality marketing strategy. He brought a fresh analytical mind to conventional practices.

I invited him to return to Jamaica to host a series of presentations to groups and classes in the hospitality industry business on segmentation and target marketing. He unfolded his extraordinary scope as a lecturer and writer, and provided new insights into the global hospitality industry. The Jamaican hospitality industry benefited greatly from his knowledge and presentations.

I am therefore not surprised at the excellent quality and content of his new book, *Segmentation Strategies for Hospitality Managers*. It not only walks us through the vital corridors of hospitality marketing but opens many doors leading to important strategies and practical procedures in identifying and winning new markets. When added to his two decades of experience as a CEO, writer, consultant, and MBA marketing faculty, this book irrevocably stamps him as an authority in the field of hospitality marketing. The substance, clarity, and logic of this state-of-the-art text should make it a must read for all hospitality managers and students worldwide.

I salute Ron on a text that is insightful, painstakingly documented, and persuasively articulated.

Desmond Henry
Former Director of Tourism
for the Government of Jamaica

Segmentation Strategies for Hospitality Managers
Published by The Haworth Press, Inc., 2007. All rights reserved.
doi:10.1300/5716_a

Preface and Acknowledgments

Yoram Wind (1978), in his seminal article on segmentation, observed that companies too often relied on a priori segmentation design and that there was a gap between academic segmentation research and business segmentation practices with little creativity in design or analysis. This gap still exists today, twenty-eight years later.

It is clear that an effective segmentation strategy results from segmentation based on empirical research into target markets, an optimal combination of segmentation bases, and the effective use of technology in leveraging the firm's segmentation strategy. Hospitality companies that continue to rely on a priori product segmentation alone (such as the traditional a priori product segments of business, leisure, and corporate travelers) are vulnerable to competitors who use a segmentation strategy that makes use of research-based segmentation bases and more effective statistical segmentation techniques.

The purpose of this book is to help fill this gap, by providing a primer on hospitality segmentation strategy for worldwide hospitality students and managers. A chapter on the impact of technology on hospitality marketing is included due to the critical nature of technology that has transformed the face of twenty-first-century hospitality management and strategy.

Any credible text on hospitality marketing strategy must build on the contributions of the most important experts and researchers in the field. Therefore, I am pleased to acknowledge my debt to the following, albeit partial, list of hospitality experts and authors who have made major contributions to the subject matter of hospitality marketing strategy and/or segmentation strategy: Philip Kotler, James Makens, Michael Olsen, Robert Lewis, Richard Chambers, Alistair Morrison, Michael Porter, Art Weinstein, Yoram Wind, James Meyers, Joseph West, and John Bowen.

I would also like to thank my editor, Professor Art Weinstein, of Nova Southeastern University. Although any errors and omissions

Segmentation Strategies for Hospitality Managers
Published by The Haworth Press, Inc., 2007. All rights reserved.
doi:10.1300/5716_b

xi

are my responsibility, what is good in this text owes much to his expertise in the areas of marketing and segmentation, and to the invaluable advice, encouragement, and editing he has provided over the past two years.

My thanks also go to my colleague and mentor, Dr. Joseph Balloun, professor and former Research Director of Nova Southeastern University, for his comments on the final draft of this book.

Finally, I would like to thank my wife Bixia, and my children, Mindy and Lenny, for their patience in putting up with a husband and dad whose time was too often occupied with the research and writing of this book.

Chapter 1

Introduction to Segmentation

OVERVIEW OF THE HOSPITALITY INDUSTRY

The hospitality industry is the world's largest industry (US$545 billion in annual revenues) and is the second largest employer in the United States. Components of the hospitality and related travel industry include hotels, resorts, airlines, restaurants, cruise lines, rental car agencies, travel agencies, tour companies, theme parks, and tourism. Travel and tourism employers, globally, pay more than $1.6 trillion in wages and salaries and create an estimated 12.5 million new jobs annually ("The hospitality Industry—A natural in Arkansas," retrieved January 20, 2005, from www.arjobs.com/stories/hospitality.html).

Restaurants

Restaurant sales growth has stabilized at about 5 percent annually. There are 8 million restaurants in the world which are traditionally segmented into full-service restaurants and fast-food restaurants. The full-service segment includes family restaurants, dinner houses, and buffet-type restaurants. The fast-food sector includes sandwich shops, and hamburger, Mexican, pizza, and chicken restaurants. Faced with a decline in the supply of restaurant workers (sixteen- to twenty-four-year-olds), restaurants are hiring more immigrants and retirees, and making greater use of automation. The restaurant industry represents a mature market with fierce competition and low profit margins. Restaurant sales have been fueled by the rise of two-income families and Americans having less time and energy to prepare their own meals (Hoover's Industry Overview, Restaurants [premium report], retrieved June 20, 2004 from premium.hoovers.com/subscribe/ind/overview.xhtml?HICID-1442).

Segmentation Strategies for Hospitality Managers
Published by The Haworth Press, Inc., 2007. All rights reserved.
doi:10.1300/5716_01

Airlines

The airline industry has been whipsawed by high labor costs, the doubling of oil prices, and security problems related to the aftermath of the September 11, 2001, terrorist attack. This attack was accompanied by a dramatic increase in security costs and a significant reduction in revenues from apprehensive travelers. Medical travel bans (e.g., SARS, bird flu) further reduced air travel to some countries. Although major airlines have recovered from this onslaught (as of 2006), they are faced with what is clearly a more serious long-term threat from discount regional carriers such as Southwest Airlines and JetBlue, who have greatly reduced costs of labor and infrastructure. Only Southwest Airlines has been consistently profitable in recent years, and several other airlines (e.g., United and Delta) are either flirting with bankruptcy or emerging from bankruptcy proceedings as of this writing.

Increasingly, members of the more lucrative business traveler segment are seeking to reduce their high ticket prices and have turned to online booking, sophisticated search engines, and relying more on video and teleconferencing technology (Hoover's Industry Overview, Airlines [premium report], retrieved June 20, 2004, from premium.hoovers.com/subscribe/ind/overview.xhtml?HICID-1600).

The Lodging Industry

The lodging industry is generally divided into the price/value segments of budget and economy motels, extended-stay hotels, mid-priced hotels and motels, resorts, and upscale and luxury hotels. In recent years, the budget and economy and the extended-stay segments have expanded to meet increased demand. Business travelers have relied more on the all-suite product segment, which offers amenities such as cable TV, a separate bedroom, and a full kitchen.

Decreased travel related to the 9/11 attack and a struggling economy had negatively impacted hotel revenues until recently, when this sector began participating in the general economic recovery characterized by strong sales and rising prices (as of 2006). Hotel chains initially countered with special deals and heavy discounts to lure guests back. Many hotels attempt to obtain a competitive edge with business travelers by offering new technology such as high-speed and wireless Internet access, and accelerated check-in and check-out. During the

industry downturn, hotels followed expectations by consolidating and using acquisitions to save money instead of building new properties (Hoover's Industry Overview, Lodging Industry [premium report], retrieved June 20, 2004, from premium.hoovers.com/subscribe/ind/overview.xhtml?HICID-1436).

Cruise Lines

Cruise lines were the only bright spot in the industry after the downturn following 9/11. According to cruise industry news, the industry had nearly $13 billion in revenues by the end of 2003, which represented an 8 percent increase over 2002 (see Table 1.1) (Retrieved from Hoover's Industry Overview, Cruiselines [premium report], retrieved June 20, 2004, from premium.hoovers.com/subscribe/ind/overview.xhtml?HICID-1218).

Overall, the hospitality industry is has now recovered from the economic downturn that began after 9/11 and was related to travelers' increased concerns associated with terrorism, the doubling of oil prices, a sluggish economy, and SARS-related risks. Another factor

TABLE 1.1. 2002 Cruise Passenger Statistics and Cruiseline Market Share

Cruiseline	Passengers (thousands)	% of Total Passengers
Carnival Cruise Line	2,548	33.3
Royal Caribbean International	1,980	25.9
Princess Cruises	776	10.2
Norwegian Cruise Line	739	9.7
Celebrity Cruises	639	8.4
Holland America Line	445	5.8
Disney Cruise Line	401	5.2
Costa Cruise Lines	54	0.7
Cunard Cruise Line	43	0.6
Crystal Cruises	21	0.3
Total	7,645	100.0

Source: U.S. Department of Transportation, 2002, www.dot.gov.

which will have a negative long-term effect on profits is the impact of the Internet. Cost-conscious travelers and companies are increasingly turning to search engines, shopping robots, and online travel companies such as Cheaptickets.com. These Web sites, Internet search engines, and software allow travelers to swiftly find the cheapest fares and hotel accommodations on a global basis and compare the value of competitive offerings. Consequently, they have dramatically increased the buying power of both corporations and individuals.

MARKET SEGMENTATION: AN OVERVIEW

Market segmentation is the process of classifying customers and prospects into groups with similar needs and purchasing behavior (Weinstein, 1994). Effective segmentation allows your company to select those groups that can be served most profitably and positions your firm to effectively service the needs of those groups.

Let us suppose that you have a sizable global coin collection. You can then organize, or *segment,* this collection by such variables as country, denomination, condition (mint, good, poor, etc.), date, and so forth. Which variables you use to organize your selection depend on what is most useful or convenient for you. You may end up, for example, segmenting your coin collection according to country, denomination, and date, in that order. This is called a *multistage segmentation* because there are three stages of segmentation used in this case.

But suppose that you later decide to start selling coins and open up a storefront in your neighborhood using your coin collection as your initial inventory. It occurs to you that you need to organize your display using a different organization, or segmentation, which more closely fits your customers' needs and preferences. Since you do not have your own customers yet to survey, you research how other coin stores segment their inventory. You then end up segmenting your inventory according to country, denomination, and condition. This is an example of *product segmentation.*

But then, you find that business is slow, so you decide to change your segmentation variables to reflect more closely the preferences of the customers in the neighborhood of your coin store, including your own customers. Market research, using surveys of coin customers in your neighborhood, finds, for example, that they are most interested in American coins at least 100 years old in mint or good condition.

You then devote a special section of your store to these coins and customize your advertising to communicate this specialty to your target customers using direct-mail and ads in specialized magazines. Business gets much better as a result.

But we have also seen, by this example, that segmentation is often a three-stage process:

1. Partitioning your market into groups or segments with similar characteristics (e.g., coin collectors, stamp collectors, etc.)
2. Selecting the segments that your company is best able to serve (e.g., coin collectors within a five-mile radius of your store)
3. Researching their needs and preferences which are used to customize your offerings to this target market (e.g., American coins at least 100 years old in mint or good condition)

Criteria for Effective Segmentation

It is not enough to segment your market if your chosen segments are of little use to you. Suppose you selected an airline segment called "Last-minute travelers" (i.e., those who prefer to use standby travel instead of making reservations). This segment choice may not be helpful if this segment does not represent a stable homogenous group of people with whom you can easily communicate. In fact, this may not be a true segment at all if this group represents most people at various times and for various reasons. Several criteria are used for effective segmentation in hospitality markets; the following are generally viewed as important for effective segmentation:

1. *Measurability.* Marketers need to know metrics such as size, frequency, growth rates, and profitability of selected segments in order to select target segments.
2. *Substantiality.* The segment must be large enough and the demand large enough to be profitable.
3. *Accessibility.* Members of the target segment must be reachable by your marketing communications (e.g., people who prefer to fly at the last minute may not be a reachable segment).
4. *Defensibility.* The marketer should be confident that the selected target segment(s) can successfully be defended against competitors. But sometimes being located in a cluster of competitors

can be a positive, such as a fast-food restaurant located in the food court of a mall, or a car rental outlet located in a rental car cluster at the airport.

5. *Durability.* The selected segment must be stable over time. Some segments are short-lived or have little growth potential. For example, the segment of attendees of nonrecurring events would not have sufficient durability.

6. *Competitiveness.* The segment must be evaluated to ensure that your company has some unique or differentiated product or service that will enable you to position your company to best serve this segment.

7. *Homogeneity.* A true segment will be different from other segments while maintaining similarity within the segment. Thus, U.S. female business executives may be an example of a growing hotel segment which has these qualifications and whose members may have special needs and preferences that are different from their male counterparts. Marriott, for example, targets the female business traveler with their Springhill Suites brand. The suite configuration has a tendency to attract more leisure and female travelers than the similarly priced Courtyard brand.

8. *Compatibility.* Managers must be careful not to try to serve incompatible segments at the same time. Your selected target segments must be compatible with your other target segments or your "customer mix." Incompatibility problems may arise from such differences in age, values, politics, and usage characteristics. For example, resort customers who prefer booming rap music may not be compatible with seniors that appreciate softer classical or pop music.

The Advantages of Market Segmentation

The underlying assumption of market segmentation is that you cannot be all things to all people and that the best way to achieve a competitive advantage is by specializing in a particular attractive segment which you are better able to serve due to both your superior knowledge of the needs and preferences of this segment(s) and also the resources, goals, and core competencies of your company.

Thus, Southwest Airlines targets the economy traveler who prefers no-frills, low-cost, one-class air transportation. They are better able to serve this group due to their low cost structure which uses only one type of plane (low maintenance costs), nonunionized employees, and "short haul" structure. On the other hand, Hong Kong's Cathay Pacific Airlines targets upscale travelers for its first-class and business-class tickets which feature elegant service by highly trained Chinese flight attendants, gourmet meals, individual monitors for TV, news, and first-run movies, and comfortable accommodations. Business travelers may pay more than US$14,000 for a first-class ticket from JFK to Hong Kong on their new nonstop service.

Some advantages of market segmentation are as follows:

1. *More effective use of marketing dollars.* Marketing is more expensive but the *net cost* of marketing is less. This is due to the fact that targeting customers most likely to purchase your services usually means that the net cost of your ads and promotions are less on a per customer basis. For example, targeting business travelers and investors using ads in *Wall Street Week* and *Forbes Magazine* for luxury hotels such as Marriott and Westin makes more sense than marketing in the mass media for customers. Even more effective would be to directly market to corporations and individuals who frequently stay at your luxury hotel. Marketing to segments who are more likely to patronize your company is cheaper in the long run because you are able to attract more of your target customers for your marketing dollar.

2. *A better understanding of the needs and preferences of your target customers.* This is due to your specialization in your target segment(s), and understanding your customers better means that you may be better able to satisfy and even anticipate their needs. Thus, the manager of an authentic Japanese sushi bar in San Francisco, Manhattan, or Hong Kong is intimately familiar with the requirements, preferences, and needs of his or her customers, many of whom may be Japanese. By serving the freshest and best cuts of local, sushi-quality fish, using highly trained Japanese sushi chefs and waitresses, and stocking the best quality brands of sake (a Japanese rice wine) distinguishes this restaurant to target Japanese and gourmet customers who demand an authentic Japanese sushi bar and are willing to pay a pre-

mium for this opportunity. For example, a dinner at one of the world's best sushi bars at the Shangri-La hotel in Hong Kong can cost more than $100 per person excluding tips and drinks.

3. *More effective targeting of communications.* For example, your resort hotel may advertise in *Conde Nast* magazine, whose target readership of wealthy world travelers is more likely to patronize your luxury hotel on Paradise Island, Bahamas, than the readership of *The New York Times.*

4. *Better positioning.* By specializing in a small number of segments, you have a better chance of achieving a competitive advantage with these segments, especially if you have chosen these segments well and there is a good "strategic fit" with the external environment, resources, and core competencies of your hospitality company. For example, it may make sense for a pastry chef to open up a pastry shop that specializes in selling gourmet cheesecakes to local corporations and families rather than to steakhouses (e.g., the Celebrity Cheesecake Company in Plantation, Florida). However, changing marketing trends, such as the present trend toward low-carbohydrate foods, fueled by the popular low-carb Atkins diet, may compel adjustments to this service, such as the addition of low-carb desserts.

Limitations of Market Segmentation

- Market segmentation is more expensive to implement (although the end results may be more sales per your promotions dollar). This is because of the high cost of market research, customer databases, and targeted communications used to facilitate the segmentation process.
- It is difficult to select the best base for segmenting a market. For example, a hotel may have to choose among geographic, demographic, psychographic, benefit, and usage segmentation bases. Which segmentation base or combination of bases should be used? No magic formula or computerized decision process exists that will automatically select the optimal segmentation bases. This requires research, analysis, experience, and some creativity.
- It is difficult to know how broadly or narrowly to segment. For example, should a cheesecake company target only its town, an

entire city, or the whole country? Should it target corporations, housewives, families, or children? World- famous Juniors restaurant of Brooklyn, New York, ships its gourmet cheesecakes to corporations and individuals across the entire country.
- There are risks related to choosing unviable segments. This may arise from the simultaneous serving of incompatible segments. For example, many restaurants offer early-bird specials to seniors, who often eat earlier. But this also has the advantage of keeping senior customers separate from younger guests, who may prefer a different sort of ambiance with their dinners.
- Some products and services appeal to a mass market, which may be more profitable than marketing to a specific segment. For example, McDonald's appeals to a mass market of fast-food customers. Even here, though, McDonald's uses market segmentation to target families with children with their colorful mini-playgrounds and Disney novelty promotions. Recently they have substantially increased revenues by offering healthier options to the health-conscious customer segment who buy salads and chicken burgers.

A Priori Segmentation versus Post Hoc Segmentation

A priori segmentation refers to segmentation that is created without the benefit of primary market research. Managerial intuition, analysis of secondary data sources, analysis of internal customer databases, and previously existing segments are used to group customers into different segments. The traditional hotel customer segments of corporate, leisure, and groups are examples of a priori segmentation. Other examples of a priori segmentation are the following:

- Heavy users versus moderate or light users
- ValsII or Prism (lifestyle) clusters (see Chapter 2)
- Replication of previous post hoc segmentation categories
- First class, business class, and coach airline travelers

This is in contrast to *post hoc segmentation,* which is based on primary (original) research into the preferences and purchase behavior of your target market. Segments are not defined until after the collec-

tion and analysis of all the relevant information collected in that research. Multivariate analysis of these data is used to define each segment (Market Segmentation, White Papers Library, DSS Web site, retrieved January 20, 2005, from www.dssresearch.com/toolkit/resource/papers/SR01.asp).

One form of a priori segmentation, product segmentation, revolutionized the U.S. hotel industry in the 1980s. Major U.S. hotel chains used this strategy to counter the competitive effects of a mature market and to more effectively serve diverse economic segments. This was the first attempt by the hotel industry to create hotels to mirror consumers' lifestyles rather than to be all things to all people. The most successful of these new products were the all-suite hotels and the limited-service hotels. Hotel chains developed *segmentation portfolios* to reach critical mass. Five hotel segments reported by one hotel marketing analyst for eleven major hotel chains are the following:

1. *Traditional:* Full service, comfort, food service, moderate price: Holiday Inn, Marriott
2. *Economy:* Limited to no-service, no-frills facilities, low price: Fairfield Inn, Sheraton Inns
3. *All-suites:* The most successful of the segments offering homelike accommodations and moderate price: Embassy Suites, Residence Inns
4. *Casino:* Moderate prices, subsidized by casino operations: Harrah's, Tropicana
5. *Upscale:* High service level, comfort, food service, expensive: Crowne Plaza, Ritz-Carlton (Dev & Hubbard, 1989)

Post hoc segmentation is the result of primary market research into industry markets. A good example of post hoc segmentation is the story of Marriott's Courtyard hotels. Marriott surveyed self-employed, independent, restricted, or non–expense account customers, a subset of the business traveler market. They asked these customers what they wanted in a relatively low-cost hotel room and what trade-offs they would accept in return for a lower price. The result is history: the very successful Marriott Courtyard chain, which has many competitors in the hotel industry (Lewis & Chambers, 2000).

SEGMENTATION PRACTICES IN THE HOTEL INDUSTRY

Until recently, all segmentation in the U.S. hotel industry used a priori product segmentation. The industry has evolved from a priori price segmentation to a priori product segmentation and is now involved with market segmentation and niche marketing. Niche marketing strategy is often accompanied by a "relationship marketing" strategy whereby hotels utilize integrated customer databases to build one-to-one long-term, interactive, mutually profitable relationships with their best customers.

Stage 1: Price Segmentation

Until the 1980s the U.S. hotel industry had developed with four basic price segments: luxury, convention, first class, and economy. Usually a hotel company specialized in one of these categories (Schultz, 1994).

Stage 2: Product Segmentation

Product segmentation revolutionized the U.S. hotel industry in the 1980s. Major U.S. hotel chains used this strategy to counter the competitive effects of a mature market and to more effectively serve diverse economic segments. This was the first attempt by the hotel industry to create hotels to mirror consumers' lifestyles rather being all things to all people (Schultz, 1994).

NICHE MARKETING

Niche marketing is perhaps best defined by Dalgic and Leeuw (1994):

> We may conclude that niche marketing could be defined as positioning into small, profitable homogeneous* market segments

---

*It is important that homogenous be interpreted as "having similar needs and purchasing behavior." This explains why firms can profitably target such niches as Europeans, women, and mature travelers that are not homogenous in the conventional sense of that word.

which have been ignored or neglected by others. This positioning is based on the integrated marketing concept and the distinctive competencies the company possesses.

Examples of niche marketing in the hospitality industry include the following:

- Residential lifestyle communities for seniors. Marriott used psychographic (lifestyle) segmentation to develop senior communities.
- Hotels for nature lovers and ecotourists. Ecotourism is a form of specialty travel defined by the International Ecotourism Society (TIES) as "responsible travel to natural areas which conserves the environment and sustains the well-being of local people." The World Tourism Organization (WTO) estimates that nature tourism generates 7 percent of all international travel expenditures. And while tourism overall has been growing at an annual rate of 4 percent, nature travel is increasing at an annual rate of between 10 and 30 percent (Ecotourism Statistical Fact Sheet, retrieved January 20, 2005, from www.ecotourism.org/research/stats/files/stats.txt).
- Cultural tourism. Club Med has "villages" in Mexico and Asia that promote tours of famous archeological sites.
- Japanese sushi bars that specialize in raw fish.
- Airlines that specialize in special charters for executives.
- Cruise lines that specialize in affluent seniors desiring luxury accommodations. One example is the Holland America line, whose guests have an average age of fifty-five (Kotler, Bowen, & Makens, 1998).

ADVANTAGES AND RISKS OF NICHE MARKETING

Advantages

1. If you have selected carefully, there may be little competition for your target niche. For example, the niche may not be large enough to be feasible for larger companies or it may not be synergetic with their other products.

2. Your company should be able to achieve a competitive advantage in servicing this niche, since it was selected because it fits the resources and key competencies of your hospitality company
3. You can serve this niche better than the competition by specializing in understanding and serving its needs.
4. Niche marketing is associated with higher margins since servicing niches better meets the needs of niche customers (decreased price elasticity).
5. There is greater opportunity for cross marketing because of the intimate knowledge of niche customers contained in your customer database (Morritt, 1997).

Risks

Here, as elsewhere, greater market returns are associated with greater risks. The following are risks and limitations associated with niche marketing (NM):

1. There is more investment risk for hospitality companies targeting a single niche since you are not diversified. For example if you are an airline that specializes in U.S. charter flights for top executives to the Caribbean region and there is an economic recession or a medical epidemic that restricts travel to that region, you will experience a loss of profitability during this time.
2. NM is more expensive; it requires more information about your customers, and greater customization of products and services. NM also requires the use of expensive customer databases to continuously track, monitor, evaluate, and control operations to profitably maximize service, quality, and perceived value.
3. NM will not make up for fundamental hospitality problems (poor service, low quality, poor facilities, lack of operational controls).
4. NM is more dependent on customer retention (since it is a small market). Therefore, perceived service and value must be excellent.
5. NM requires the full support and commitment of both upper management and employees who must buy into the marketing concept that NM implies.

6. If you are targeting more than one segment, you will need to be concerned about possible interactions, conflicts, and synergies between segments (Morritt, 1997).

Hospitality companies that successfully pursue niche markets include Ritz-Carlton hotels (wealthy customers and executives who demand the highest quality), Rent-a-Wreck car rentals (people who prefer low-cost rentals), and Taco Bell restaurants (people who desire low-cost Mexican fast food).

USING MARKET SEGMENTATION TO DEVELOP A COMPETITIVE ADVANTAGE

Market segmentation is used to select the most attractive customer segments for your hospitality firm. These segments, ideally, have a good strategic fit with the resources and core competencies of your firm and external environmental factors. Thus, JetBlue Airlines is able to achieve superior perceived value from its target customers who are looking for low-cost, luxury air service in an environment of high airline costs and prices due to expensive legacy infrastructures, rising fuel and safety costs, and declining customer service.

1. Market segmentation is then used to discover the needs and preferences of your target customers. This enables your firm to modify its marketing mix (products, promotions, price, and distribution) to offer superior value to your target customers. For example, Holland America cruise line (since acquired by Carnival Corporation) has a reputation for excellent food and service to its senior segment with an average age of fifty-five.
2. Segmentation research is used to discover how your customers perceive your products and services relative to your competition and substitute product and services. Thus, McDonald's and other fast-food chains recently started offering healthier meals and salads due to this type of market research.

Using this information, hospitality companies position their products and services to achieve a unique differentiation and/or superior value in the minds of their target customers. This in turn leads to sustainable competitive advantage (Porter, 1985). Companies such as

Southwest Airlines, Marriott Courtyard hotels, Taco Bell restaurants, and Carnival cruise lines are examples of hospitality companies that have differentiated themselves in the minds of their target customers in a way that is highly valued by these customers. They have achieved a competitive advantage in their target markets by delivering what is perceived as superior value by these target customers (Weinstein & Johnson, 1999).

SUMMARY

In this chapter we have provided an overview of the hospitality industry and described how segmentation strategies are used by many companies in the hospitality industry. Primary and secondary segmentation research is used to gather data from present customers, markets, including competitive intelligence, and environmental factors. These data are analyzed and converted into information used to select the optimal target customer segments with a good "strategic fit" with the resources, core competencies, and goals of the firm. These segments should also fit the criteria for effective segmentation. Research into the needs and preferences of target segments is then used to customize the firm's offerings and promotions to best satisfy these target customers.

Finally, the firm uses this information to position itself by the strategies of differentiation and/or superior value (Porter, 1980). The firm then communicates this positioning using a marketing mix customized for these target customers. Additional market research is required to continuously monitor the needs and preferences of target segments, marketing trends, and the internal and external environments to ensure that the firm continues to be the vendor of choice for its target segments.

EXERCISES

1. Explain what customer segments are targeted by your company (or a selected hospitality company). Is this selection justified or does it need to be improved?

2. What are the advantages of post hoc customer segmentation (segmentation based on market research) over a priori or product segmentation?

3. Select a major target segment for your company (or a selected hospitality company) and explain what special needs and preferences lead them to choose your company (or a competitive company) over the competition.

4. Explain which segmentation bases (benefit, usage, psychographic, demographic, etc.) are used by your company and why.

5. Explain how your company (or a selected hospitality company) gathers internal data from its customers. Where are these data stored?

6. Explain the target advertising media and message for a selected product or service of your company. Explain why this is appropriate or not for the target segments of this offering.

7. Does your company (or a selected hospitality company) serve any niche markets that are underserved by the competition? Do you have any suggestions for serving a new niche market that has a good "strategic fit" with the resources, competencies, and goals of this company?

8. Explain how you would respond to recent safety concerns from target customers in one of the following hospitality and travel sectors: airlines, hotels, cruise ships, casinos, theme parks, travel agencies, tour companies. What targeted communications would you use?

9. Explain how your customer's use of hospitality search engines on the Internet can reduce the pricing power of your company (or a selected hospitality company). How would you maintain your margins in the light of this trend?

10. Explain the types of market research performed by your hospitality company, or its marketing firm, that supports your company's segmentation plan. How often is this research updated?

Chapter 2

Segmentation Bases: Panning for Gold in the Hospitality Sector

In the last chapter we discussed the importance of segmenting your hospitality market in order to achieve a competitive advantage with your target customer segments. In order to select the most attractive segments it is necessary to first select the optimal combination of segmentation bases.

A *segmentation base* is a screen or filter that allows a company to target the most appropriate or desirable segments of its potential market. This is like the screen that 1848 Gold Rush prospectors used to filter out the larger chunks of gold from the grains of sand and dirt in the California streams.

Hospitality companies may use *single-stage, two-stage,* or *multistage* segmentation, which depends on the number of segmentation bases used. Multistage segmentation (the use of more than one segmentation base) is generally viewed by marketers as superior to single-stage segmentation, with the proviso that the primary segmentation base should be the variable with the greatest influence on the purchase behavior of customers (Morrison, 1996).

For example, using the single-stage (purpose of trip) "business traveler" segmentation base allows the targeting of the lucrative business traveler segment who are less price sensitive. This segmentation strategy often leads to increased profitability for hospitality companies. This is because these companies have a fixed capacity, thus replacing less profitable customer segments with more profitable customer segments usually leads to increased profitability.

The "filter" metaphor allows us to think in terms of multiple filters and also of optimal filters. For example the filter of "business travelers" may be too coarse, allowing too many subsegments to be tar-

Segmentation Strategies for Hospitality Managers
Published by The Haworth Press, Inc., 2007. All rights reserved.
doi:10.1300/5716_02

geted. For example, a hotel in Peoria whose guests usually come from a twenty-five-mile radius may not want to include the subsegment of Chinese business travelers in their segmentation base. A more optimal two-stage filter may be "business travelers within a twenty-five-mile radius." Thus, the optimal filter for a particular hospitality company usually involves two or more filters. In this case we first use geographical segmentation to isolate potential customers within a twenty-five-mile radius of the hotel. Next, we filter out the business customers from the previous group. We may also wish to use the "heavy user" segmentation filter to further screen our customers and end up with the multistage segment "frequent business travelers within a twenty-five-mile radius of our hotel."

SEGMENTATION BASES USED IN THE HOSPITALITY INDUSTRY

Geographical Segmentation

This is the most widely used segmentation base for hospitality companies. It has the advantage of being low cost with readily available data. This process involves segmenting customer groups into separate geographical locations. Geographical segments can be very small (e.g., a one-mile radius of the business) or as large as countries. Thus, a small midtown Manhattan pizzeria may target customers within a one-mile radius, while a destination hotel like Sandals of Jamaica may target the United States, Japan, and Europe.

For another example, fast-food companies often vary their menus according to regional tastes. McDonald's introduced a Texas Burger (large burger with lettuce, tomato, and condiments of choice) for Texans, and KFC in Jamaica serves a very spicy fried chicken geared to Jamaican tastes. This modification was later introduced into the U.S. market and was a hit with Hispanics and people of Caribbean extraction who prefer spicy food.

The U.S. government defines large metropolitan areas in terms of supposed economic boundaries called standard metropolitan statistical areas (SMSA), such as the New York City SMSA. The government produces voluminous data on these areas, including population, ethnic mix, household size, occupations, and so forth. These data are useful when the market is being segmented by demographic bases.

Another geographical division is the designated market area (DMA) developed by the A.C. Nielsen research company. These areas are based on geographic areas served by TV stations. Their data include demographic characteristics used for reaching specific audiences by TV. These designations are used by fast-food chains such as McDonald's and Burger King.

However, geographical segmentation does not help in determining the diverse needs and wants of target customers and usually needs to be supplemented by other segmentation bases, such as demographic, usage, psychographic, and benefit segmentation.

Demographic Segmentation

Demographic segmentation involves segmenting potential markets according to such population statistics as age, gender, household income, family size, and ethnicity. It is common for demographic segmentation to be used in conjunction with geographical segmentation. This is a two-stage segmentation process called *geodemographic* segmentation.

Two of the more useful demographic variables for the hospitality industry are *age* and *family life-cycle stage*. For example, McDonald's targets children who will bring their parents. And the growing and affluent senior citizens market is an attractive one for many hospitality companies. For example, American Express focuses on the "mature" market, which accounts for an estimated 70 percent of the tour industry's bookings.

Stages of the family life cycle include single person, married couple, married with children, married with grown children, "matures," and the widow(er). Other nontraditional segments include couples without children and who are both working, single parents and nonparents, second and third marriages, and gay travelers. Marketers that have tapped into this segmentation bases include resorts that target singles (e.g., Club Med), restaurants with early-bird dinners, and cruise lines that target gay travelers. Online travel agencies, such as Oceanvoyager.com and Orbiz.com, book gay cruises for large cruise lines such as Carnival, Royal Caribbean, and Norwegian Cruise Lines.

The Marriott and Hyatt hotel chains have developed senior living centers for seniors. This is a growing industry, with an estimated 12

percent of people over age sixty-five needing assistance with transportation, shopping, meal preparation, and medical treatment. Dining room food sales alone in this segment are in excess of $2.5 billion (Kotler et al., 1998.

Trend analysis (the analysis of current marketing trends) is an important part of demographic segmentation. For example, marketing trends in the past decade include a significant increase in minority populations (African-American, Asian, Hispanic), significant increases in household income, and an increased number of women business travelers (Weinstein, 1994). The point here is that it is better to target segments that are supported by positive marketing trends.

Purpose of Trip Segmentation

Hospitality companies have traditionally used "purpose of trip" segmentation as the primary base to segment their customers into the "business travelers" and "leisure travelers" groups. This is because the needs, preferences, and price sensitivities of these two groups are usually quite different. According to Holiday Inn, purpose of trip and experience desired (benefit segmentation) are the prime reasons for choosing a hotel (Lewis & Chambers, 2000).

An estimated 80 percent of urban hotel occupancy in the United States derives from business travelers, and the business expense account customer is an important segment for upscale restaurants as well. Business travelers are typically less price sensitive (especially if traveling on expense accounts), more time constrained, and require business services such as fax and computer facilities, meeting rooms, high-speed Internet access, and secretarial services. Business purpose for hotel companies can be further broken down into the subsegments of conventions, associations, corporate, expense account, and non–expense accounts.

A variant of this type of segmentation is occasion-based segmentation. Caesar's Pocono's Resort targets honeymooners with a champagne glass in-room spa and heart-shaped tubs. Also, convention hotels subsegment the business convention market into annual conventions, chapter meetings, board meetings, educational seminars, and sales meetings.

Product Segmentation

Hospitality companies have traditionally used price/product segmentation to initially segment their market. Examples are budget hotels, fast-food restaurants, all-suite hotels, luxury hotels, gambling casinos, and so forth. This type of product segmentation is usually put in place without the benefit of market research and is derived from managerial intuition or what is common practice in the industry. Thus, hotel chains feature a number of price/product lines such as budget, economy, all suite, middle tier, and upscale properties. Hotel chains such as Marriott use product segments which include budget hotels, business hotels, family hotels, luxury hotels, ultra luxury hotels, all suites, extended stay hotels, and destination hotels (see Table 2.1). Airlines use product segments of coach, business class, and first class travel accommodations.

TABLE 2.1. Marriott Price-Segmented Hotel Brands

Brand name	Price range ($) (double occupancy)	Market segment
Fairfield Inn	50-70	Economy class business and leisure travelers
TownPlace Suites	60-75	Moderate income travelers with weekly stays
Springhill Suites	80-100	Business and leisure travelers desiring moderate accomodations
Courtyard	80-105	Designed for independent business travelers and those on a budget
Residence Inn	90-115	A residential-style hotel
Marriott Hotels and Resorts	95-255	For the upscale traveler
Renaissance Hotels and Resorts	95-255	Ultra-luxury hotels for the rich and powerful
Ritz-Carlton (acquired by Marriott)	185-310	

Source: Adapted from Lewis & Chambers, 2000.

Hotels are also segmented within price segments. For example, in the United States today we have low budget, middle budget, upper budget, and luxury budget hotels.

More effective post hoc segmentation is based on market research and develops products for target customer groups. Marriott hotels evolved from a priori product segmentation to post hoc segmentation when they identified and targeted the attractive segment of self-employed, independent, non–expense account customers (a subset of the business traveler market) and asked these customers what they wanted in a relatively low-cost hotel room, what tradeoffs they would make, and what they would give up to pay less. The result was the highly successful Marriott Courtyard product line which is often copied by other competitors.

Usage Segmentation

Usage segmentation segments customers based on frequency of use or share of total demand. This is related to the Pareto principle, a rule of thumb that states that 80 percent of sales volume is usually generated by 20 percent of the customers. Thus, hospitality firms target "frequent travelers" who are usually business travelers. Although frequent travelers are only 11 percent of business travelers, they make up about 45 percent of all business trips. One survey of frequent travelers found that they took an average of 16.9 business trips per year. Thus, the marketing dollars spent on frequent business travelers may produce a higher return than the same dollars spent on other hospitality segments (Morrison, 1996).

However, there is often intense competition for this lucrative segment, which may mitigate the effectiveness of targeting this segment. The frequency segmentation filter is often part of a multistage segmentation process which may combine frequency segmentation with purpose of trip and geographical segmentation. Thus, a convention hotel may target annual business convention business from within their city.

Another form of frequency segmentation groups customers according to their usage status. For example, customers can be grouped into nonusers, former, regular, and potential users. Different marketing promotions are often used for these different usage groups. Low frequency can also be an important segment to the hospitality indus-

try if it occurs with regularity. A restaurant, for example, may have a few hundred customers who come only once a year for an anniversary date. These customers would be an important segment that deserves special marketing attention (Lewis & Chambers, 2000).

Another type of segmentation relates to purchase size. This may refer to travelers who have high checks in restaurants or buy expensive wines. The high purchase customer in the hotel industry may be the segment that uses the more expensive rooms/suites, eats in the hotels restaurants, and orders expensive room service. Or they may be the "high rollers" or "whales" of the casino industry.

Brand Loyalty Segmentation

This segmentation process groups customers according to their level of brand loyalty. Four brand loyalty segments are hard core loyals, split loyals, shifting loyals, and switchers. In the hotel industry, a hard core loyal is someone who almost always stays at the same hotel; a split loyal is someone who always stays at two to three hotels; a shifting loyal switches loyalties periodically to different brands; and a switcher has no loyalty to any brand and may purchase based on best price or value (Morrison, 1996).

Hospitality firms build brand loyalty through relationship marketing programs. They develop guest databases and use this information to customize offerings and communications to their target segments. For example, some upscale restaurants keep files on frequent VIP customers, with their preferred captain, wines, table choice, last visit, and preferences noted.

Also, it is common practice for hospitality firms to offer rewards to frequent guests in order to retain their most profitable customers. However, at this time there has been little research on the effectiveness of these rewards and which ones are best at retaining target customers.

Benefit Segmentation

A powerful form of segmentation is known as *benefit segmentation*. This process groups customers according to similarities in the benefits sought. This is because customers do not just buy services but rather purchase a package of benefits believed to be part of the

specific service chosen. For example, when I purchase a ticket to Hong Kong on Cathay Pacific Airlines, my previous experience with this airline leads me to expect a consistent service package which includes gourmet meals, superb service from highly trained and friendly stewardesses, individual seat monitors (providing news, TV shows, and first-run movies), and nonstop service from New York City.

Research conducted by American Express found that demographic variables (except household income) were not effective in describing tourism markets. However, eleven benefit variables, including scenic beauty, shopping, rest and relaxation, cuisine, history/culture, accommodations, sports facilities, water sports facilities, attitudes of people, entertainment/nightlife, and airfare cost, were useful in profiling three benefit segments: passive entertainers, sports types, and outdoor types (Weinstein, 1994).

Studies of the lodging industry have determined that three of the most desired benefits are related to location, cleanliness, and price. Other benefits sought by hotel guests are comfort, prestige, recognition, attention, romance, quiet, and safety. Convention planners highly value food quality and service. Convenient schedules, on-time departures, and lower fares are highly valued by business airline travelers.

In a study of family, atmosphere, and gourmet restaurants, it was found that there are five major appeal categories for restaurant customers. The relative importance of food quality, menu variety, price, atmosphere, and convenience were studied. It was found that patrons of family restaurants sought convenience and menu variety. Patrons of atmosphere restaurants sought food quality and atmosphere most. And patrons of gourmet restaurants most valued food quality (Kotler et al., 1998).

Benefit segmentation can be used to focus on other segmentation variables such as the singles or mature market. These markets are not homogeneous in their needs and preferences, but identifying benefits sought within these markets can result in unique subsegments which can be uniquely targeted (Lewis & Chambers, 2000).

Channel Segmentation

In addition to marketing directly to customers, hospitality companies also market to intermediaries (e.g., travel agents and tour compa-

nies). Different marketing approaches are appropriate for these two types of customers. Channel segmentation involves segmenting travel intermediaries by function and by common characteristics shared by functional groups. Examples of such groups are travel agents, incentive travel planners, and tour wholesalers and operators. Hospitality organizations using the services of these intermediaries must decide which channel segments match the profile of their target markets. Segmenting distribution channels follows customer segmentation in a multistage segmentation process (Morrison, 1996).

The Internet is a channel of distribution in wide use by hospitality companies today. In 2004 there were over 228 million Internet users in the United States, with more than half having home access. This represents a 110 percent increase in internet usage since 2000. The Internet is used by airlines, hotels, rental car companies, cruiselines, restaurants, and travel agencies for online sales, companywide communications, and advertising. Research shows that wealthy consumers now dominate hospitality Internet purchases (see Chapter 10).

Lifestyle Segmentation

Lifestyle segmentation, or *psychographics,* involves the development of psychological profiles of customer groups and psychologically based measures of lifestyles. A lifestyle is a way of living and includes the activities, interests, and values of different lifestyle segments. The VALS™ technique is one popular segmentation scheme based on market research. Statistical techniques such as factor or cluster analysis are used to identify specific lifestyle segments based on the answers of respondents to survey questions. The following are eight consumer lifestyle segments identified by the VALS™ typology (*Source:* SRI Consulting Business Intelligence [SRIC-BI]: www .sric-bi.com/VALS [Weinstein, 2004]):

1. *Innovators:* take-charge, sophisticated, curious
2. *Thinkers:* reflective, informed, content
3. *Achievers:* goal oriented, brand conscious, conventional
4. *Experiencers:* trend setting, impulsive, variety seeking
5. *Survivors:* nostalgic, constrained, cautious
6. *Believers:* literal, loyal, moralistic
7. *Strivers:* contemporary, imitative, style conscious
8. *Makers:* reasonable, practical, self-sufficient

One user of psychographic segmentation has been Club Med, which for years promoted a hedonistic and singles lifestyle for the "swinger" singles segment. Not surprisingly, this all-inclusive vacation company has followed recent marketing trends and repositioned itself for families and conferences.

For example, nightclubs often target young singles wanting to meet other young singles. The Kapinski Group of German hotels selected market segments based on social class and lifestyle for hotels in New York, Boston, and Washington, DC. They targeted a market niche of upscale business travelers who appreciate and can afford smaller European-style hotels with "old world" service (Kotler et al., 1998).

However, there are limitations to psychographic segmentation that include the following (Weinstein, 1994):

1. This type of segmentation involves complex primary research and data collection, and analysis can be problematic due to the complex statistical techniques and the voluminous amount of data involved.
2. This type of segmentation study is relatively expensive (>$50,000) and may not be an appropriate use of limited marketing funds.
3. There is no uniform approach toward this type of segmentation; it should not be used on its own but should be part of a multistage segmentation approach (e.g., used with geographical and demographic segmentation to identify target segments) (Morrison, 1996).

Organizational Segments

In addition to individual customers, hospitality companies often sell to organizations such as corporations, schools, and associations. The decision-making unit of an organization is called the buying center and is comprised of all those individuals and groups who participate in the purchasing decision process, and who share common goals and the risks arising from the decisions (Kotler et al., 2003).

The organizational segments of the hotel industry include corporate business travelers, corporate meetings, associations, tour groups, incentive travel, airline crews, government agencies, and the SMERF

group (social, military, educational, religious, and fraternal organizations).

Organizational buyers are typically viewed as rational purchasers who emphasize rational considerations such as price and functionality. Organizational buyers are generally grouped into the following categories:

1. *Gatekeeper:* Gathers the information for the buying decision process and controls the information flow. Gatekeepers have the power to prevent sellers or information from reaching members of the buying center. This may be a secretary who controls access to the main decision maker.
2. *Influencer:* Are consulted about the buying decision. Influencers help define needs and set selection criteria. This could be an executive secretary or spouse.
3. *Decider:* Has the formal power to make the final decision. The decider is also often involved in setting the terms of the purchase and negotiations
4. *Approvers:* Authorizes the proposed actions of deciders or buyers. For example, this may be a corporate vice president.
5. *Buyer:* Has the formal authority for selecting suppliers and arranging the terms of purchase. Buyers may also play a role in selecting vendors and negotiations.
6. *Users:* Actually use the hospitality services, such as the students of a university who use the services of a food service company.

Hotels may also segment organizations by size and number of room nights generated, and offer special packages to their best customers.

Different business segments have different needs and require different marketing strategies. Large hospitality companies may have separate marketing strategies for different elements of the buying center, while smaller hospitality companies with limited funding may just target the decision maker in the sales process. Exhibit 2.1 indicates some common ways of targeting these segments; however, it should be noted that effective targeting of the firm's best organizational customers requires long-term relationship marketing techniques rather than the short-term transactional tactics (see Chapter 5).

EXHIBIT 2.1. Targeting Hotel Organizational Segments

Segment	Contact and Characteristic
Individual business travelers	Contacting local or distant referral source; discounts for high-use organizations
Corporate meetings	Customized facilities
Business and Professional Associations	
Large	Facilities for conventions, trade shows
Small	Small quiet meeting rooms
Tours, incentive houses	Special pricing for wholesalers, tour groups
Airline crews	Quiet rooms
Government	Special rates
SMERF	Special rates

Source: Adapted from Hsu & Powers, 2002.

YOUR CUSTOMER DATABASE

One of the best sources of information for choosing segmentation bases is your present customer database. If this database is structured correctly and kept up to date, analysis of this database should reveal

1. which customer segments have the highest lifetime value (estimated total profit to the firm over the average number of years that you will keep this type of customer);
2. which customer segments spend the most money (on average);
3. who your most frequent customers are;
4. which customer segments are most loyal;
5. which customers contribute the highest margins; and
6. why frequent guests patronize your hospitality company.

For example, you may find that your best customers (using the metrics above) may be female business travelers living within a twenty-mile radius of your company. This would indicate the use of a multi-stage segmentation base of geography, gender, and purpose of trip. This segmentation filter can then be used to identify new customers with similar profiles from your group of potential customers.

SUMMARY

In this chapter, we have reviewed ten major segmentation bases used by hospitality companies and explained how they are used as part of a single or (preferably) multistage segmentation process. It is generally recognized that there is no universal formula or well-defined procedure for choosing the optimal combination of segmentation bases. The right choice for a given hospitality company comes out of a combination of market research (segmentation studies and competitive intelligence studies) analysis, creativity, and managerial intuition. It also involves learning from the successes and mistakes of competitors.

It is also important to realize that target segments need to be monitored on a regular basis. The purpose is to ensure that changes in target customer segments (e.g., demand, growth, and profitability), the current resources of your company (e.g., being acquired by a new chain or brand), new marketing trends and technology, or new competitive offerings do not require changes in your company's segmentation variables or your ability to offer superior value to your target segments.

For example, McDonald's restaurants recently rejuvenated an extended period of lagging sales and profits by adapting its fast-food meals to a new generation of health-conscious consumers who were moving away from unhealthy high-calorie meals toward low-calorie and low-carb meals. McDonald's adapted to this trend by offering fresh salads and chicken and fish sandwiches. By monitoring its customers on a regular basis, these changes could have been anticipated much earlier, with huge savings for this company.

EXERCISES

1. Explain and justify at least two segmentation bases that are used by a selected well-known hospitality company (such as Southwest Airlines, Polle Tropical Restaurants, Carnival Cruise Lines, or Motel 6). How could this company improve its segmentation?
2. Which segmentation bases are used by your hospitality company? Justify or critique this choice.

3. Many hospitality companies (such as hotel chains and airlines) use product segmentation. Is this best used alone or with the addition of customer segmentation? Explain.

4. It is not unusual for companies to target and encourage their most profitable segments, such as high-margin business travelers, and to discourage less profitable segments. This often translates into service quality discrimination. Do you view this as an unethical business practice? Explain why or why not.

5. What segmentation base(s) would you select for a new midtown restaurant that specialized in organic health food. Why?

6. What segmentation base(s) would you select for a new hotel that was designed for women executives and located in San Diego? Why?

7. How would you advise the owner of a new casual dining, moderately priced, seafood restaurant in Manhattan, who decided to use the segmentation bases of income and geography to target people within fifty miles whose annual household income was at least $100,000?

8. Explain how an analysis of your guest database would help in your choice of segmentation bases.

9. Do segmentation bases ever need to be changed for a given company? Explain.

10. Explain the role of market research in deciding on the optimal combination of segmentation bases. Should this be primary or secondary research?

Chapter 3

Niche Marketing

INTRODUCTION

In the highly fragmented and intensely competitive global hospitality industry of the twenty-first century, niche marketing has been an important marketing strategy for both survival and increased profitability. We are all aware of niche marketers such as Ritz-Carlton hotels, Southwest Airlines, Japanese sushi bars, and Rent-a-Wreck car rentals. However, the area of niche marketing has been largely neglected in the hospitality and marketing literature and there are multiple definitions of this term.

We will use here a modification of the definition of niche marketing provided by T. Dalgic: "the positioning of a firm specializing in small, profitable, market segments with shared special interests which are underserved by others" (Dalgic & Leeuw, 1994). This definition agrees with such clear examples of niche marketing as a magazine that targets Hispanic teenage girls *(Quince)*, an insurance company that sells insurance to high-risk drivers (Progressive), and a company that sells ostrich meat online. Examples of niche marketing in the hospitality industry are cited later in this chapter.

Characteristics of an attractive niche cited by Kotler (1991) are as follows:

1. The members of a niche have a set of similar, but unmet or underserved, needs.
2. The members of a niche will pay more for a product that satisfies these needs.
3. The niche is not attractive to many competitors.
4. The nicher can best satisfy these needs by their distinctive set of resources.

Segmentation Strategies for Hospitality Managers
Published by The Haworth Press, Inc., 2007. All rights reserved.
doi:10.1300/5716_03

5. The niche has sufficient size, disposable income, growth, rate, stability, accessability, and profitability (Kotler, 1991).

Relationship marketing skills are critical for success in niche marketing (see Chapter 5, "Relationship Marketing"). This is because niche customers are really purchasing a package that includes the brand, product, and services provided. Niche customers will pay a premium, and be loyal, to the company that best satisfies their special needs. Thus, it is only through effective relationship marketing that niche companies are able to erect barriers to potential competitors, build their brand, and retain loyal customers and key suppliers (Davis & Davidson, 1991).

Niche Marketing and Segmentation

Niche marketing has been differentiated from segmentation in the marketing literature. For example, niche marketing is said to be a "bottom-up approach," whereas segmentation marketing is held to be a "top-down approach." In a bottom-up approach, the marketer starts from the needs of a few customers and gradually builds up to a larger customer base. In the top-down approach, the marketer starts with the mass market and divides it into micromarkets. Niches are viewed as different from segments because of the smaller size of the market and the increased level of specialization required by the organization (Shani & Chalasani, 1992) (see Exhibit 3.1).

EXHIBIT 3.1. Segmentation versus Niche

Segment	Niche
Top-down approach	Bottom-up approach
Division based on differences	Aggregating based on similarities
Undifferentiated homogenous groups	Emphasis on individuals
Larger in size	Smaller in size
Emphasis on division into smaller groups	Emphasis on fulfilling a specific need

Source: Adapted from Shani & Chalasani, 1992, p. 45.

Again, niche marketing is held by Weinstein (1994) to be a form of concentrated target marketing as opposed to segmentation, which discerns and responds to customer needs through concentration or differentiation. Weinstein (1994) defines *market differentiation* as segmenting the market based on different customer needs, and includes product, promotional, price, and distribution strategies for different segments. An example would be Snapple soft drinks, differentiated by their high juice content. A concentration strategy, on the other hand, implies that a firm has decided to serve one of several potential segments of the market. The example given by Weinstein is a computer dealer who concentrates on the home-user segment of the market.

Kotler (2003) differs from the view that niche marketing is a bottom-up approach, explaining that "[m]arketers usually identify niches by dividing a segment into sub-segments." The example given is dividing the smoker segment into those that want to stop smoking and those that do not. However, Kotler does differentiate niches from segments, noting that niches are generally smaller and normally attract only one or two competitors. He also cites the vulnerability of larger companies such as IBM who often lose pieces of their market to nichers.

But clearly niches are a special kind of segment and niche marketing strategy is one kind of segmentation strategy. They are not to be differentiated by the process of segmentation, since that may vary with different companies and products and within the same company at different times. Neither can they always be differentiated by the current size of the segment, since niche markets sometimes grow into larger markets (e.g., health-conscious fast-food consumers which have been recently exploited by fast-food companies such as McDonald's). Rather, they are differentiated by being a certain type of market segment which sometimes changes radically with time but which usually begins as a small underserved market consisting of customers with a special shared need and the willingness and ability to pay a premium to the company that best satisfies that need.

Classic examples of niche marketing in the hospitality industry are Enterprise car rentals (specializing in car owners who have their cars under repair), airlines that provide luxury jet charter services to top executives, hotels and cruise lines that cater to ecotourists, restaurants

that specialize in health foods, and spa hotels that cater to the health conscious.

Advantages of Niche Marketing

The advantages of niche marketing include the following:

1. There is limited competition from other firms, who may ignore this segment because of its size or perceived marginal profitability.
2. The company has superior knowledge of the needs and preferences of the target niche due to specialization in this niche.
3. Individuals in this niche will pay more for specialized services. For example, Ferrari commands a high price for its cars because of the niche it has carved out among high-end performance-driven sports car drivers. Likewise, Ritz-Carlton hotels command a premium price due to their reputation of personalized, unmatched customer service along with elegant facilities for the discriminating guest.
4. Companies earn brand equity (including brand recognition and high perceived quality) derived from the superior ability to serve their niche market.

THE RITZ-CARLTON HOTEL CHAIN: WORLD-CLASS NICHE MARKETING

Perhaps no other hospitality company has received more awards or is so widely recognized for their expertise in niche marketing as Ritz-Carlton hotels (now owned by Marriott International). Ritz-Carlton received 121 quality awards in 1991 alone, including "Best Hotel Chain in the United States" by the Zagat Travel Survey (Partlow, 1993). In this section, we will discuss the niche marketing practices of this hotel that has earned a loyal following in the *ultra-luxury niche* of the luxury hotel market. We will see that Ritz-Carlton uses elements of market orientation, relationship marketing, and total quality management (TQM) to provide exceptional quality and value to its target niche of the wealthy and powerful.

Ritz-Carlton, a 1992 Baldridge Quality Award winner, displays its service strategy on a small plastic card given to all employees; the card contains the company's service credo and motto: "We are Ladies

and Gentlemen serving Ladies and Gentlemen." It also lists the Ritz-Carlton "Basics" (see Exhibit 3.2).

Ritz-Carlton also develops a market-oriented company culture by starting each new hotel with a seven-day orientation process led by top management and which utilizes a TQM program to ensure the highest quality of customer service. All employees are told that the highest mission of the hotel is "the genuine care and comfort of our guests" (see company credo in Exhibit 3.2).

Best practices of Ritz-Carton Hotels include these:

1. Emphasis on employee selection, training, and satisfaction. New employees are chosen for such qualities as a willingness to serve others, drive, enthusiasm, sincerity, and an optimistic friendly attitude ("Ritz-Carlton Employees Go for the Gold," 2002). All workers receive 100 hours of quality training annually. In addition, each department of the hotel goes through a ten- to fifteen-minute "line-up" where employees review the credo and the service steps they employ each day (Eisman, 1993).
2. Development of a corporate culture of excellence in service and quality standards.
3. Employee empowerment. All employees are given the authority to spend up to $2,000 to solve guest problems without the approval of management ("Ritz-Carlton Employees," 2002).
4. Use of total quality management techniques (TQM) to ensure this excellence, including employee empowerment, quality deployment procedures, quality teams, daily quality production reports for each of 720 work areas, and supplier certification processes. Quality production reports are compared with predetermined customer expectations to improve services and maintain high levels of customer satisfaction. These quality performance standards are set by employee quality teams.
5. All guest data (including likes and dislikes) are entered into a computerized guest database that is used to personalize services at all Ritz-Carlton hotels. From the moment a guest books in, his or her guest history profile begins and is available to all Ritz-Carlton hotels. Ritz-Carlton employees carry guest preference pads and take every opportunity to note and record guest preferences, which are later added to the guest history database. The guest history also provides the opportunity to reward frequent guests in more effective ways than providing a free breakfast,

fruit basket, or reward points. For example, a frequent guest with a preference for ocean views may be given a free upgrade to a deluxe room with that view, while the staff member thanks the guest for his or her patronage. Ritz-Carlton's success in building relationships is apparent in its 60 percent repeat business company wide (Conlon, 1996).

PORTER'S STRATEGY OF FOCUS

No discussion of niche marketing strategy would be complete without including the "Focus" marketing strategy (part of the four generic strategies of Michael Porter, 1980). Surprisingly, Porter's contribution is rarely discussed in the niche marketing literature.

> Although the low cost and differentiation strategies are aimed at achieving their objectives industry wide, the entire focus strategy is built around serving a particular target very well and each functional policy is developed with this in mind. This strategy rests on the premise that the firm is thus able to serve its narrow strategic market more effectively or efficiently than competitors who are competing more broadly. As a result, the firm achieves either differentiation from better meeting the needs of the particular target, or lower costs in serving this target, or both. (Porter, 1980, p. 38)

The Focus marketing strategy is about targeting particular narrow segments or *niches* (Table 3.1), which can be more effectively targeted using either a differentiation strategy, a low-cost leadership strategy, or both (Porter, 1980). Thus, Motel 6 is an example of a niche strategy using the low-cost Focus leadership generic strategy, while Ritz-Carlton is an example of the Focus differentiation strategy.

Porter adds that the Focus strategy has the potential to earn above-average returns for its industry and also provide defenses against each of Porter's competitive "five forces" that in total determine the level of industry attractiveness. According to Porter (1980), the state of competition within a particular industry depends on the five forces: power of new entrants, power of suppliers, power of buyers, threat of substitutes, and competitive rivalry. The collective strength of these five forces determines the profit potential or ROI. In addition, a Focus

EXHIBIT 3.2. The Ritz-Carlton "Gold Standards"

The Ritz-Carlton Credo

The Ritz-Carlton is a place where the genuine care and comfort of our guests is our highest mission. We pledge to provide the best service and facilities to our guests who will always enjoy a warm, relaxed, yet refined ambiance. The Ritz-Carlton experience enlivens the senses, instills well-being, and fulfills even the unexpressed wishes and needs of our guests.

The Ritz-Carlton Motto

"We are Ladies and Gentlemen, serving Ladies and Gentlemen." Practice teamwork and "lateral service" (i.e., employee-to-employee contact) to create a positive work environment.

Three Steps of Service

1. A warm and sincere greeting. Use the guest's name if and when possible.
2. Anticipation and compliance with guest's needs.
3. Fond farewell. Give guests a warm goodbye and use their names, if and when possible.

The Ritz-Carlton "Basics"

1. The Credo will be known, owned, and energized by all employees.
2. The three steps of service shall be practiced by all employees.
3. All employees will successfully complete Training Certification to ensure that they understand how to perform to The Ritz-Carlton standards in their position.
4. Each employee will understand their work area and hotel goals as established in each strategic plan.
5. All employees will know the needs of their internal and external customers (guests and fellow employees) so that we may deliver the products and services they expect. Use guest preference pads to record specific needs.
6. Each employee will continuously identify defects ("Mr. BIV": Mistakes, Rework, Breakdowns, Inefficiencies, and Variations) throughout the hotel.
7. Any employee who receives a customer complaint "owns" the complaint.

(continued)

(continued)

8. Instant pacification will be ensured by all. React quickly to correct the problem immediately. Follow up with a telephone call within 20 minutes to verify that the problem has been resolved to the guest's satisfaction. Do everything you possibly can to never lose a guest.
9. Guest-incident action forms are used to record and communicate every incident of guest dissatisfaction. Every employee is empowered to resolve the problem and to prevent a repeat occurrence.
10. Uncompromising levels of cleanliness are the responsibility of every employee.
11. "Smile. We are on stage." Always maintain positive eye contact. Use the proper vocabulary with our guests. (Use words like "Good morning," "certainly," "I'll be happy to," and "my pleasure.")
12. Be an ambassador of your hotel in and outside of your workplace. Always talk positively. No negative comments.
13. Escort guests rather than pointing out directions to another area of the hotel.
14. Be knowledgeable of the hotel information (ours of operation, etc.) to answer guests' inquiries. Always recommend the hotel's retail and food and beverage outlets prior to facilities outside the hotel.
15. Use proper telephone etiquette. Answer the phone within three rings and with a "smile." When necessary, ask the caller, "May I place you on hold?" Do not screen calls. Eliminate call transfers when possible.
16. Uniforms are to be immaculate; wear proper and safe footwear (clean and polished), and your correct name tag. Take pride and care in your personal appearance (adhering to all grooming standards).
17. Be certain of your role during emergency situations and be aware of fire and life-safety response processes.
18. Notify your supervisor immediately of hazards or injuries and of equipment or assistance that you need. Practice energy conservation and proper maintenance and repair of hotel property and equipment.
19. Protecting the assets of a Ritz-Carlton Hotel is the responsibility of every employee.

Source: www.ritzcarlton.com

TABLE 3.1. Niche Marketing in the Hospitality Industry

Company (by industry)	Place	Niche
Hotels		
Kalahari Resort and Convention Center	Wisconsin	Indoor water parks serving families with children
Ritz-Carlton Hotels	Chain	Ultra-luxury accommodations for the rich and powerful
Wyndham Hotels	Chain	Targets airline employees
Monaco Hotel	Chicago	Caters to sports athletes more that 6 feet tall (10' ceilings, 86" beds)
Sutton Hotel	New York City	Caters to upscale extended-stay guests
Campamento Camani Hotel	Venezuela's Amazon jungle	Targets nature lovers (ecotourism)
Restaurants		
El Pollo Loco	California	Hispanic market (chicken)
IHOP	Chain	Targets seniors and Hispanics
La Madeleine	Texas	Single female diners
Su Shen	Florida	Japanese sushi bar
Sublime	Florida	Vegetarian, macrobiotic, organic foods
Airlines		
JetBlue	New York City, Long Beach (23 cities)	Comfortable low-cost flights to a few select destinations
AirRoyale	New York City, Los Angeles, London	Charter jet service for executives
Seaborne Airlines	San Juan, St. Croix, St. Thomas	Seaplane service to the Virgin Islands and Puerto Rico
Cruiselines		
SeaEscape Cruises	U.S.	Targets the gay cruise market
Festival Cruise Lines	U.S.	Designed for European tastes
Royal Caribbean	U.S.	Targets conventions and meetings

strategy can be used to select targets less vulnerable to substitutes, where competitors are the weakest. A Focus (niche) strategy usually implies limitations with regard to market share achievable. This is due to the inherently smaller size and specialization involved in Focus marketing (Porter, 1980).

Risks Involved in a Niche Marketing Strategy

Porter (1980) lists three risks of a focus strategy:

1. The risk that the cost differential between broad-range competitors and the focused firm widens to eliminate the cost advantages of serving a narrow target or to offset the differentiation achieved by focus. An example would be a local grocery that targets the health food niche market not being able to compete with larger firms, such as Whole Foods supermarkets, because of the economies of scale held by the latter due to high volume sales and superior purchasing power. The same risk would apply to a local health food restaurant that is similarly vulnerable to the market entry of a large chain health food restaurant.
2. The risk that the difference in desired products or services between the strategic (focus) target and the market as a whole narrows. This would be the case if the low-cost airlines lost their niche because most of the major airlines converted to the low-cost airline model or had brand extensions that used this model. Examples include Song and Ted airlines, which are products of Delta and United airlines, respectively. This is what is happening to Southwest airlines, as many new discount airlines enter the market (see Exhibit 3.3). The competitive advantage once held by Southwest airlines, due to their low-cost, comfortable air travel in a friendly, no-frills economy travel market niche, is gradually being eroded by these new entrants into the discount airlines market. Thus, the problem for this high-flying niche marketer now becomes how to differentiate themselves from these new competitors who offer similar products for similar prices.
3. The risk that competitors will find submarkets within the strategic target and outfocus the focuser. The conference-center resorts in the hotel industry that focus on a subniche of the corporate meeting market offer a good example. The developers of

this subniche identified the need for specialized resorts exclusively serving smaller corporations and association meetings and conferences. Unlike the major hotel chains, these resorts do not pursue the major corporate and association conventions. Rather, they rely on sophisticated audio visual equipment, customized meeting facilities, attractive destinations, and high levels of personal service in arranging and coordinating their clients' meetings (Morrison, 1996).

SUMMARY

Niche marketing is defined as the positioning of a firm specializing in small profitable market segments with shared special interests which are underserved by others. There has been an increased use of branded niche marketing (using different brands for different niches) by different sectors of the hospitality industry, including restaurants, hotels, and airlines. Examples of this trend include Song airlines by Delta, Courtyard by Marriott, and On the Border (Mexican grill) by Brinker International. The use of a separate branding strategy for companies launching new niche market offerings avoids the risk of brand dilution while building brand equity with customers of the selected target niche.

EXHIBIT 3.3. The Proliferation of U.S. Discount Airlines

Airline Name	Web Site
JetBlue	www.jetblue.com
Song (owned by Delta)	www.flysong.com
Ted (owned by United)	www.flyted.com
Airtran	www.airtran.com
Southwest	www.southwest.com
America West	www.americawest.com
Spirit	www.spiritair.com
Delta Express (owned by Delta)	www.deltaexp.com/
Frontier Airlines	www.flyfrontier.com
American Transair	www.ata.com
Gulfstream International Air	www.gulfstreamair.com

Porter (1980) has made a significant contribution to niche marketing strategy by the development of his two generic FOCUS strategies of low-cost leadership and differentiation (which may be used separately or in combination). He has also articulated the three major risks of a Focus or niche marketing strategy: (1) the risk that the cost differential between broad-range competitors and the niche marketer widens to eliminate the cost advantages of serving a narrow target or to offset the differentiation achieved by marketing to this niche; (2) the risk that the difference in desired products or services between the niche market and the market as a whole narrows; and (3) the risk that competitors will find submarkets within the niche market and out-focus the niche marketer, thus eroding the niche market selected.

Ritz-Carlton hotels has exemplified some of the best practices in niche marketing. They have done this through the development of a market-oriented corporate culture, the integration of TQM procedures, and the setting of new standards for service quality that are both friendly and anticipate the needs of their customers. Service has also been supported by customer-database-assisted relationship marketing which makes their influential guests feel like royalty at any Ritz-Carlton location.

EXERCISES

1. Select a niche marketing company in the hospitality sector. Critique the effectiveness of their niche marketing strategy, including their ability to command premium prices, loyal customers, and offer superior value to these target customers.
2. Critique the level of "strategic fit" of this company (fit of resources, skills, competencies, corporate goals, environmental factors, marketing trends) with their choice of target segments.
3. What is the role of hospitality employees in the niche marketing process? Give an example to illustrate your point.
4. What is the role of the customer database in the niche marketing process? What information should be entered into this database for a selected hospitality company?
5. What are the advantages and liabilities of serving more than one niche at the same time?
6. What is the role of market research in effective niche marketing?

7. Product/service positioning usually involves distinguishing your offering (in the mind of your customers) from the competition in a way that is valued by these target customers. With this in mind, is it possible for a hospitality company to effectively niche market a commodity product/service? Explain.
8. Can you have a niche market of one? Explain.
9. How would you advise a travel company that niche marketed vacation packages to World War II veterans and their families to sites that were significant in that war, such as Normandy and North Africa?
10. What can other hotels (who target different segments) learn from the niche marketing strategy of Ritz-Carlton hotels?

Chapter 4

Niche Marketing for Hotel Managers

INTRODUCTION

The focus of this chapter is niche marketing as a strategy for smaller resort hotels (<125 rooms), but the strategic implications of niche marketing are generalizable over the entire hotel industry. The tourist hotel market in Jamaica, not unlike the global market, is involved in a consolidation phase associated with the later stages of a mature market. There is fierce competition, commoditization of products, price cutting, declining revenues, and declining profitability. Morritt (1995) has argued that that the key to smaller hotels competing successfully in this market is related to three programs: market segmentation, strategic alliances, and automation. This chapter will focus on niche marketing as a segmentation strategy for smaller hotels.

A Brief Overview of Niche Marketing Strategy

Market segmentation is the process of classifying customers and prospects into groups with similar needs and purchasing behavior (Kotler, 1991; Weinstein, 1995). Effective segmentation allows the firm to select those groups that can be served most profitably and positions the firm to effectively service the needs of those groups. Niche marketing is perhaps best defined by Dalgic and Leeuw (1994):

This chapter appeared previously as R. Morritt, "Niche marketing for hotel managers," *Journal of Segmentation in Marketing, 1*(2), 103-119. © 1997 by The Haworth Press, Inc. Reprinted with permission.

Segmentation Strategies for Hospitality Managers
Published by The Haworth Press, Inc., 2007. All rights reserved.
doi:10.1300/5716_04

45

We may conclude that niche marketing could be defined as positioning into small, profitable homogeneous market segments which have been ignored or neglected by others. This positioning is based on the integrated marketing concept and the distinctive competencies the company possesses.

How does NM lead to competitive advantage?

- If you have selected carefully, there is little competition for your target niche. (The niche may not be large enough to be feasible for larger hotels or it may not be synergetic with its other products.)
- Your hotel should have a competitive advantage in servicing this niche, since it was selected because it fits the resources and key competencies of your hotel.
- You can serve this niche better than other hotels by specializing in understanding and serving its needs.
- Niche marketing is associated with higher margins, since servicing niches better meets the needs of niche customers (greater price elasticity).
- Niche marketing is a strategy that hotels like Marriott and Club Med have already proven to be effective in competing in the hotel industry.
- There is greater opportunity for cross marketing because of the intimate knowledge of niche customers contained in your customer database (e.g., tours, rental cars).

Here, as elsewhere, greater market returns are associated with greater risks. The following are risks and limitations associated with NM.

- There is more investment risk for smaller hotels targeting a single niche, since you are not diversified.
- NM is more expensive; it requires more information about your customers and greater customization of products and services.
- NM requires the use of customer databases to continuously track, monitor, evaluate, and control operations to profitably maximize service, quality, and value, *as defined by the customer.*

- NM will not make up for fundamental hotel problems (poor service, low quality, poor facilities, lack of operational controls).
- NM is more dependent on customer retention (since it is a small market). Therefore, perceived service and value must be excellent.
- NM requires the full support and commitment of both upper management and employees who must buy into the marketing concept that NM implies. What is this marketing concept? It is that the mission of your hotel (and every employee) is to provide facilities and service which *best serve the needs and expectations of your target customers.*
- Segment selection needs to be supported by periodic market research and analysis of guest histories and surveys. A priori selection of segments without empirical support exposes the hotel to the risk of selecting a segment that is of not sufficient size, that will not be attracted to the hotel product being offered, or is not sufficiently profitable.
- If you are targeting more than one segment, you will need to be concerned about possible interactions (conflicts/synergies) between segments.

Segmentation Practices in the Hotel Industry

Writers often distinguish between "a priori" (predetermined) and "post hoc" segmentation. The latter identifies market segments based on market research, including market surveys. A priori segmentation, on the other hand, usually involves selecting groups from the market and then finding out if they are true segments, often without new research (Struhl, 1992). Until recently, all segmentation in the U.S. hotel industry was a priori segmentation, which was an attempt to appeal to different customer segments by a differentiation strategy. The U.S. hotel industry has evolved from price segmentation to product segmentation and is now experimenting with market segmentation and niche marketing. Niche marketing strategy is often accompanied by a "relationship marketing" strategy whereby hotels utilize integrated customer databases to build long-term, interactive, mutually profitable relationships with their best customers.

Stage 1: Price Segmentation

Until the 1980s, the U.S. hotel industry had developed with four basic price segments: luxury, convention, first class, and economy. Usually a hotel company specialized in one of these categories (Schultz, 1994) (Exhibit 4.1).

Stage 2: Product Segmentation

Product segmentation revolutionized the U.S. hotel industry in the 1980s. Major hotel chains used this strategy to counter the competitive effects of a mature market and to more effectively serve diverse economic segments. This was the first attempt by the hotel industry to create hotels to mirror consumers' lifestyles rather than to be all things to all people (Schultz, 1994). The most successful of these new products were the all-suite hotels and the "limited service" hotels ("super segments," in Kotler's terminology). Hotel chains developed *segmentation portfolios* to reach critical mass. Five hotel segments

---

EXHIBIT 4.1. Marriott's Use of Price Segmentation

Marriott trio: Call it marketing genius or marketing madness, but Marriott International is turning its 782 room Miami Airport Marriott into three hotels. When an $8.5 million renovation project is completed there in October, guests will find a 365 room full service Marriott, a 125 room Courtyard by Marriott, and a 285 room Fairfield Inn. The original hotel built in 1972 as a Marriott Motor Inn, had a main tower and two low rise buildings in the back. The hotel became outdated in the 1980's as guests began to prefer the main tower with its indoor corridor to the hotels two smaller buildings with their outside corridors. Eventually the two smaller buildings housed mostly airline crews. Each of the remodeled hotels will have a lobby, front desk, hotel staff, and complementary airport shuttle. The three hotels will share a pool and a fitness center. The Courtyard and the Marriott will each have a restaurant that will serve guests of all three hotels. Rates will differ: Marriott is $99 a night; Courtyard, $79; Fairfield Inn, $49 to $59. Travelers call the regular Marriott number to make reservations for any of the three. "When travelers call us for this location, we can give them a menu of choices" says Joseph Brown, the hotel's director of marketing. "It used to be you didn't have a choice."

Source: USA Today, August 17, 1995.

reported by one hotel marketing analyst (Dev & Hubbard, 1989) for eleven major hotel chains are as follows:

1. Traditional: Full service, comfort, food service, moderate price (Holiday Inn, Marriott)
2. Economy: Limited to no- service, no- frills facilities, low price (Fairfield Inn, Sheraton Inns)
3. All-suites: The most successful of the segments, moderate price, offering homelike accommodations (Embassy Suites, Residence Inns)
4. Casino: Moderate prices, subsidized by casino operations (Harrah's Tropicana)
5. Upscale: High service level, comfort, food service, expensive (Crowne Plaza, Towers)

Stage 3: Special Interest Segmentation

In recent years hotels have experimented with special interest segmentation, a niche marketing strategy that targets special interest groups. However, too often this is a priori strategy and does not take advantage of the benefits of market research and the new technology afforded by computerized guest databases. Special interest groups targeted by hotels include the following:

- Hotels that cater to the fifty and over traveler (Choice Hotels). Customized features include lever handles on doors, brighter lights, larger buttons on remote controls, and ground-floor locations ("Rodeway Seniors Get Choice Rooms," 1993).
- Hotels targeting nature lovers (ecotourism) (e.g., Campamento Camani, which has thirteen units). David Anderson (president of the Anderson Group) noted that ecotourism is the world's fastest growing tourist market (Deneen, 1993).
- Hotels that accommodate guests needing to stay for a month or more (the eighty-four-room Sutton in New York City). These hotels often offer maid service, a free health club, fully equipped kitchens, and apartment- sized closets and storage areas (Selwitz, 1992b).
- Hotels that cater to culture buffs (e.g., Boston's Four Seasons hotel, which markets arts packages tied to the city's cultural institutions and theater) (Selwitz, 1992a).

- Hotels that exploit historical attractions of places (e.g., Club Med has "villages" in Mexico and Asia that promote tours of famous archeological sites) (Club Med, 1995).
- Hotels that cater to gourmets and aficionados (e.g., Omni Hotels just launched its cigar dinner program, the first in a series of "smoking soirees" that will take place at seventeen Omni properties in the United States ("Omni Hotel Plays Up to Aficionados," 1995).
- Hotels that cater to the auto/truck traveler (e.g., Forte Hotel's Thrift Lodge brand) (Escalera, 1994).
- Hotels that have created a new business: residential life-care communities targeting the growing mature (sixty-five and older) population (e.g., Marriott made use of psychographic segmentation which has been found to be useful in segmenting the elderly market for the development of these communities) (Camacho, 1988).
- Hotels that have entered the time-share and condominium markets. Time-sharing and condominium conversions stabilize revenues and bring in capital (you sell your units but still get to rent and manage the property). Jamaica's upscale Half Moon resort has a time-share plan on rooms and villas where the units revert back to the owners in twenty-five years (GMS, 1993).

Niche marketing allows a hotel to successfully compete against larger hotels by its superior ability to service its chosen niche. This is no accident but is based on a niche marketing strategy that includes market research and transformation of the hotel culture to deliver what this niche considers to be value in an unexcelled and profitable manner.

For example, as a small resort hotel in Jamaica, you would not want to compete with an international chain like Wyndham for business clients. They have superior resources and are better positioned for this market, having conference facilities, marketing departments, relationships with many large corporations, and large marketing budgets. You do not have the resources to compete effectively in this segment, so you need to target a different, perhaps smaller, niche that is more consistent with your abilities and resources. An example of one small hotel in Jamaica that successfully targeted a niche market is Negril Gardens in Jamaica,[1] which has achieved excellent occupancy

rates as well as insulating themselves from a recent U.S. recession by targeting the European market (16 percent of present market) (Exhibit 4.2).

Stage 4. Integrated Niche Marketing

We define *integrated niche marketing* as niche marketing that is integrated with customer databases, quality control systems,[2] and relationship marketing[3] to achieve a sustained competitive advantage by offering unequaled services to target customers. This entails knowing how your target customers define value and consistently delivering this value in an efficient, friendly, and cost-effective manner.

Integrated niche marketing enables you to better achieve and maintain a competitive advantage in servicing your target niche. Customer databases, relationship marketing, and quality control systems are tools to customize, optimize, and control all hotel operations in servicing your target niche.

A recent example of successful integrated niche marketing is Baldridge award winner Ritz-Carlton hotel (Mene, 1994; Partlow,

---

EXHIBIT 4.2. Negril Gardens Jamaica

Negril Gardens is a small hotel in Negril which has achieved unusually high occupancy rates during a recent recession by managing to effectively serve middle class European tourists who were comfortable with a small moderately priced beach front hotel which adopted European standards of higher service quality, continental cuisine, nightly entertainment, clothing optional beaches, relaxed atmosphere (no scheduled activities), multilingual staff, and no phones in your room. An added and unique benefit of this hotel was that Europeans associated with other European guests who had similar values and lifestyles. Thus you avoided the "ugly American" problem. Negril Gardens has achieved a competitive advantage in the Jamaican tourism market. If you wished to vacation in Negril (world famous 7 mile long beach) and desired a European type ambiance in a moderately priced hotel, Negril Gardens was one of the few places to achieve these goals. While most small hotels were struggling with marginal occupancy rates during this period, Negril Gardens had occupancy rates exceeding 90 percent and due to its target niche, was relatively immune to the US recession at the time.

Source: GMS, 1993.

1993), which services the super luxury market niche. Hotel manager Mene used Baldridge award criteria, guest databases, and guest surveys to improve hotel processes and personally build teamwork and transform the culture of the hotel to deliver what guest surveys indicated as constituting value[4] to their guests and significantly minimized service defects (reported by committed employees as well as guests). Employees were trained, empowered, evaluated, and rewarded according to satisfaction of corporate goals of service quality and teamwork, and the conversion process was "top down," led by the general manager.

Use of a computerized guest database that provides detailed information on 240,000 frequent guests was used to help employees become familiar with individual guests likes and dislikes resulting in personalized quality service and long-term relationships with customers.

SEGMENTING YOUR LOCAL MARKET

Market Structure

It is important to first understand the major forces, trends, and critical success factors that drive the present market. The market for Jamaican resort hotels, for example, has a particular structure at any given time that can be uncovered by researching the answers to the following questions:

- Who are the major and minor buyer groups?
- Where do they come from?
- What are the competing products?
- How successful are these products?
- What are the major reasons why buyers purchase this product?
- How is this product used?
- What are the standard variations of this product?
- What is the industry life cycle stage in this market?

The major geographical sources of tourism for Jamaica have been the United States, Canada, and Europe. Thus, in the first two months of 1995 the United States constituted 66 percent of the market; Canada, 13 percent; Europe, 16 percent; and all others, 6 percent.

The Jamaican market has long been divided into the all-inclusive hotel group (which, following the Club Med model, offers a package that includes meals, entertainment, sports, and often an air package) and the non-inclusive hotel group (which charges separately for meals and often does not have entertainment or a great variety of sports). In recent years the most profitable hotels have tended to be of the all-inclusive variety, of which Sandals is representative (GMS, 1993).

The major tourist centers in Jamaica are Montego Bay, Negril, and Ocho Rios. The average occupancy rate over the past five years has been below the break-even point of 65 percent for most non-inclusive independent hotels. This situation, associated with the mature life cycle stage of the tourism market, was exacerbated by recessions in the United States and Europe, the Gulf War, overbuilding in major tourist areas like Montego Bay, competition from other Caribbean tourist destinations, and government cutbacks on tourism marketing. Adding to this problem was an image problem associated with an inaccurate perception of inadequate security propagated by the media and travel agents.

The major hotel products in the Jamaican market consist of all-inclusive hotels (Sandals, Cibboney) which target the younger active couples segment, hotels that target businesspeople and convention business (Wyndham, Holiday Inn), hotels that cater to the affluent market (Half Moon, Lido), a few hotels that target the European market (Negril Gardens), condo hotels that offer suites (Montego Bay Club, Turtle Towers, Seawind), and many small hotels on or near the beach which do not target anyone.

The larger and international hotel chains have decided competitive advantages in this market, not only by economies of scale and deeper pockets but also by expert management teams, computerized hotel systems, staff training programs, and international brand equity. Many of these larger hotels are operating sophisticated database marketing systems and frequent guest programs. Smaller hotels have an increasingly difficult time competing in the latest shakeout phase of the Jamaican tourist market, and most have occupancy levels below the break-even level of 65 percent (Morritt, 1995).

Tourists come to Jamaica for many different reasons that can be exploited by a niche marketing strategy for smaller hotels (benefit segmentation). Some of these reasons are the tropical climate, lush vege-

tation, magnificent beaches, water sports, snorkeling and scuba diving on the coral reef, sports fishing, world-class golf and tennis facilities, the Jamaican culture (including reggae music and the historical celebrity of Port Royal), and the friendly people. Affluent tourists often come to be pampered in elegant spalike surroundings and never get to see Jamaica or its people at all outside the sterile surroundings of an "international hotel." There are presently no casino hotels. Jamaica, famous for bauxite and tourism, is the home of some of the best coffee (Jamaican Blue Mountain) and cigars (Royal Jamaican) in the world.

MARKET TRENDS

The Dominance of the All-Inclusive Vacation Package

Many tourists prefer an all-inclusive package that includes airfare, meals, accommodations, drinks, airport transportation, and tips. This successful pricing strategy, originated by Club Med, is now used successfully by Sandals and other all-inclusive hotels worldwide. Statistical data continue to show that all-inclusive hotels, as a group, have higher occupancy rates than noninclusive hotels in Jamaica.

Ecotourism: The Fastest Growing Sector of Tourism

Ecotourism describes the recent trend toward concern for environmental protection, cleanliness, safety, and health as well as the enjoyment of natural beauty including parks, forests, lakes, rivers, oceans (whale watching), coral reefs (diving and snorkeling), and wildlife. A growing number of tours are springing up in countries like Jamaica for tourists to see the unspoiled backcountry as well as activities like glass-bottom boat rides, scuba diving, and snorkeling at the coral reefs.

According to the Travel Industry Association of America, 43 million Americans are self-proclaimed "eco-tourists," willing to pay 8.5 percent more for environmentally friendly travel suppliers. The potential is cited by one travel writer for the creation of the "Eco-hotel" (or Ecotel), which would tap this huge market (Rushmore, 1993).

Health Tourism

Health tourism is defined by Goodrich (1994) as the attempt on the part of a hotel or destination to attract tourists by promoting its health care services and facilities in addition to its regular amenities. Goodrich cites such services as medical treatments and special diets. I shall use the term "health tourism" in a wider sense to include spa and fitness facilities as well as special meal plans to target health-conscious tourists.

It is common for Caribbean tourists and locals to enjoy ocean-bathing as a healthy activity, and islands such as Bahamas, Cuba, and Jamaica have mineral springs and health resort facilities such as the Silver Reef Health Club in Grand Bahama Island and Doctors Cave Beach in Jamaica (Goodrich, 1994).

Furthermore, certain all-inclusive hotels in Jamaica (Cibboney and Swept Away) have successfully differentiated themselves by offering spa and/or fitness facilities, including "spa food" (fresh ingredient, low-calorie gourmet meals), to tourists who are health conscious.

The targeting of health-conscious tourists is an especially interesting niche for smaller hotels in the Caribbean, since it is a relatively inexpensive modification of facilities and also has the potential to bring in local business for restaurants and fitness facilities. Several hotels in Jamaica have special diet facilities such as Lady Diane (macrobiotic food) and Cibboney (spa food).

The Rise of Cultural Tourism

Cultural tourism refers to tourism that promotes the local culture, history, or archeology of the tourist destination. Kotler, Haider, and Rein (1993) speak of a major trend in place revival called *heritage development:* "the task of preserving the history of places, their buildings, their people and customs, the machinery, and other artifacts that portray history" (p. 209). I mentioned previously that Club Med has special archeological villages in Mexico and Asia and also that Boston's Four Seasons hotel offers cultural packages tied to local cultural institutions like theaters, concerts, and opera. One example of the attraction of historical landmarks is New York's famous South Street Seaport (including the museum, restored giant clipper ships, and concerts of old sea chanteys.)

The Implications of U.S. Population Trends for the Hotel Industry

Demographic trends indicate a significant increase in the sixty-five-and-over population (baby boomers maturing), the distribution of wealth skewing toward the older population, the increase in the percentage of dual-income families, increased business travel for women, and an increased proportion of nontraditional households (Yesawitch, 1991). Other trends include

- the trend toward shorter, more frequent vacations. U.S. vacations average four to five days, in contrast to European vacations, which average ten to fourteen days.
- greater numbers of visitors from Germany and the Pacific Rim, as higher disposable incomes in these countries and currency devaluation make countries such as the United States, Mexico, and Jamaica a better value.
- greater competition from cruise ships (growing twice as fast as the hotel industry in the Caribbean) and strong competition from other Caribbean Islands such as Puerto Rico, Dominican Republic, Cuba, and the Bahamas.

SEGMENT SELECTION

A rigorous segment selection procedure involves using primary and secondary market research using selected segmentation criteria (e.g., benefits, lifestyles, usage, and "special interests") and market structure analysis to target potentially profitable segments in your local market. The subsequent short list of potentially most profitable segments is further screened by the criteria of "fit" with the resources, culture, and special competencies of your hotel, actionability (how reachable are your customers?), and structural attractiveness (growth of, profitability of, and competition for this segment).

Risks of Segment Selection

Since most hotel segmentation at the time of this writing consists of an a priori segmentation strategy called "product segmentation," any discussion of hotel industry segmentation must address the risks of "product segmentation" that is not supported by market research,

as opposed to pursuing a market-based segmentation strategy such as niche marketing.

It is possible to attract true segments by both methods (defined as a group with similar needs and purchasing behavior). However, a priori segmentation, *unless supported by market research,* involves special risks:

- The selection of a product segment that is not a true segment because the group does not exhibit similar needs and purchasing behavior (e.g., the decision to specialize in mature tourists [fifty-five and older], Europeans, or affluent tourists does not address a true market segment since these groups are not homogenous).[5]
- The risk that the size, growth, needs, preferences, and/or purchasing behavior of your target segment may change. Without ongoing market research, you will be flying blind with your marketing and strategic planning.
- You may miss significant new market trends, competitive threats, unidentified neglected market niches, and major differences between competitive products in a product class (Struhl, 1992).

SEGMENTATION BASES

We have cited earlier the following segmentation variables used by the hotel industry to target segments and niche markets:

- price segmentation (traditional, economy, upscale, ultra)
- benefit segmentation (limited-service hotels, business hotels)
- psychographic/lifestyle segmentation (VALS; younger, active, health-conscious couples)
- demographic segmentation (country, age, income, family size)
- usage segmentation (frequent guest programs, RFM analysis used in relationship marketing)
- special interest segmentation (ecotourists, culture tourists, health tourism)

Once we have examined our market structure and determined the available segment options, we must then undertake the task of seg-

ment selection. What criteria should be used for selecting your target segment?

SELECTION METHODS
FOR SMALLER RESORT HOTELS

Smaller resort hotels are typically limited in their financial, technical, physical, and marketing resources. Therefore, there is an acute need for practical segmentation methods that are market based, relatively inexpensive, cost-effective, require a minimum of technical sophistication, and can be fitted to the resources and culture of the smaller resort hotel.

1. Adapt Another's Successful Target Segmentation

Modify a successful target segmentation that fits the selection criteria cited above (including empirical support) and modify it to your particular market and hotel. Then position your hotel, services, promotions, and communications to this segment. The success of the Sandals chain in targeting young, active couples who prefer all-inclusive pricing is an example of selecting a niche that was built on an already proven model. This was the model responsible for the success of the French international resort hotel company known in the United States as Club Med., which originated the all-inclusive vacation concept. This "bundling" strategy has the added advantage (in addition to cost savings) that guests do not have to worry about money or extra costs during their vacation. This concept was modified to include couples only, and Sandals is now the most successful hotel company in Jamaica (Club Med, 1995).

2. Use Your Guest Database to Profile Your Best Customers

Use your guest register or customer database to profile your best customers. Major service companies use database marketing to target their best companies by RFM[6] and "lifetime value" analysis, and then build a one-to-one relationship with these customers by tracking relevant guest data such as demographics, usage, family members, guest preferences (room type, child care facilities, business facilities, etc.), when they travel, their needs or requests, as well as their credit and

purchase history. Larger hotels like Marriott, and Holiday Inn offer "frequent guest" programs that provide special benefits to guests for frequent stays. Furthermore, the achievement of successful niches allows for additional revenues from cross marketing and the sale of additional products to your target niche.

3. Find a Gap in the Present Industry Segmentation of the Market

Find a gap in the present industry segmentation of the market that meets the segment selection requirements outlined above. By creating a perceptual map of your local market using such attributes as product segmentation, price, target profile, customer preferences, and purchase intentions (Struhl, 1992), you may find unsatisfied gaps in your local market. Some of these gaps may be empirically verified to represent significant demand for your current or projected resort hotel offering. For example, there seems to be (at the time of this writing) few resorts in Jamaica targeting singles, culture-oriented tourists, health and ecotourists, and tourists who would like clean and comfortable economy hotels (e.g., Motel 6)[7] while touring the country.

4. Hire a Market Research Consulting Firm to Do a Segmentation Study

Hire a reputable consulting firm to research the attractive segments available in your particular market that conform to the selection criteria cited above, confirm attractive a priori segments, and analyze your current customer base. Although a segmentation study can be relatively expensive, it is an investment that often pays off over the long term in future revenues. And you can inexpensively purchase secondary (previous) research from trade associations such as the Caribbean Tourism Association and the Jamaican tourist board. You can also reduce costs by strategic marketing alliances with other smaller hotels (Dev & Klein, 1993). You may also want to talk to your consultants about setting up a guest database that is integrated with the reservation system.[8] This will enable you to automatically obtain guest information critical to selection of target niches, monitoring their profiles, needs, preferences, attitudes, and perceptions of your hotel.

This information is used to continuously improve and customize your promotions, services, and facilities to the needs and desires of your target niche.

NOTES

1. I am indebted to Pat Morgan, the European manager of this hotel, for her assistance.

2. Includes monitoring and continuous improvement of customer satisfaction, employee performance and morale, and hotel processes.

3. Building long-term, mutually profitable, interactive relationships with customers, usually with the help of a customer database and personalized marketing.

4. Speedy, friendly, personalized service, long-term relationships with customers, responsiveness to complaints, a "good feeling" associated with the hotel.

5. For a contrary view that treats the affluent as a homogenous segment of the hotel market despite an opening remark dividing the affluent into three segments, see Mann (1993).

6. Recency, frequency, and monetary value: This is a procedure adapted from direct marketing firms who, which considerable success, identify their best customers (using a weighted algorithm) related to the recency of their visits, the frequency of their visits, and their total amount of purchases. Lifetime value analysis computers the lifetime value of each customer (estimated revenue less costs over a five- to ten-year time frame), then all marketing projects are evaluated on their incremental effects on the lifetime value of the targeted customers (Hughes, 1994).

7. For an interesting account of the Motel 6 success story based on solid marketing research using focus groups and guest surveys, see Cunningham and Dev (1992).

8. For information on database marketing strategies for hotels, see Durr (1989), Francese and Renaghan (1990), Durocher and Niman (1991), Nowakowski (1991), Mondy and Hollingsworth (1984), and Sparks (1993).

Chapter 5

Relationship Marketing

INTRODUCTION

A loyal customer is one who values the relationship with the company enough to make the company a preferred supplier. Loyal customers don't switch for small variations in price or service; [instead] they provide honest and constructive feedback, they consolidate the bulk of their category purchases with the company, they never abuse company personnel, and they provide enthusiastic referrals. (Reichheld, 2002, p. 126)

Relationship marketing (RM) is a marketing program designed to achieve a competitive advantage by building a long term, one-to-one, interactive, mutually profitable, relationship with the firm's best customers. This includes guests, employees, and suppliers. It is based on the assumption that building good long-term relationships with your customers will result in increased customer loyalty and retention, an increased share of the customer, and increased profitability.

RM has been called a refocus of traditional marketing with a greater emphasis being placed on the creation of "customer value" (Payne, Christopher, Clark, & Peck, 2000). Thus, it is important the hospitality firms understand (using target segment research) what constitutes superior customer value (SCV) for their target segments. This SCV is defined by your customers and will vary with different segments. Thus, the customers of Motel 6 budget hotels will have a different view of what constitutes superior value (low-price, clean, comfortable accommodations) than the target customers of a luxury hotel chain like Westin hotels. And the child customer segment of McDonald's will have a different view of what is most important to

them in a dining experience than the customers of a fine dining restaurant in a five-star hotel.

Trust and Commitment

In a 1994 research study, it was found that the variables of trust and commitment were two determinants of successful business relationships (Morgan & Hunt, 1994). This was no great surprise, since the literature on business relationship marketing is based on similar studies in individual personal relationships. These variables also represent important factors in personal relationships such as marriage and friendship. Most people would rather do business with people or companies they know, trust, and who are committed to the relationship.

Theodore Levitt compares the relationship between the seller and the buyer to that of a marriage. (This is understandable, since much of the literature on RM is derived from the psychology and sociology of human relationships.):

> The sale merely consummates the courtship. Then the marriage begins. How good this marriage is depends on how well the marriage is managed by the seller. That determines whether there will be continued or expanded business or troubles and divorce, and whether costs or profits increase. . . . It is not just that once you get a customer, you want to keep him. It is more a matter of what the buyer wants. He wants a vendor who will keep his promise, who'll keep supplying and stand behind what he has promised. The age of the blind date or the one night stand is gone. Marriage is both more convenient and more necessary. . . . In these conditions success in marketing, like success in marriage, is transformed into the inescapability of a relationship. (Levitt, 1983)

Since much of what we know about relationships is derived from research into relationships between individuals (e.g., marriage relationships), it is not surprising to learn that researchers have found that good service relationships involve *trust and commitment* (Lewis & Chambers, 2000; Morgan & Hunt, 1994).

A study by Bowen and Shoemaker (1998) of luxury hotel guests found the following antecedents that affect trust and commitment:

- Fair costs: perceived fairness of prices
- Switching costs: costs and efforts in going elsewhere
- Benefits: what the customer received from the service firm that supports value and trust
- Understood values and goals: beliefs and values in common regarding the expectations of both parties

Positive consequences that are related to trust and commitment or lack of it are

- increased product usage and
- voluntary actions on behalf of each other (such as word-of-mouth advertising).

Negative consequences related to trust and commitment were found to be

- uncertainty or lack of trust in service delivery and
- opportunistic behavior where one party takes advantage of the other (hotel raising rates according to demand).

Commitment is defined by Bowen and Shoemaker (1998) as "the belief that an ongoing relationship is so important that the partners are willing to work at maintaining the relationship and are willing to make short-term sacrifices to realize long term benefits" (p. 15). Their study found that benefits and trust are the most important antecedents to customer commitment. Perceived value and switching costs had a positive but lesser impact on relationship commitment. Benefits that determined loyalty in this study of luxury hotel guests included hotel upgrades, flexible check in/out times, customized service, and involved employees who were perceived as caring about their guests. It was noteworthy that only a small percentage of guests (27.8 percent) viewed frequent guests programs and affinity cards as having a significant effect on their loyalty. This is evidence for the evolving nature of relationship marketing benefits where many benefits that start out as offering superior perceived value to target customers evolve into benefits that are only significant in their absence since they have become a routine part of expected services.

RM and Database Marketing

RM, a form of *database marketing,* is usually an integrated form of marketing which makes use of customer databases and quality control procedures, to consistently deliver superior value to target customers and suppliers. Information in customer databases is used to customize services, promotions, and frequent buyer programs for target customers. Thus, hospitality and travel companies like Cathay Pacific Airlines and the Aladdin Casino hotel keep extensive guest databases which keep track of target customer personal information, preferences, and activities and which are used to create special promotions and incentives for the firm's best customers. For example, Cathay Pacific e-mails preferred customers who travel to Hong Kong their latest promotions to Hong Kong every month, including their promotions for nonrestrictive fares for senior travelers and their new nonstop service from New York to Hong Kong. Exhibit 5.1 illustrates some of the differences between RM and traditional marketing.

Disney's Magic Kingdom Club

A good example of a database marketing program was Disney's Magic Kingdom Club. The program went national in 1971 when Dis-

EXHIBIT 5.1.
Relationship Marketing versus Traditional Marketing

Relationship Marketing	Traditional Marketing
Focus on LT relationships	Focus on single transactions
Many interactive contacts	Contact only on sale
Focus on superior perceived value	Focus on benefits or brand name
Long-term focus	Short-term focus
Emphasis on excellent service quality	Minimal service
Meeting and exceeding customer expectations	Emphasis on company goals
Quality is everyone's business	Quality is limited to a specific department

Source: Adapted from Bowen & Shoemaker, 1997.

ney World opened in Florida. Disney offered group travel packages, discounted car rentals and hotel rates, and a club newsletter, among other benefits. Club members also received a 10 percent discount on Disney stores, discounts on Disney videos, and a Disney mail-order catalog.

Disney has enrolled more than 6 million members in more than 30,000 companies. A sophisticated computer program allows Disney to select the most promising companies for each park location. The Disney RM program was then exported to Europe for use at Euro-Disney outside Paris. The returns from this program were reported to be huge by the national director of this club (Jones, 1998). (This program was eventually replaced with the present Disney Visa Card, which allows members to accumulate points for benefits.)

Benefits of RM

The benefits of relationship marketing derive from increased customer retention, increased purchases from loyal customers, reduced marketing costs, decreased price sensitivity of loyal customers, and word-of-mouth referrals from loyal customers. Reduced marketing costs derive from the fact that far fewer marketing dollars are required to keep a customer than to create one. And new customers are created from loyal customers through word-of-mouth advertising. And targeted promotions are less costly based on ROI.

The importance of customer retention is supported by a study by Reichheld and Sasser (1990), who found that a 5 percent increase in customer retention resulted in a 25 to 125 percent increase in profits in the service groups they studied.

Limitations of RM

But relationship marketing is not deemed appropriate for all customers. It is designed for those individuals or companies whose value to the company exceeds the costs of targeting them in an RM program. This may include customers or companies who are most profitable for the company, have the greatest lifetime value for the company, or who have the greatest influence on the company's success. A transactional approach may be appropriate for some other customers,

while unprofitable customers may often be encouraged to exit by an RM program.

For example, it makes good business sense for hospitality companies like Ritz-Carlton Hotels and Aladdin Casino to build partnership-like RM programs with their most profitable frequent customers who are the major source of profits for these companies. However, it may not be appropriate for a McDonald's restaurant that serves mostly transients in Midtown New York to have a similar program with its customers. Here a transactional basis is more appropriate because the investment in the RM program is not justified by its returns to the company.

Hospitality Company Relationships

Relationship marketing in the hospitality industry is exemplified

1. between hospitality companies (hotels, airlines) and marketing intermediaries (tour companies, travel agencies, online travel companies);
2. between hospitality retailers and key business customers, such as large corporations and government agencies;
3. between food service retailers (McDonald's, Burger King) and other organizations (universities, bus terminals, airline terminals);
4. between hospitality companies and key suppliers (food and security suppliers);
5. between hospitality companies and their employees;
6. between hospitality companies and their marketing agencies, banks, and law firms; and
7. between hospitality companies and their best (most profitable) individual customers (Kotler, 1999, pp. 12-13).

One way that hospitality companies create and maintain relationships with target customers is through direct marketing, including Internet marketing. One tool for retaining preferred customers is through frequent buyer programs, which increase customer retention by increasing switching costs for these customers.

For example, Sheraton Club International offers members special rates, upgrades, their own floors, and often their own lounge with complimentary beverages. American Airlines offers its Advantage

members free air miles, upgrades, and priority seating. And hotels like Aladdin Hotel and Casino of Las Vegas offer a myriad of exclusive benefits for their best casino customers, who are nicknamed *whales*. These "comps" often include free benefits such as plane tickets, exclusive suites, gourmet restaurants, limousines, and special facilities dedicated to these special customers, who may lose up to millions of dollars in one visit to this casino. Clearly there is intense global competition for these "high rollers," and their extravagant perks are evidence of their high value to casinos and casino hotels.

A 1998 study of 892 luxury hotel business guests supported the position that RM can increase customer loyalty and retention. Benefits that received strong support from these guests were room upgrades, flexible check-in and out, personalized services, and expedited registration. Trust was found to be an antecedent of guest loyalty. This study found that hotels can build trust by ensuring guest safety, providing consistent quality service, seeing that employees follow through promptly on guest requests, and being truthful and communicating accurately (Bowen & Shoemaker, 1998).

But the customer retention goals of RM are seldom satisfied by the existence of frequent buyer programs or a Web site alone. These tools are usually of a transactional nature, and the proliferation of these tools has meant that they are benefits that are now expected by the customer rather than benefits that provide superior value to target customers and thus successfully differentiate the firm with respect to its competitors. The early stages of RM, which were characterized mostly by such tools as frequent buyer programs, Internet Web sites, and targeted promotions, has given way to a more comprehensive view of RM that is influenced by the asset view of the customer.

The Asset View of the Customer

Relationship marketing programs currently view the customer as an *asset*. RM manages customer assets by attracting, maintaining, and enhancing target customer relationships. Thus, RM develops customer-specific objectives and strategies unique to each customer segment. Customers are segmented according to their value to the firm, and different relationship levels are selected for different value levels of customers. Measures such as *lifetime value of the customer* and *RFM analysis* are used to measure the relative value of hospital-

ity customers. Marketing programs and promotions are evaluated by the rate of return from the customer and segments targeted by these programs.

STRENGTHS AND LIMITATIONS OF RM

Strengths of RM

The following is a list of some of the strengths of a relationship-marketing program:

1. Increased retention of the profitable subgroup of loyal customers, who are less price sensitive.
2. Reduced marketing costs due to the lower costs of maintaining existing customers (who spend more money).
3. Greater sales volume from target segments due to increased customer retention, cross selling, the introduction of new products and services (that fit your "best customer" profile), customized offerings, and increased customer bonding. Getting close to your customer and being an expert on his or her needs, preferences, and lifestyle puts you in the position of being able to sell your satisfied target customers new products that they need.
4. Produces switching costs to target customers, which makes it more difficult for them to switch vendors or relationships.
5. Increases satisfaction of target customers, employees, and suppliers.
6. If designed correctly, it creates a culture of *market orientation.*
7. Uses technological innovation (including databases, software, Web sites, wireless, and integrated communications) to achieve a competitive advantage with the firm's best customers. This is because a successful RM program is difficult to copy and expensive to create. This is especially important with mature industries like the hotel and airline industries in which there is little differentiation in the same product segment (Bowen & Shoemaker, 1998).

8. Acquisition of new customers based on present customer profiles. By analyzing customer databases and profiling the firm's best customers, the company has the opportunity to target other market segments which are similar to the one now targeted. Thus, if a life insurance company finds that their best customers are senior citizens living in an upscale area of Florida, it might target seniors in other upscale areas of the Sunbelt with their ads and promotions.
9. Marketing to more profitable customer segments. Generally people are willing to pay more for products and services that are customized to their needs.
10. Increased competitive intelligence through interactive dialogues with target segments (use of personal and mail surveys and focus groups).
11. RM is an important source of new customers. Happier guests are an important source of word-of-mouth advertising. It has been estimated that a happy customer tells about six people on average of his or her experience, but an unhappy customer tells about nine people on average.

Limitations of RM

1. RM will not work without the full support, commitment, and example set by upper management, including necessary resources for at least three years and authority to proceed with the entire strategic plan associated with RM
2. There must be a viable marketing plan which explains how RM will increase profits, in a cost-effective manner, with timetables, controls, and specific operational measures of effectiveness and profitability.
3. The initiation of an RM strategy is usually capital intensive. It involves a high initial cost for the hardware and software involved, training programs, specialized personnel, and increased customer service.
4. It is subject to all the problems and liabilities of any program that involves a *change of organizational culture*. This includes the resistance to change from existing departments, managers, and employees, lack of support from upper management, and inadequate use of training, incentives, and controls.

Loyal Customers Are Not Always Profitable

A recent study by Reinhartz and Kumar (2002) revealed revolutionary findings that contradicted previous assumptions about loyal customers—that they were usually more profitable, cheaper to serve, and provided more word-of-mouth advertising. These assumptions were all contradicted by this study. The researchers recommended that companies develop more accurate measures of profitability and purchase probabilities. And companies need to target their most profitable customers (rather than their most loyal), which includes both some loyal customers (with loyal attitudes) and also some short-term customers (who are very profitable). This article was revolutionary in its finding that the previous assumptions about loyal customers were not confirmed for the three companies studied. The lesson here is that firms need to be careful about spending large amounts on marketing to loyal customers who are not profitable.

RM As Part of the Organizational Culture

RM works best when it is combined with an organizational culture of market orientation and TQM as exemplified by Ritz-Carlton, a hotel company targeting the ultra-luxury segment of the hotel market. Guests must believe that hospitality employees can be trusted, are concerned about what is important to them, and are empowered to correct any problems in a timely and friendly manner to the full satisfaction of the guest. Every employee represents the hotel, and thus the job of RM marketing is the job of all employees, who work together to provide a consistent market-oriented image of the hospitality firm. Integration of a quality control program ensures a program of timely problem identification, monitoring, and controls designed for continuous improvement of services.

Every employee represents the company in the eyes of its customers. The attitude of the housekeepers, janitors, and busboys in a hotel are just as important as those of the managers, waiters, and front desk people in influencing guest satisfaction and retention.

Do your hospitality employees exceed the expectations of your customers by their appearance, attitude, and excellent, cheerful service? Do they go the extra mile to ensure that customer needs are met promptly and to the customer's satisfaction? These are your best hospitality employees. It is the goal of employee relationship manage-

ment that the hospitality company obtains, trains, and retains the best employees by providing the best environment, training, and benefits for these valuable internal customers.

MEASURING CUSTOMER RELATIONSHIPS

Lifetime Value of the Customer

Lifetime customer value (LCV) can be defined as the net present value of the profits that you will realize on the average new customer during the expected life of the customer. LCV may change from month to month, due to changes in the variables that determine this number (Hughes, 1994).

Here the goal is to assign to each customer an estimate of the net profitability of that customer to the firm. Past profitability is calculated using the sum of all margins of products/services purchased less the costs of reaching that customer (which does not include mass advertising). Then the estimated average life of this type of customer relationship is entered into the calculations. LCV is then calculated by adding forecasts for the major parameters and discounting future profits for net present value. The customers with the highest LCV would then be classified as preferred customers and provided with enhanced services and other special treatment designed to keep them happy and also to expand their individual purchases (Winer, 2001) (see Table 5.1 for an example of a LCV).

RFM Analysis

RFM analysis uses a combination of three measures to segment the firm's most attractive customers. These measures are *recency* of purchase, *frequency* of purchase, and *monetary* value.

Recency of Purchase

Recency is the most powerful single measure. Someone who has just purchased your product/service is much more likely to purchase again than someone who has not purchased for several months or years.

TABLE 5.1. Customer Lifetime Value Table

Revenue	Year 1	Year 2	Year 3	Year 4	Year 5
Customers[a]	1,000	400	180	90	50
Retention rate[b]	40	45	50	55	60
Avg. yearly sales[c]	1,500	1,500	1,500	1,500	1,500
Total revenue[d]	1,500,00	600,000	270,000	135,000	75,000
Costs					
Cost percent[e]	50	50	50	50	50
Total costs[f]	750,000	300,000	135,000	67,500	37,500
Profits					
Gross profit[g]	750,000	300,000	135,000	67,500	37,500
Discount rate[h]	1	1.2	1.44	1.73	2.07
NPV profit (NPVP)[i]	750,000	250,000	93,750	39,020	18,120
Cumulative NPVP[j]	750,000	1,000,000	1,093,750	1,132,770	1,150,890
Lifetime value (NPV)	750	1,000	1,093.80	1,132.80	1,150.90

Source: From A. Hughes, 1994, *Strategic Database Marketing* © 1994, McGraw-Hill. Reprinted with permission of The McGraw-Hill Companies.

[a] This line shows a specific group of hospitality customers obtained in one year over a five-year period. It does not include new customers obtained in the following years.

[b] In Year 1, these 1,000 customers purchased from this company. But the retention rate is poor, as this firm retained only 40 percent of those same customers in the second year. But the retention rate gradually improves over the five-year period.

[c] This is the average amount spent by each customer per year. This is computed by dividing the total sales by the total number of customers. However, this chart shows that sales per customer are not growing during this five-year period. It is one of the goals of RM to increase the amount of sales from each customer every year.

[d] This line shows the total revenue derived from customers still remaining from the original group. It is computed by multiplying the customers by the average yearly sale.

[e] Costs can be computed in many ways depending on the particular company. In this case, the company computes costs as 50 percent of revenues. This is multiplied by total revenue to get total costs.

[f] This is the total costs as computed in the previous note.

[g] Gross profits are total revenue less total costs.

[h] This is the most complex part of the analysis. It is necessary to discount future revenues to obtain their net present value (NPV). Here the 1994 market rate of interest is used (10 percent). If this chart were made today, the market rate would be closer to 2 percent. That 10 percent interest rate is doubled to 20 percent to include risks. Risks include the risk of interest rates increasing, competition stealing customers, economic risks, etc. The formula used here to obtain the discount rate is $D = (1 + I)n$ where D is the discount rate, I is the interest rate, and n is the number of years you have to wait. Thus, the discount rate in Year 4 is computed as $D - (1.20)3 = 1.728$.

[i] NPV profits = profits divided by the discount rate. Thus, the NPV of the 135,000 profits expected in Year 3 is 93,750 (the result of dividing 135,000 by the discount rate of 1.44).

[j] This line adds the NPV of all the profits in the present year and each previous year. Thus, the NPV of the profits realized by Year 3 is equal to the NPV of the profits in Year 1 + Year 2 + Year 3.

Frequency of Purchase

The second most powerful measure is the frequency of purchase by the customer. Customers are often divided into quintiles based on the frequency of purchases. Frequency measures are often combined with recency measures to properly understand customer purchase behavior. The most dependable group is the "55" quintile group, which are the customers who are the most recent and the most frequent purchasers (Hughes, 1994).

Monetary Value

Monetary value is calculated by adding up the total dollar amount of purchases since the customer entered your database. Then customer records in the database are sorted according to monetary value. Customers are divided into quintiles according to monetary value, and the quintile becomes part of the customer record (Hughes, 1994).

Clickstream Analysis

Clickstream analysis refers to patterns of mouse clicks from responses to Web sites. The goal is to increase "conversion rates," or the percentage of browsing customers who become actual buyers. Companies such as Blue Martini and Net Perceptions sell software that allows Web-based stores to customize Web sites in real time to the type of customer that is visiting the site based on the previous purchase patterns of the customer (Winer, 2001). Previous customer patterns are derivable from Web site visits recorded on company databases and hidden "cookies" left on customer computers.

CUSTOMER RELATIONSHIP MANAGEMENT (CRM)

CRM is the current stage of relationship marketing. It strives to optimize target customer relationships through computer-based tools (e.g., databases and CRM software). The goal is to maximize long-term return on investment (ROI) by selective customer retention. Potentially profitable customer relationships are enhanced by CRM, while unprofitable relationships are discouraged. A typical CRM in-

stallation costs between $500,000 (medium-size company) to $3 million (large company). Sales of CRM software were $25 billion dollars in 2001 and were expected to double by 2006 (Bergeron, 2002). The critical importance of high customer retention and the managing of customer expectations, the basis of high customer satisfaction ratings, is well known. CRM is designed to increase customer retention of the firm's best customers, manage expectations of different categories of customer, and support a consistent integrated image of the firm.

Customer Contact Points

Common customer contact points included in CRM systems include

- e-mail,
- fax,
- mail,
- advertisements and promotions,
- personal contacts,
- telephone,
- Web sites, and
- wireless versions of phones, the Internet, e-mail, and Web sites.

Customers expect the company to remember them and their past communications, orders, complaints, and preferences when they interact with any of these contact points. CRM allows the firm's customer contact people to access this data through either the firm's databases or a central database. The goal is to increase the firm's long term ROI by increasing the loyalty of the firm's best customers. This means

- increasing the perceived value of the firm's services for its target customers.
- increasing target customer share of business.
- increasing switching costs.
- Increasing bonding with target customers.
- increasing convenience of purchase.
- minimizing problems with the purchase process (Bergeron, 2002).

Managers often talk about the 80/20 rule. For example, 80 percent of the firm's profits often come from only 20 percent of its customers. And 20 percent of the firm's salespeople usually account for 80 percent of sales revenue. This is especially true in the service areas of banking and brokerage services (financial services), and casinos where 20 percent if the firm's customers may account for 80 percent of the firm's profits. Thus, it is especially important to increase perceived value for these preferred customers. This may be accomplished by enhanced customer services to these preferred customers. Banks and casinos often assign specially trained personal account representatives to handle their most profitable customers. CRM systems allow firms to provide customer service in proportion to the profitability of the customer.

It does this by using CRM systems to categorize the customer in terms of profitability. Then more profitable customers obtain increased levels of customer service. Unprofitable customers and unprofitable transactions are discouraged by CRM systems. This may translate into higher fees for these unprofitable customers and transactions or it may mean greatly reduced customer services.

Vendors and Practices

Leading vendors of CRM systems include Siebel Systems (the industry leader) as well as Peoplesoft, Oracle, SAP, and Nortel Networks. These companies provide software tools for increasing employee productivity in sales, service, and customer management. For example, customers entering AT&T's call centers are automatically identified by phone number and market segment and routed to the appropriate customer segment personnel before the calls are even answered. The same software allows the employees to view the entire account history of the caller at the same time as the incoming call. Call center personnel have access to the various plans and options offered by AT&T and are empowered to make decisions, answer questions, and resolve problems. By customizing and improving service to different customer segments, the CRM system allows employees to be more productive, build relationships with customers, and retain the firm's best customers. See Exhibit 5.2 on the Guestware CRM system for the hotel industry.

EXHIBIT 5.2. The GuestWare CRM System for Hotels

Scope	Found in over 700 hotels including Marriott, Kimpton, Hyatt, and Starwood hotels.
Purpose	Increased guest satisfaction, lower operating costs, and increased revenues.
Features	
Rapid response	Streamlines service delivery for hotel call canters. Provides fast way for staff to long on, dispatch, and follow up on guest requests and problems while building a guest database of incidents.
Incident tracking	GuestWare Incident tracking automates and organizes log book information, collects data, and improves follow up.
Guest recognition	GuestWare guest recognition system helps retain valued guests by recording guest information and segmenting and tracking the firm"s best customers.
Comment card tracking	Using guest surveys, the GuestWare comment card system manages customer feedback and follow up.
Two-way messaging	The GuestWare communications server integrates two-way messaging with GuestWare's rapid response system. This allows staff to receive and close requests and maintenance issues from a pager or cell phone. This software uses existing e-mail systems or a modem to communicate with messaging devices.
Improvement analysis	GuestWare"s improvement analysis provides you with information to eliminate recurring problems which lowers operating costs and improves customer satisfaction and retention.

Source: Adapted from www.hotel-online.com/Trends/GuestWare/.

Limitations of CRM

However, these results cannot be expected to be derived from the purchase and use of a CRM system alone. A CRM system is only as good as the customer contact employees who use the system. Unless the culture of firms like AT&T and Verizon are changed to that of a

customer-oriented firm with the full example, support, and funding of top management, the rollout of an expensive capital and training intensive CRM system is doomed to be an exercise in futility.

For example, what good does it do to provide state of the art CRM technology to call center personnel who are rude, do not listen, and think that that customers are adversaries. (Some of you may remember having such experiences with call center personnel who are not averse to hanging up on customers.) And the current trend to outsourcing call centers and customer service centers to third world countries such as India is often a case of trading short-term savings for long- term losses. This is due to decreased customer satisfaction associated with call centers that do not have access to current records, Intranet downtimes, language problems, and inexperienced personnel.

Employee Relationship Management

Industry expert Robert Lewis (1989) has noted that the most important element of the hospitality industry (confirmed by ongoing customer surveys) is the manner in which the "goods" of the hospitality industry are delivered, sometimes called *service quality.* But service quality is dependent on the relationship between the service providers (hospitality employees) and their customers. The goal of employee relationship management is to create happy, motivated, and empowered employees who have excellent attitudes, believe in their company, and are customer focused. These employees are better able to deliver high standards of service to your customers.

There is no more important aspect of relationship marketing in the hospitality industry than employee relationship management, often called *internal marketing.* You cannot expect your customers to be very happy or satisfied if they are being served by unhappy employees. Sheraton Hotels found that its employee satisfaction index accounted for 50 percent of the variance in its customer satisfaction index, which in turn accounted for 50 percent of the variance in its revenue per available room (REVPAR) (Lewis & Chambers, 2000). And good relationships with employees, suppliers, and guests involve trust and commitment.

Research by Bowen and Shoemaker (1998) found that the leading responses to the question what does it mean to have trust in a luxury hotel were as follows:

1. The hotel does things as promised
2. The feeling that my personal property is safe in my room
3. If I receive a fax, I know it will be delivered to my room
4. Employees provide quick and correct answers
5. I trust a hotel that trusts its employees.

Internal marketing is not restricted to customer contact people but refers to all the employees of a service firm. For example, Marriott empowers all its employees to spend up to $2,000 each to repair problems with its guests. The employee may be a housekeeper or janitor who is engaged in repairing this problem to the guest's satisfaction.

I remember staying at the San Francisco Marriott for a marketing convention a few years ago. During the night the small refrigerator in the room developed a gas leak, which emitted a toxic odor. The front desk was notified and the assistant manager promptly visited my room and personally supervised a work crew who pumped the air out of this room. My room was immediately changed during this process. They also refunded half of the price of my entire stay and asked me to send them any doctor bills, saying that they would pay any medical expenses related to this incident. They also paid for a free breakfast. It was my impression that they did all that was possible to rectify this unfortunate situation to my satisfaction. And it was done quickly and cheerfully by the manager on duty. This is one example of how happy and motivated employees influence customer satisfaction and retention at their hotel. It is also an example of how effective complaint management can strengthen the relationship between your company and its customers. It is also an example of how to convert critical incidents to a positive bonding experience with your customers.

Hospitality firms must compete for the best internal customers (employees) as well as the best external customers. This may include offering better compensation, more attractive amenities (benefits and awards), flexible schedules, better training, and more opportunities for advancement, or more challenging positions (Lewis, 1989). But it also may mean better communications with employees. Employees need not only to be fully knowledgeable about their products, but also

believe that they are highly valued by the firm and that the company is doing its best to meet their needs and expectations. This knowledge, along with employee empowerment, will enable hospitality employees with the right attitude to better serve their customers.

But employee relationship management cannot be successful without a change in corporate culture of the hospitality firm. RM programs require employees and managers to be customer oriented and require the strong support and example of upper management in order to develop a culture that supports this program. This may require radical changes in the attitude and behavior of all employees and policies, training, and compensation that encourage and are consistent with this change.

For example, Ritz-Carlton, known for its legendary customer service, is famous for opening up new hotels with all new employees and providing them with a ten-day orientation by top management and an extensive training program prior to opening. They use principles of TQM (total quality management) and internal marketing to create an optimal service culture for the hotel that is monitored daily. The success of this program is related to the fact that they avoid many of the problems of cultural change by starting with a new hotel and new employees. And Outback steakhouse is well-known for its hiring of servers who are people oriented (who get along well with people, like their job, and who are liked by their customers).

I am reminded of the well-known anecdote about the president of Marriott hotels, Bill Marriott, who asked potential hotel managers about what was the most important part of their job. Most of them talked about keeping the customers happy. But the "correct" answer was about keeping the hotel employees happy and satisfied, because the president of Marriott was well aware that a precondition of customer satisfaction is employee satisfaction.

Transactional Marketing versus Relational Marketing

Many firms have found transactional approaches to be more appropriate for marginal and transient customer types and relationship marketing approaches to be more appropriate for high profitability and loyal customers. The differences between the two marketing approaches are illustrated in Table 5.2.

TABLE 5.2. Differences Between Transactional and Relational Marketing Programs

Dimension	Transactional Perspective Transaction marketing	Database marketing	Relational Perspective Interaction marketing
Purpose	Economic transaction	Information and economic transaction	Interactive relationships between a buyer and a seller
Nature of communication	Firm to mass market	Firm to individual	Individuals with individuals (across organizations)
Type of contact	Arms length, impersonal	Personalized (yet distant)	Face to face, interpersonal (close, based on commitment, trust, and cooperation)
Duration	Discrete (yet perhaps over time)	Discrete and over time	Continuous (ongoing and mutually adaptive, may be short or long term)
Formality	Formal	Formal (yet personalized via technology)	Formal and informal (i.e., at both a business and social level)
Managerial intent	Customer attraction (to satisfy the customer at a profit)	Customer retention (to satisfy the customer, increase profit, and attain other objectives such as increased loyalty, decreased customer risk, etc.)	Interaction (to establish, develop, and facilitate a cooperative relationship for mutual benefits)
Managerial focus	Product or brand	Product/brand and customers (in a targeted market)	Relationships between individuals
Managerial investment	Internal marketing assets (focusing on product/service, price, distribution, promotion capabilities)	Internal marketing assets (emphasizing communication, information, and technology capabilities)	External market assets (focusing on establishing and developing a relationship with another individual)
Managerial level	Functional marketers (e.g., sales manager, product development manager)	Specialist marketers (e.g., customer service manager, loyalty manager)	Managers from across functions and levels in the firm

Source: Adapted from Coviello, Brodie, & Monroe, 1997.

SUMMARY

RM has been defined as a marketing program designed to build a long-term, one-to-one, interactive, mutually profitable relationship with the firm's best customers. The goals of this program include increased customer retention, increased revenues from these customers, increased customer satisfaction and loyalty, and increased profitability for the firm. The key to building successful customer relationships lies in developing a greater understanding of the specific needs of target customers and the delivery of superior perceived value to these customers while enhancing profitability.

Hospitality firms have developed relationships not only with their employees and customers but also with their suppliers, travel agencies, tour companies, corporations, government agencies, banks, law firms, and other travel companies. Frequent guest programs are common in the hospitality industry but are usually transactional in nature. They are now viewed as an expected benefit by customers rather than a unique benefit that supports a competitive advantage for the firm. Nevertheless, they do involve switching costs for customers and thus have a positive impact on customer retention. For example, an Advantage account with American Airlines leads to free flights based on total accumulated miles. As an Advantage customer, I would usually prefer to book my flights on American Airlines to build my frequent flyer points. However, frequent flyer programs are now available for all major carriers and price, convenience, and service quality will often decide which airline is chosen for frequent flyers.

Casinos are clearly the most extreme example of RM, where extravagant perks are distributed to customer *whales* (the most lucrative casino customers who may spend up to millions of dollars in a few days at the gambling tables). Extensive data about these whales are recorded on company databases, including their gambling activities, preferences, dislikes, special requests, personal information, and so forth. These perks include free first class airline tickets and limousines, free accommodations in expensive hotel suites, gourmet meals in exclusive restaurants, and a special host whose job it is to keep these lucrative customers happy.

CRM, which involves software automation of RM functions, is now an industry of over $32 billion per year. CRM enables employees, when interacting with guests at customer contact points, to ac-

cess personalized data through the firm's customer database. The goal of CRM is to increase return of investment (ROI) by increasing the satisfaction and loyalty of the firm's best customers. This is accomplished by allowing firms to provide customer service in proportion to the profitability of the customer. Examples of leading CRM vendors are E.Piphany, Siebel Systems, Peoplesoft, Oracle, SAP, and Nortel Networks.

It is important to note that RM programs are not appropriate for all customers or all hospitality firms. RM is usually a capital-intensive program that requires expensive customer databases, employee training, a market-oriented company culture, and the full support of upper management. It should be justified on an ROI basis for selected customer segments and is usually reserved for the most profitable customers of the firm. RM programs are not considered appropriate for transactional customers, unprofitable customers, or where the program's expense is not justified by increased returns from target customers over the first few years.

EXERCISES

1. Relationship marketing is not appropriate for all hospitality customers. Do you agree with this comment? Explain why or why not.
2. Is it more important to build mutually beneficial relationships with hospitality customers or the employees of a hospitality company? Explain.
3. What metrics would you use to measure the success of an RM program for a given hospitality company? Justify your answer.
4. Explain what is meant by "lifetime value of the customer" and how this is related to RM.
5. You are the new marketing director of a luxury hotel that is trying to implement an RM program. You find that the company culture is not market oriented and not supporting this program. What do you do and why?
6. Explain how the company's Web site can be used to support the company's RM program. Be specific and use concrete examples.
7. Which is more important to having better employee support of your RM program: regular employee training or having an em-

ployee screening process that results in the hiring of only friendly, customer-oriented contact people? Justify your answer.

8. RM programs are expensive to implement, and it may take years to accomplish the retention and revenue goals of such a program. At a meeting of management of your hospitality company, it is proposed that your RM program target only the firm's most lucrative external customers, excluding employees, suppliers, and banks. What is your reasoned response to this suggestion?

9. Customer relationship marketing existed hundreds of years ago when the owner/managers of town general stores knew personally their customers and their needs and developed long-term relationships with these customers. What is different about the RM practices of hospitality managers today, and is this a positive or negative change? Explain.

10. It has been said that hospitality services are undergoing increasing commoditization and that today's hospitality customers are opportunistic and will choose whatever company gives them the best deal at a given time. If this is true, then why not save money by not investing in an expensive RM program that may not be supported by your employees and that will take years to show any returns even if it is successful? Explain and justify your position.

Chapter 6

Special Issues in Segmentation

Once your firm has completed its marketing research, you must determine with whom you compete, forecast demand, and decide from which segments you intend to cultivate business and to what degree. All of this information—and more—should be part of your marketing plan. (Yesawich, 1988)

THE EVOLUTION OF MARKET SEGMENTATION

In this chapter we will discuss the integration of segmentation strategy with the special issues of yield management, branding strategy, and information technology (IT).

Market segmentation has evolved from the a priori *product segmentation* that was evidenced in the hotel industry in the 1980s and to a large extent continues today. Table 6.1 shows a listing of U.S. hotels by price/quality segments, which is an example of product segmentation.

But many hotels now use *market research based segmentation* to enhance existing properties and build new brands that are customized to target segments. For example, the Marriott's *Courtyard* brand was a hotel created on the basis of market research into the needs of independent business travelers who were not on an expense account and therefore needed reasonably priced accommodations with business facilities, clean comfortable rooms, and an in-house restaurant. Its subsequent success with this segment has fostered several copy-cat brands.

Due to the increased emphasis on research-based segmentation, major hotel chains are now pursuing the high growth segments of the

Segmentation Strategies for Hospitality Managers
Published by The Haworth Press, Inc., 2007. All rights reserved.
doi:10.1300/5716_06

TABLE 6.1. Hotel Brands by Price/Quality Segment

Upper upscale	Upscale	Midscale+F&B	Midscale-F&B	Economy
Embassy Suites	Adam's Mark	Best Western	Comfort Inn & Suites	Econolodge
Four Seasons	Courtyard	Clarion	Fairfield Inn	Knights Inn
Hyatt	Doubletree	Holiday Inn	Hampton Inn & Suites	Microtel
Marriott	Homewood Suites	Howard Johnson	La Quinta	Motel 6
Ritz-Carlton	Radisson	Quality Inn & Suites	Springhill Suites	Super 8
Weston	Residence Inn	Ramada		Travelodge
	Wyndham			

Source: Adapted from O'Neill, 2006.

ultra-luxury and value-oriented business travelers with new brands and sometimes the enhancement of existing brands.

But, as mentioned earlier, the targeting of market segments has evolved into relationship marketing (RM). RM is a program designed to maximize revenues and customer retention by building a one-to-one, long-term, mutually beneficial, interactive relationship with the firm's best customers. This involves segmenting the firm's customers so that different relationships are designed for different types of customers. The most desirable customers are targeted for more involved and beneficial relationships (which are more expensive), while the least profitable customers may be targeted for transactional, or minimal, relationships and may be encouraged to leave.

Hospitality Company Segmentation Strategies

U.S. hospitality companies commonly engage in one or more of the following segmentation strategies:

1. Price-quality segmentation (see Table 6.1)
2. Standard product segmentation (business, leisure, and transient travelers)

3. Small niche marketing (hotels that cater to tall people, the environmentally conscious, cigar smokers, vegetarian restaurants, culture seekers, etc.)

4. Large niche marketing (companies that cater to families with children [McDonald's], business women, business travelers [Hilton Hotels], vacation travelers, seniors, health-conscious customers)

5. Segmenting by profitability measures (such as RFM analysis and lifetime value of the customer) for special relationships.

6. Geodemographic segmentation.

7. Lifestyle segmentation (Sandals resorts, Jamaica's *Hedonism* resorts, hotels that cater to gays, etc.)

8. Low price customer segments (discount airlines)

9. Ultra segment focus: Ultra-rich and powerful, high margin, prestige conscious, customers (Ritz-Carlton, Aladdin's London club casino)

10. The use of all-inclusive product segments providing insulated, no stress environments (spas, all-inclusive destination resorts, cruise ships .)

11. Segmentation based on statistical techniques such as cluster analysis and data mining.

12. A multisegment strategy designed for extended market coverage. Some hospitality companies serve many segments at once such as Marriott International (through its many brand extensions which are usually segment specific), Holiday Inn (which services the major product segments of business travelers, families with children, and leisure travelers), restaurants that service a large variety of food catering to many segments such as diners, and online travel agencies that service customers for many hospitality travel needs such as airline reservations, hotel accommodations, rental cars, and packages (Expedia, Travelocity).

But the process of segmentation is not insulated from other marketing and functional processes of the firm and does not stand alone from these processes. An understanding of the interactions with these other processes leads to an integrated view of segmentation for the hospitality firm. Segmentation issues that will illustrate this point follow:

Integration of Segmentation Strategy
with Yield Management

Yield management is form of discriminatory, demand-based, pricing, used by hospitality firms. It is usually aided by a software system which suggests prices based on variation in the price elasticity of different customer segments, inventory, and actual and forecasted demand. A yield management system results in different prices for different customer segments, depending on demand factors and inventory status. In the hotel sector, yield management is an attempt to maximize room rates when demand exceeds supply, and maximize occupancy when supply exceeds demand (Jones & Hamilton, 1992).

Yield management is the process of allocating the right type of capacity to the right kind of customer at the right price, yielding maximum revenue. However, this creates a trade-off situation. In the hotel industry, the 2 separate yet related parts of yield management are pricing and room-inventory management. Although this process started in the airline industry, its techniques are appropriate in other industries as long as certain characteristics are met. A good yield-management system depends upon an excellent information system to track critical aspects of the business, such as: 1. booking patterns, 2. demand patterns by market segment, and 3. effects of price changes. When hotels use yield-management techniques, they must address additional problems of multiple-night stays, the multiplier effect of rooms on other hotel functions, and decentralized information systems. A firm commitment from top management is essential in making the system work. Issues dealing with employee problems and incentives along with alienated customs must be attended to by management. (Kimes, 1989)

Hospitality firms have traditionally used price segmentation to charge different prices for different customer segments to increase overall revenue and profitability. What makes the airline and hotel industry ideal for yield management are the properties of having relatively fixed capacities, where demand can be segmented into clearly identified segments, where the product is perishable, where inventory is sold in advance, where demand fluctuates substantially, and when

marginal sales costs are low but marginal production costs are high (Hanks, Cross, & Noland, 1992; Kimes, 2003).

Let's consider the airline industry, which was the first hospitality sector (in 1977) to initiate yield management. Charging different prices to business travelers and leisure travelers based on the customer's ability to book in advance and their price sensitivity is an example of yield management. This isolates the business traveler customer (whose timetable is less flexible and who is less price sensitive) from the more time-flexible and price-sensitive leisure traveler. Hotels often have different rates for corporate customers, senior citizens, and those able to book weeks in advance.

Another example of yield management in the airline sector is the special "Worry Free" senior tickets currently issued by Cathay Pacific airlines (from New York to Hong Kong and points east). These tickets are not only discounted but have no restrictions and are fully refundable without penalties. Passengers must show proof of age (fifty-five and older). Tickets are good only for economy class service and only for one destination beyond Hong Kong. Worry free senior fares do not qualify for frequent flyer miles.

These restrictions make it difficult for most of the other higher margin airline segments to take advantage of this service. Most business travelers are either not old enough, would not be traveling coach class, or would need to make stops not covered under the restrictions of this ticket. On the other hand, this promotion increases occupancy rates using seats that are not already booked by passengers paying higher rates. However, requiring the time-flexible seniors to book at least thirty days in advance may go further to minimize segment leakage from the business traveler segment.

But the problem in the hotel industry is that rates are often negotiable and loopholes exist that allow higher priced segments access to lower rates which are intended for low price segments. This practice defeats the purpose of price segmentation and yield management, which are designed to maximize revenues by charging higher prices to segments which are less flexible and price sensitive.

The following is an illustrative example of a typical call placed to the reservations office of an independent hotel.

CALLER: I'd like to book a room for next Friday.
FRONT DESK: Our single rate is $250.

CALLER: That is a little high for me. Anything less expensive?

FRONT DESK: You could try our corporate rate of $200.

CALLER: Still beyond my budget!

FRONT DESK: What about our special promotions rate this week for $150?

CALLER: Do you have anything less than that?

FRONT DESK: Well, we do have one room left at the super value rate of $105.

CALLER: I'll take it.

This problem of negotiable rates has only been exacerbated by the advent of Internet travel sites like Expedia, Travelocity, Hotwire, and travel search engines, which compare the prices of multiple online travel sites for a given reservation. These search engines leverage the customer's ability to find the cheapest reservation, thus reducing margins for hospitality companies.

Price Fences

It is generally agreed that "price fences" are needed to ensure the integrity of yield management goals of price segmentation (Hanks et al., 1992). For example, airlines impose restrictions that minimize the ability of high price segments, like business travelers, to access the low price fares reserved for low price segments like leisure travelers. These restrictions, or "fences," include advance booking requirements, nonrefundable fares, and Saturday-stay bookings.

Taking advantages of these differences, hotels expand the price elastic segments through innovative packages and promotions while protecting the inelastic segments by restrictions that are guided by the differences above. The idea is to replace the somewhat arbitrary negotiation practice of hotel pricing with a rational pricing system supported by restrictions designed to separate the price elastic from the price inelastic segments. This system should be perceived to be fair to customers and nonnegotiable for individuals. This pricing system can be adjusted for demand and inventory factors but is the same for everyone at a given point in time. This is more like the pricing plans of various cell phone companies, which have different plans for different segments but which are not negotiable for individuals.

Using a rational pricing system with price fences not only supports the yield management goals of price segmentation but also addresses perceived fairness problems when guests learn that others have obtained the same accommodations and services at greatly reduced prices.

Another example of yield management in the restaurant sector is the "early bird special" used by many restaurants to discount meals for those that eat between five and six o'clock. This appeals mainly to the senior citizen segment of the market who is more price sensitive and usually eats at an earlier time. It also has the advantage of increasing occupancy during what would normally be low occupancy periods, allowing greater capacity for later times when the prices (and margins) are higher.

Successful price discrimination is associated with the following conditions (Hanks et al., 2002; Kimes, 2003; Kotler, 2003):

1. Relatively fixed capacity, perishable inventory, advanced reservations, low variable costs, high fixed costs, variable demand patterns, ability to forecast future demand, and ability to segment customers on their varying needs behavior, and willingness to pay.
2. Different customer segments must have different price elasticities for the same service product.
3. The different segments must be identifiable and a mechanism must exist to price them differently based on inventory, demand, and price sensitivity.
4. There should be fences established that prevent customers in higher priced segments from obtaining the prices reserved for lower priced segments.
5. The segments should be large enough to make the use of the system cost-effective.
6. The costs of the system should not exceed the incremental revenues obtained by using the system.
7. The system should be easy to understand and fair, so that customers will not be confused or upset by the differential pricing.
8. Prices should be consistent across distribution channels. Thus, the customer should ideally obtain the same price for a given reservation whether booking directly or using a travel inter-

mediary. It is customary to receive a small discount for the cost savings to the hospitality company for booking on the Internet, but prices should be the same for a given reservation from all Internet sites.

9. There should be an accurate system of demand forecasting.
10. Reservations personnel need to be well trained in revenue management policies and conflict resolution.

Branding Strategy

A brand is a collection of promises in the consumer's mind represented by a name or symbol. It is about the perceived quality, value, and service expectations associated with the name of your company, product, or logos. A strong brand is perhaps the most potent weapon in the hospitality marketer's arsenal in the battle for competitive advantage. Ten advantages of having a strong brand are as follows:

1. A strong brand is a powerful incentive for potential customers who have never purchased your services.
2. It allows customers to feel more secure about purchasing your services because of the consistency, dependability, and reputation of your brand.
3. It is helpful in attracting and managing different customer segments.
4. It allows for more successful brand extensions and new product development.
5. Customers will often pay a premium for a strong brand, especially if it is a prestige brand.
6. It facilitates profitable chains, franchising, and cross marketing opportunities with a global scope.
7. It facilitates target customer retention (brand loyalty) and positive word-of-mouth advertising.
8. It offers some protection against switching behavior. Customers are less likely to switch because of lower prices or a special promotion.
9. It enables the creation of a unique identity that is well-known to target segments.

10. It enables more effective and cost-effective advertising, promotions, purchasing (associated with economies of scale), and hiring of key personnel (attracted to successful brands).

Examples of strong hospitality brands in the U.S. hospitality market include

- McDonald's (represented by the ubiquitous Golden Arches logos), which targets families with children by providing low price, tasty fast food, in a clean environment;
- Motel 6 (which is "keeping the light on" for us), which provides basic, low-cost services for its low-budget, price-sensitive customers);
- Ritz-Carlton, a prestige hotel chain that targets the ultra rich and powerful, with unsurpassed facilities and service quality;
- Southwest Airlines, which targets the leisure, price-sensitive traveler with friendly, low-priced service; and
- Princess Cruise Lines, which has distinguished itself by offering flexible and affordable cruises to the most destinations of any cruise line. This includes multiple dining locations, flexible entertainment, and affordable private balconies. Among their multiple target segments are honeymooners, entrepreneurs, corporate executives, adventure traffic, and families.

Integration of Segmentation Strategy with Branding Strategy

Segmentation strategy should be well integrated with branding strategy, including the use of brands which target specific, actionable, customer segments that have favorable forecasted market trends and demand projections. The firm must deliver on the *drivers* of target customer segments (which provide the firm with its competitive advantage) while at the same time providing those features *(satisfiers)* expected by target customers but do not differentiate the firm from its competition (McKinsey & Company, 2002).

For example, recent research indicates that that the relatively new segments of the value-driven business traveler and the luxury-driven business traveler (Table 6.2) will erode segments related to the traditional service-oriented business travelers. This would threaten brands

TABLE 6.2. Business Travelers versus Leisure Travelers

Leisure Travelers	Business Travelers
Are time flexible	Are time inflexible
Focus on price and value	Focus on quality, comfort, and prestige
Destination flexible	Destination inflexible
Location flexible	Location inflexible
Price elastic	Price inelastic
Longer stays	Shorter stays
Do not need business facilities	Need business facilities (fax, Internet access, secretarial service, interpreters, etc.)

Source: Adapted from Hanks et al., 1999.

such as Marriott, Sheraton, and Hilton (McKinsey & Company, 2002).

A McKinsey & Company study found that existing brands were appropriate for the value-driven business segment by offering value rates or by offering value subbrands such as Four Points by Sheraton or Courtyard by Marriott. But it was also found that the luxury business segment demanded specifically targeted brands such as Ritz-Carlton, W Hotels, or Four Seasons (McKinsey & Company, 2002).

The cornerstone of this new brand strategy (aligned with segmentation strategy) is the fusion of customer insights, demand projections, and an analysis of organizational capabilities. This strategy allows companies to focus on segments that will thrive as industry economics evolves, and develop superior selling propositions with distinctive benefits and drivers while over delivering on triggers that matter (McKinsey & Company, 2002).

The Integration of Segmentation Strategy with IT Strategy

It is well-known that segmentation applications in the hospitality industry (e.g., relationship marketing, customer relationship management [CRM], yield management, and Internet marketing) require expensive integrated databases and computerized operational systems.

These include decision support systems, property management systems, yield management systems, global distribution systems, integrated reservation systems, and integrated customer databases. But these different systems need to be integrated with one another (so they can communicate and share information), and also with segmentation strategy, to maximize effectiveness and cost-effectiveness in the achievement of segmentation and IT goals.

Effective implementation of segmentation-based marketing strategies requires integration of segmentation marketing strategy with IT strategy on several levels:

1. Revenue and yield management systems need to be integrated with segmentation strategy. This is because yield management is based on differential pricing for different segments who have different price elasticities and different priorities. Changes in target segments, their price elasticities, and priorities, as well as competitive offerings, need to be continuously updated into the enterprise-wide database system which supports the yield management system. Otherwise your hospitality company will be vulnerable to competitive companies whose yield management systems offer better value to your firm's most profitable customers (Clemons & Weber, 1994).

2. Relationship marketing and customer management programs (CRM) are in wide use in the hospitality industry and are especially important to high value, return guests (i.e., usually business guests who are less price sensitive and more quality sensitive) and who represent a disproportionate share of the profits to the company. RM programs are designed to provide personalized interactive services to these most profitable customers and deliver superior value leading to higher customer retention. CRM software is designed to automate parts of this process.

3. This personalized service to target customers requires up-to-date, enterprise-wide, comprehensive, guest databases which enable customer contact personnel and hospitality managers to input, measure, and anticipate guest preferences. This in turn allows the company to exceed target customer expectations by the delivery of services customized to their current needs and preferences Thus, the segmentation strategy of relationship marketing requires integration with IT strategy (including effective

development and implementation of enterprise-wide guest databases and information gathering procedures) to leverage the ability of customer contact personnel and managers to effectively deliver these personalized services to target customers.

4. Thus, I may be pleasantly surprised to find that a hotel property that I have never before visited asks me about my children by name and has sitters available for them, has my favorite paper delivered to our room, has the honor bar stocked with our favorite items, and delivers flowers and a complimentary bottle of champagne to our room on the occasion of our anniversary! They are also aware of my dietary restrictions in the hotel restaurant and provide me with a special low-carb menu. Needless to say, our room is equipped with high-speed Internet, a portable computer, and a fax machine used in the past for business communications while I am on the road so I do not need to carry any equipment (except a portable hard drive) with me.

5. The effective use of IT systems to effectively leverage service to target segments on an enterprise-wide basis. This is associated with current problems in the hospitality industry related to the efficient and cost-effective use of technology.

Problems with IT Implementation

Between 1995 and 2002, the U.S. hotel industry spent US$8 billion on information technology, which was more than 9 percent of the pretax profits during this period. Expenditures during this period have increased by 15 percent every year, increasing IT investment per room by almost 75 percent. The majority of spending was on property management systems, central reservation systems, and the interface between the two, with the remainder including customer relationship management systems, revenue management, and electronic distribution through the Internet. IT facilitates most activities in selling room nights, including marketing, pricing, inventory control, booking, and guest check-in and checkout (Brown & Stange, 2002).

However, it was found that companies are required to spend on IT just to maintain competitive parity, and there was little evidence that this massive spending has increased industry profitability. Furthermore, it does not seem to have been a source of competitive advantage either. Companies designated as IT champions by Cornell Uni-

versity (the number one ranked hospitality school in the United States) and *Internet Week,* on average, were not able to increase REVPAR (revenue per available room) faster than competitors. The major reasons cited for these suboptimal returns from IT were found to be the following (Brown & Stange, 2002):

1. The difficulty experienced by hotels in charging for the consumer benefits associated with technology improvements. Although segmentation strategies like relationship marketing, CRM, and yield management have been built on capital intensive data warehouses, property management systems, and enterprise computer networks, these capital-intensive investments have often not resulted in greater profits or even a competitive advantage for these hospitality companies. But they have certainly increased customer expectations. Thus, world class services provided by IT improvements are now viewed as mandatory satisfiers to compete in an increasingly competitive hospitality market. However, these improvements provide no sustainable competitive advantage or increased profitability.
2. A highly fragmented industry structure that hampers efficient investment. Approximately half of the total IT spending by hotels is fixed and independent of the number of properties in the chain. Each chain designs and builds proprietary or highly customized IT systems to address the same functional needs. Thus, decades of redundant IT investment have resulted in financial inefficiencies.
3. IT systems are often not well integrated with PMS systems and legacy systems (Brown & Stange, 2002).

In addition, training and implementation costs in a highly distributed IT environment, like the hotel or restaurant industry, can often equal or exceed the cost of software development. Given that the majority of hotel IT investment is currently spent on PMS (property management system) and CRS (central reservation system) enhancements, the training and implementation costs for these projects is expected to remain a significant investment in the foreseeable future (Brown & Stange, 2002).

Finally, there is the problem related to current outsourcing of IT-related projects to cheap labor countries like China and India, in the

attempt to lower costs. But recent evidence indicates that there are also significant risks in outsourcing IT projects. Shoddy quality, security problems, and poor customer service often wipe out any intended benefits from this practice. Published studies by Forrester Research Inc. and Gartner Inc., based on a study of 219 clients who outsourced projects offshore and domestically, indicate that half of the projects undertaken in 2003 will fail to deliver the anticipated savings. And this does not take into account the negative effects of word-of-mouth advertising by dissatisfied customers which expand geometrically. (One study found that dissatisfied customers complain, on average, to nine other people).

But there are also problems with the intranet communications with the outsourced companies, language problems with outsourced customer service systems, lack of adequate training by outsourced personnel, and erosion of perceived service quality and brand name as a result of such problems.

Recommendations

1. Prioritize and target against highest potential customer segments. Understanding the relative value that customers place on different services is critical when considering hotel technology investments. For example, business travelers will not place the same value on high-speed Internet access as vacation travelers. Hotels are advised to prioritize in which properties technology services should be employed and what types of rooms (club floor, business rooms) within each property. This is related to cost-effective distribution of IT for serving the most profitable customer segments (Brown & Stange, 2002).
2. Avoid the enterprise launch of new IT systems and wait until new IT systems have been properly tested and used in a pilot study first. This will minimize problems related to bugs and incompatibility problems with existing (legacy) systems, as well as problems with the acceptance of this system with hospitality employees. Also, systems should be adopted which are appropriate to individual properties and the needs of their target segments and the employees at this property.
3. Ensure that yield management systems are integrated with the segmentation strategy of your property. Yield management sys-

tems should be supported by an enterprise database that is continuously updated with regard to the current behavior of target segments, including their preferences, price sensitivity, and current demand forecasts.

4. Measure and monitor the effectiveness of your IT systems in delivering the stated goals of your company's segmentation strategy. Your IT systems should be leveraging the segmentation goals of your company, including target customer and employee needs, and increasing target customer retention. Ensure that they are focused on these needs and are cost-effectively delivering the segmentations goals of your company. This can be done only by regular measurement and monitoring the effectiveness of these systems in satisfying these goals. This includes periodic market research into the perceived value of these systems to these target segments and employees.

5. It is recognized that IT issues are not within the "comfort zone" of most hospitality managers and they tend not to ask hard questions or challenge answers. But the basic business questions remain the same: What is the project ROI (return on investment)? Where will the projected cost savings/revenue increase come from? How will you measure success? What IT companies are signing up to deliver these numbers? What are the risk factors? And (perhaps the most important) what are your alternative courses of action? Perhaps the most fatal error in the IT capital allocation process is the failure to assign accountability for delivering the expected benefits (cost reduction/revenue increases) (Brown & Stange, 2002).

6. Carefully research the proposed outsourcing of IT projects and critical customer service access to companies in third world countries. Understand the high risks involved in third world outsourcing, which often ends in the trading of short-term financial gains for larger long-term losses in customer and employee satisfaction and retention. This is of special concern to the hospitality industry, which is more sensitive to customer service.

7. Engage a cross-functional team to review and help modify the system. It is good practice to engage a cross-functional team guided by a market segmentation consultant to develop the requirements and modifications needed for a specific IT project. This will ensure that the needs of different functional depart-

ments (e.g., marketing, finance, operations, HR, IT, etc.) are met and that the system is focused on the needs of your target segments, your corporate goals, and the needs of your employees.

SUMMARY

It is clear that an effective segmentation strategy cannot stand alone but must be integrated with other functional departments and elements of the company's strategic plan. This means the segmentation strategy should be integrated with yield management (finance and operations), marketing trends, branding strategy, new product development, quality assurance programs (marketing), IT strategy and implementation (IT), and integrated into the company culture including employee buy-in, training, compensation systems, and internal marketing (marketing and HR).

Integration with yield management will yield higher overall profits by maximizing revenues when demand is high and occupancy when demand is low. This creates low price opportunities for price-sensitive segments when demand is low and ensures availability for price-insensitive segments when demand is high. The implementation of price fences ensures that price-insensitive segments like business travelers cannot take advantage of opportunities designed for price-sensitive segments like leisure travelers on a budget.

Integration with brand strategy will strengthen segmentation strategy by aligning the brands of the company with specific target segments that have favorable forecasted market trends and demand projections. This will exploit the added advantages of well-positioned strong brands, which include protection from price switching, greater customer loyalty, more effective advertising and promotions, and facilitated development of chains, franchises, and cross marketing opportunities.

Finally, the integration of segmentation strategy with IT strategy allows for the leveraging of the selection and customized servicing of target segments. This is accomplished by the use of cost-effective technology and information systems customized to the needs of target segments as well as company and employee needs. Effective integration of IT strategy with IT strategy and systems will minimize current problems associated with ineffective IT systems which either do not justify their cost or are not focused on the needs of target seg-

ments, managers, and employees, or cannot be integrated with existing legacy systems.

Like other programs designed to change the company culture, an effective segmentation program requires the full financial and moral support of upper management, who need to be actively involved and lead in the implementation of their company's segmentation and positioning plan. It also requires appropriate, cost-effective, and user-friendly information systems and databases, and the regular measurement, monitoring, and control systems to effectively implement your integrated segmentation plan.

EXERCISES

1. What are the main product and customer segments for your own or a selected hospitality company? Explain the justification for this segmentation.
2. Many hospitality companies use demand-based pricing, which is a form of price discrimination since it favors certain customer segments over others. (For example, leisure air travelers who book in advance may pay much less than business travelers for the same seat). What is the justification for this kind of discrimination and is it fair?
3. What is the effect of the Internet (including travel Web sites and travel search engines) on hospitality pricing? What can be done to prevent further price erosion and the commoditization of services for your hospitality company?
4. Justify the importance of accurate forecasting for branding and pricing decisions. How does this affect segmentation selection and monitoring?
5. What are the advantages of a strong brand in achieving a competitive advantage with your target customers?
6. What is the role of company information systems and databases in the selection and servicing of target segments? Include the advantages and liabilities of one of these systems.
7. Evaluate how well your company's segmentation plan is integrated with its yield management system.
8. What is the branding strategy of your company including its major target segments and how it attempts to achieve a competitive advantage with these segments?

9. What are the *drivers* (features regarded most important in their purchasing decisions) by your major target market segment and how well does your company's IT systems leverage your company's ability deliver on these drivers?

10. What research-based segmentation studies are performed by your company on an annual or semi-annual basis to support your company's segmentation plan? What studies are omitted but are needed for your plan?

Chapter 7

The Impact of Marketing Trends on Segment Selection

> A new product or marketing program is likely to be more successful if it is in line with strong trends rather than opposed to them. (Kotler, 2003)

Marketing trends have changed dramatically in the past three decades. It is important that hospitality companies continuously monitor marketing trends that impact their markets to ensure that target segment and marketing mix selections (customized to the needs and preferences of target segments) are supported by, rather than contrary or neutral to, these trends.

For example, McDonald's recently reversed a negative trend in sales (by acting contrary to a trend toward healthy eating) by offering its health-conscious customers new products with fewer calories and less fat. Club Med spent twenty years and millions of dollars to change its outdated image from one of singles and sex to one of family fun when the trend in the 1980s was toward more monogamous relationships and away from high-risk promiscuity (Whelihan, 1991).

MAJOR MARKETING TRENDS

Some of the major marketing trends currently affecting the hospitality industry are the following.

The Graying of America

The changing age structure of the United States has caused special attention to be paid to the "baby boomers" (aged forty-two to sixty by

Segmentation Strategies for Hospitality Managers
Published by The Haworth Press, Inc., 2007. All rights reserved.
doi:10.1300/5716_07

2006), who are estimated to be about 40 percent of the population and who regard travel as a luxury. The highest growth segment of the population are "mature travelers" (fifty-five and older) who are estimated to number 75 million by the year 2010). Examples of hospitality companies targeting these lucrative markets are Holiday Inn (Preferred Seniors Program) and the AARP travel service (which targets the 23 million members of AARP) (Morrison, 1996).

Changing Household Structure

The proportion of married couples in the United States has changed from 56 percent in 1990 to 53 percent in 2000 as more and more households are composed of single people, people living together outside of marriage, gay couples, older people living alone, divorced people living alone, and single parents. Hospitality services that have targeted this market include the French company Club Med (who pioneered vacations for the singles market), restaurants who specialize in catering to singles, and the famous Hedonism resorts of Jamaica.

The Growth of Hispanic and African-American Segments

The U.S. Census results showed that the population of the United States in April 2000 was 281.4 million. Of the total, 36.4 million, or 12.9 percent, reported their race as black or African American. This segment is projected to grow to 61.4 million (or 14.6 percent of the population) by 2050.

The same census showed that the Hispanic population in the United States increased by 58 percent, from 22.4 million in 1990 to 35.3 million in 2000 (or another 12.5 percent of the population). This segment is projected to grow to 102.6 million in 2050 (or 24.4 percent of the total population). Thus, minority groups will make up about 50 percent of the U.S. population by 2050 (see Table 7.1).

Of the 35.3 million Hispanics in the United States in 2000, the Mexican-origin population numbered almost 21 million persons and was the largest Hispanic group. Puerto Ricans were 10 percent and Cubans were 3.5 percent of the Hispanic population. Half of all Hispanics live in two states, California and Texas (http://www.census.gov/). By contrast, the white population (non-Hispanic) of the United States, which was estimated at 69.4 percent of the population in the

TABLE 7.1 Projected Population of the United States, by Race and Hispanic Origin: 2000 to 2050 (thousands)

Race or origin	2000		2010		2020		2030		2040		2050	
	Number	%	Number	%	Number	%	Number	%	Number	%	Number	%
White alone	228,548	81.0	244,995	79.3	260,629	77.6	275,731	75.8	289,690	73.9	302,626	72.1
Black alone	35,818	12.7	40,454	13.1	45,365	13.5	50,442	13.9	55,876	14.3	61,361	14.6
Asian alone	10,684	3.8	14,241	4.6	17,988	5.4	22,580	6.2	27,992	7.1	33,430	8.0
All other races	7,075	2.5	9,246	3.0	11,822	3.5	14,831	4.1	18,388	4.7	22,437	5.3
Hispanic of any race	35,622	12.6	47,756	15.5	59,756	17.8	73,055	20.1	87,585	22.3	102,560	24.4
White, not His-panic	195,729	60.4	201,112	65.1	205,936	61.3	209,176	57.5	210,331	53.7	210,283	50.1
Total	282,125	100.0	308,936	100.0	335,805	100.0	363,584	100.0	391,946	100.0	419,854	100.0

Source: U.S. Census, March 2004 (www.census.gov/ipc/www/usinterimproj/natprojtab01).

2000 census (195.7 million), is projected to decrease to 50.1 percent by the year 2050 (210.3 million). Hotels have addressed this trend by offering more ethnic foods, providing front desk personnel who are fluent in minority languages, and, in some cases, targeting minority segments with ethnic-oriented advertising.

The Growing Importance of Women Travelers

Women travelers represented 50 percent of the market in the year 2000. Hotels that target this segment should know that women have different preferences than men in their hotel selection and use criteria. Women consider security, room service, cleanliness, and low price to be important, while men were more concerned with the availability of fax machines and suite rooms with separate bedroom and office spaces (Kotler, 1996).

The Trend Toward Healthier Lifestyles

This includes eating healthier foods, exercise, losing weight, and down-aging (adopting a younger lifestyle). Many hotels and resorts offer guests low-calorie meals, spa facilities, exercise rooms, and massage services. Low-calorie meals have been available by request for years on airlines. And although many restaurants offer healthy choices on their menus, there has been in increase in restaurants that specialize in healthy foods. Subway, for example, emphasizes the health value of its sandwiches as a main advertising theme, featuring a customer (Jared) who lost more than 100 pounds eating Subway sandwiches.

The Trend Toward Increased Security for Travelers

Increased security became a critical issue for hospitality companies and their guests following the 9/11 terrorist attack on the World Trade Center and subsequent terrorist attacks worldwide. Surveys found that security was an important issue for hotel guests, and security is part of a service quality measure (SERVQUAL) introduced by Parasuraman, Berry, and Zeithaml. The authors defined security as freedom from danger, risk, or doubt (Parasuraman, Berry, & Zeithaml, 1988).

Hotels have improved their security by using smart cards, electronic security systems, closed-circuit television systems, employee training, increased use of security personnel, and emergency preparedness and crisis management plans. Security is a special concern for hotels such as Ritz-Carlton who target celebrities, dignitaries, and corporate CEOs, all of whom are at greater risk for personal terrorist attacks). The efforts of airlines in approving security after 9/11 are well-known and include luggage x-rays, personal searches, metal detectors, database screening, and bulletproof cockpit doors. These measures have involved extended but necessary delays for passengers.

Although the travel and hospitality industry was negatively impacted by the events of 9/11, travelers are increasingly taking to the sky again, heartened by increased security measures and increased airline safety.

The Trend Toward Ecology and Conservationism

Some hotels and resorts have exploited a trend toward "green marketing" and targeted guests supporting ecology and conservationism. *Ecotourism* describes the recent trend toward concern for environmental protection, cleanliness, safety, and health, as well as the enjoyment of natural beauty, including parks, forests, lakes, rivers, oceans, coral reefs,and wildlife.

A growing number of tours are springing up in destination markets, such as Jamaica, for tourists to see the unspoiled backcountry as well as engage in activities like glass-bottom boat rides and scuba diving and snorkeling at the coral reefs. Whale watching has proved to be a successful business for many seaside locations such as Santa Barbara, California.

According to the Travel Industry Association of America, 43 million individuals in the United States are self- proclaimed ecotourists, willing to pay as much as 8.5 percent more for environmentally friendly travel suppliers. The potential is cited by one travel writer for the creation of the "eco-hotel" or ecotel that would tap this huge market (Rushmore, 1993). Many hotels also advertise to their guests that they practice conservation, ranging from recycling of waste materials to water and energy conservation.

The Trend Toward Greater Speed and Convenience

Americans are working more, have less free time, and are demanding greater speed and convenience in the services they patronize. This has led to the trend of instant gratification for many consumers, which has supported the growth of fast-food restaurants, instant and priority check-ins, ticketless airline travel, and day and weekend cruises.

Hospitality companies often need to improve the speed and convenience of guests which will, in turn, improve guest satisfaction. This includes improvement in guest check in and check out, billing, and car parking for hotels and airlines. For restaurants, it may mean not waiting too long for a server, the meal, or the bill. And it may mean greater reliance on the Internet for easy booking of hospitality services.

Increased Demand for Ethnic Foods

Changes in demographic trends and increased awareness of foreign ethnic foods due to globalization have led to increased demand for ethnic foods. This includes Spanish restaurants (e.g., Polle Tropical), Japanese restaurants (Benihana), Mexican restaurants (Taco Bell), Cuban restaurants, Italian restaurants and pizzerias, Chinese restaurants, Jamaican restaurants, Thai restaurants, and Jewish restaurants. It is not unusual to enter an airport mall at a major hub and find several of these ethnic restaurants filled with travelers. Many upscale hotels and resorts have one or more of these ethnic restaurants in-house as well.

Especially popular to large upscale hotels are Japanese, Italian, and Mexican restaurants in addition to American steak and seafood restaurants. Most U.S airlines now offer meals only on long flights and have a limited choice of only vegetarian and kosher meals. However, foreign airlines such as Cathay Pacific, Singapore Airlines, and Air Jamaica offer excellent ethnic cuisine to their guests. Hospitality companies need to continuously monitor the food preferences of their target segments to ensure customer satisfaction that exceeds expectations.

The Trend Toward More Sophisticated Technology

Recent research has indicated that technology is viewed by hospitality executives as one of the most significant sources of competitive

advantage for hospitality firms. Technology can improve customer service, information management, and hotel design. This includes technology-based management information systems, decision support systems, property management systems, property yield management systems, and database marketing. The result has been increases in efficiency, effectiveness, and customer satisfaction (Lewis & Chambers, 2000).

Technological advances include computerization, central databases, high-speed Internet, smart cards, interactive TV, WI-FI connections, picture phones, all-purpose mobile phones, interactive video, videophone conferencing, and Internet booking of hospitality services.

I remember being pleasantly surprised by my first experience with electronic restroom faucets at a San Francisco hotel where I attended a marketing conference about ten years ago. They operated automatically using an electronic eye to detect the presence of hands under the faucet. On the other hand, many airport washrooms worldwide have no hot water or have broken faucets. And many Western guests are surprised to find that most restrooms in Chinese airports (and many hotels and restaurants) have no sit-down toilets.

Guests appreciate the use of high technology if it is user-friendly and brings about changes that they value. Thus, in-room checkout, using interactive TV, is valued by hotel guests because it makes the process easier and faster. Hotels and airlines that target business guests will need to provide the high-tech services that these travelers expect, such as high-speed Internet access, high-tech conferencing systems, interactive TV, and business services. Providing more than they expect in this area may provide an advantage to a hospitality company. For example, it was predicted that hotels in the future will have in-room computers, which will enable many business guests to travel without their computers.

The trend toward more sophisticated technology has provided a competitive edge for some hospitality companies. For example, there is the trend in hotels toward wireless online locking and security systems. Every time a guest or employee enters a room or controlled area, the card reader sends a record of who accessed the room, together with the time and date of entry, to the central computer's permanent storage. This system can also be interfaced with guestroom smoke detectors to alert the front desk as to which rooms have a prob-

lem, alert housekeeping to the occupancy status of a room by linking this to a motion detector, and provide heating, ventilation, and A/C control for energy management (Worcester, 1998).

The Increased Bargaining Power of the Consumer Due to the Internet

The price-searching capabilities of the Internet and the rise of powerful travel search engines (like Cheaptickets.com and Travelocity .com) combined with the increasing commoditization of the hotel and airline industry have resulted in dramatically increased bargaining power of consumers. Hospitality companies need to construct strategies to deal with this phenomenon.

One way is to target upscale market segments that are less price sensitive and are willing to pay for a customer-valued differentiation that the company offers. Ritz-Carlton and Four Seasons are examples of this strategy because of their choice of guest segments who are willing to pay premium prices for differentiated services.

I recently dined at the Kansas City Seafood and Steakhouse in Fort Lauderdale, Florida. Their prices were about 40 percent higher than other steakhouses that I had previously patronized. However, the dining experience was so good (they use prime aged beef cooked to perfection and the service was excellent) that they have made me a regular customer.

Another strategy is to engage in niche marketing and offer specialized services to a small hospitality segment willing to pay more for services that are customized to their needs. For example, my experience is that authentic Japanese sushi bars (featuring Japan-trained sushi chefs and fish caught the same day) are rare and charge premium prices (they do not seem to be vulnerable to the increased price pressures cited earlier).

Companies (like McDonald's) that pursue what Michael Porter calls the "low cost leadership" generic strategy will need to monitor the prices of competitors and maintain lowest industry costs while providing excellent value for their target customers so that they can continue to dominate their market.

The Trend Toward Fantasy and Adventure

This trend includes cruises, theme parks, remote destinations, gambling facilities, spas, safaris, white-water rafting, and so forth. A recent special broadcast on the Travel Channel featured a Russian icebreaker ship that had been converted and outfitted to take about 100 upscale global adventure travelers to extreme polar ice-locked locations not accessible by other means. They enjoyed gourmet cuisine during the trip and traveled to exotic, inaccessible locations by helicopter from the heliport on the ship. Although the accommodations were relatively spartan for these upscale guests, a high customer return rate for this type of extreme adventure travel was reported.

Hospitality companies that take advantage of the fantasy trend are companies like Disney (theme parks), hotel casinos like Aladdin of London and Las Vegas, cruise lines like the Princess Line which operates out of Fort Lauderdale, Florida, and famous hotel spas such as Canyon Ranch. Some hotels and cruise lines also offer their guests gambling facilities. Movies offered by hotels and airlines sometimes exploit this trend toward escapism and fantasy, which can take different shapes for different target segments (e.g., note the difference between segments targeted by Disney cartoon films, James Bond and Austin Powers films, and adult-rated films, all offered by hotels).

Increased Numbers of Foreign Travelers

This trend is presumed to be related to the recent devaluation of the U.S. dollar and to globalization. Hospitality companies like Sandals (which targets a global honeymoon segment) and McDonald's (which offers a consistent image and popular product internationally) take advantage of this trend by offering products that appeal to global segments.

Hospitality companies targeting global segments may need to have bilingual or trilingual employees at the customer contact level, and offer services such as authentically prepared ethnic foods, European-style service, and accommodations, activities, and facilities that are valued by these guests. Foreign exchange services, concierge services, in-house car rental services, and area maps that highlight the area's cultural and entertainment attractions are also of assistance to foreign tourists.

FUTURE TRENDS

A 1998 study by the Cornell School of Hotel Administration polled its alumni to project future trends twenty-four years into the future, when the school would turn 100. The predictions of this study included the following:

1. Customer service and satisfaction will be lodging's greatest challenge in the coming years.
2. Jobs earmarked for extinction include front desk positions, concierge service, and telephone operators.
3. Video conferencing will replace many face-to-face business meetings.
4. Computers will be standard equipment in hotel rooms.
5. Properties will get larger, with 10,000-room hotels being erected.

Room keys, newspapers, business centers, and room phones were also targeted for extinction by this survey (Whitford, 1998).

A research study by the National Leisure Travel Monitor predicted the following ten trends for the leisure travel industry:

1. *Unprecedented affluence:* 14 percent of U.S. households were correctly projected to have an annual income in excess of $100,000 by 2003.
2. *Time poverty:* Increased employment and longer hours have led to the perception that there is not enough time in a day to accomplish desired ends and not enough leisure time for vacations.
3. *The pursuit of pleasure:* Reduced savings and a trend toward self-indulgence are characteristic of current trends. This is related to increased credit card use, consumer debt, inclusive vacations and cruises, and fine dining.
4. *New and different:* Seven out of ten Americans say they would welcome more novelty and change in their lives. This is related to the increase in gambling-related travel, cultural travel, ecological travel, and theme park travel, as well as travel to new and exotic destinations such as China, Africa, and Australia.
5. *The techno-generation:* "Technos" are those who are predisposed to experiment with new technology and explore its application to both professional and personal pursuits. There is increase evidence of the relative growth of this group. The Internet

is rapidly becoming a primary source of information for consumers planning vacations. More than one-third of travelers consult the Web for their travel plans. Use of the Internet to plan and book travel services is considered to be among the most significant trends affecting the travel industry.

6. *Inclusive pricing:* Almost six out of every ten adults cited the desire for more simplicity in their lives. This has translated into increased use of packaged and all-inclusive vacations. Half of all leisure travelers use inclusively priced vacations, such as those offered by Sandals, which have one price that includes everything in the package (and often includes airfare too).

7. *Closer to home:* Most hotels cite a decline in the length of stay of their leisure guests in recent years. Of all the vacations taken by Americans in 1999, only 23 percent were for five nights or more, 21 percent for four nights or less midweek, and 46 percent for four nights or less tied to a Sunday night. Thus, more vacations are now scheduled around the demands of the workplace. The implications for marketing strategy are that travel marketing programs must have separate campaigns for these different traveler needs, and they must target fundamentally different prospects. The trend toward limiting extended travel has been accelerated since 9/11, due to both security fears and the discomfort expressed by travelers who are not willing to journey to destinations requiring more than three to four hours' travel time (Yesawich, 2000).

8. *Active vacations:* There has been a marked increase in pursuing a variety of activities while vacationing. These include shopping, exercise programs, and participation in a variety of sports programs. There is also increased interest in pursuing learning activities, including acquiring a new skill (cited by four out of ten travelers).

9. *Alternative accommodations:* There is an increasing trend toward alternatives to conventional lodging. Six out of ten travelers say they are interested in condominium- style accommodations, especially among family travelers. Three out of ten travelers are interested in all-suite hotels, and 17 percent of travelers expressed interest in purchasing some form of vacation ownership or timeshare within the next two years. Almost half of all travelers who patronize conventional accommodations ex-

pressed an interest in accommodations that featured a distinctive architectural style or service theme.

10. *Strategic control:* About two-thirds of travelers say that the Internet has enabled them to develop a greater sense of strategic control when evaluating and making purchase decisions. This is due to instant access to a wealth of pertinent information for products and services and competitive pricing for these offerings. It is likely that the next generation of electronic "bots" will be programmed to do these same searches in less time and with more efficiency. This will strengthen the current trend (mentioned earlier) of the balance of power shift to the consumer and away from suppliers. This is likely to result in more comparison shopping, commodity pricing, and downward profit margins.

ENVIRONMENTAL SCANNING

Environmental scanning is the process of observing uncontrollable forces that will affect the business environment in the future and looking for ways to benefit from or protect the business from these forces. Environmental forces include demographic, legal, economic, ecological, sociological, technological, political, cultural, and industry market forces and/or trends. The primary responsibility for environmental scanning rests with the marketing department. We have discussed many of the trends impacting the hospitality industry in this chapter. But how is forecasting accomplished in the hospitality industry?

Forecasting of Trends

Environmental and demand forecasts are usually done using either quantitative or qualitative analysis. Quantitative forecasting uses mathematical models such as times series analysis, econometric models, and multivariate analysis. Because of its complexity and cost, this method is used much less frequently than qualitative analysis in the hospitality industry (Reich, 1997).

The dominant method of forecasting for the hospitality industry is qualitative forecasting, which uses expert opinion, debates, or surveys. The following are major methods of qualitative analysis (Reich, 1997):

- *Brainstorming:* This is the open sharing of ideas by a group, usually with a leader. Ideas are recorded and everyone is involved.
- *Juries of executive opinion:* These are forecasts from various levels of management based on an analysis of the issue.
- *Delphi Method:* This is usually a series of surveys to individual experts in a specified field. The opinions are disseminated to this group and further surveys are executed until a consensus is reached.
- *Customer surveys:* This makes use of surveys (usually quantitative) to the firm's customers on their attitude on the topic of interest. This method also uses non-probability samples.
- *Scenario analysis:* This involves creating a hypothetical situation about the future. For example, if there were more airline hijackings, how would this affect airline occupancy rates and fares, and what would be the best strategy to minimize or exploit these effects?
- *Sales force or employee estimates:* This involves soliciting opinions from employees with access to key information. This may not be the most reliable method of forecasting, since managerial and sales estimates are often biased.

Six steps of environmental scanning:

1. *Watch for broad trends.* Examples are later marriages, two-income families, increased travel, increased adherence to "diets," and more vacation time and discretionary income.
2. *Determine significant trends.* You need to isolate only the trends that will have a significant impact on your company. There are no set-in-stone rules for doing this. It requires creativity, imagination, and insight.
3. *Analyze the impact.* If significant, what is the size and timing of the impact? What is the impact in terms of the service, target market, competition, cost, employee attitude, and other variables that could be affected? If it is a threat, can it be controlled, minimized, or turned into an opportunity? Get reactions from managers and employees, where appropriate.
4. *Forecast direction.* Market research is helpful in forecasting direction and is more dependable than managerial intuition. Focus groups and outside consultants are also helpful.

5. *Assess opportunity.* It is too easy to look only on the negative side of these trends. Negative trends should be assessed for opportunities. For example, the trend away from unhealthy foods was an opportunity for those companies that offered healthier fare to their customers.

6. *Relate the outcome.* Assess the outcome of this scanning process to the firm's marketing strategy and marketing plan for the short, medium, and long term. What modifications are needed? A systematic method of periodic environmental scanning is needed by all hospitality firms to avoid being negatively impacted by some of these trends and missing the opportunity afforded by others (Lewis & Chambers, 2000).

Countertrend Marketing

Although conventional wisdom rightly supports marketing plans and programs that are supported by major marketing trends, it must be noted that niche marketing opportunities exist for programs that go against these marketing trends. Examples of successful marketing of products/services that appear to go against major trends include

- super-premium ice cream with high fat content (Haagen-Dazs, Cold Stone restaurants);
- steakhouses that serve red meat (Longhorn Steakhouse, Outback Steakhouse);
- resorts for swinging young singles (Hedonism resorts in Jamaica); and
- gas-guzzling SUVs, which pollute the environment and are safety hazards due to rollover and crash risks.

But these examples of countertrend marketing can be explained by the appeal to other conflicting marketing trends such as the trend to self-indulgence. This would include the demand for faster, more powerful cars, binging on premium ice cream, and eating red meat. And since major marketing trends do not affect all people, there are often niche marketing opportunities to serve customers whose tastes are not aligned with major marketing trends.

SUMMARY

Major marketing trends that have been reviewed here include the graying of America, the growing importance of women and minority segments, the trend toward a healthier lifestyle, the concern for preservation of the environment, the desire for greater speed and convenience, and the need for greater security.

It is clear that marketing plans and programs that are supported by major marketing trends are more likely to succeed. Thus, it is important to continuously monitor marketing trends (current and future) that are likely to impact your organization. Methods of monitoring these trends have been reviewed above. They include the qualitative methods of brainstorming, Delphi Method, customer surveys, and scenario analysis. Secondary research is also available for this purpose.

It is also clear that sometimes marketers find lucrative niche marketing opportunities that appear to go against these trends. This may include attempts to deliver services and products to the needs of smaller segments which may have very different needs and preferences. And these needs may not be in alignment with major marketing trends.

EXERCISES

1. Identify two marketing trends that will significantly affect your hospitality organization (or a selected hospitality organization) and how these trends will affect marketing to one target segment of this organization.
2. How will you obtain current information on current hospitality marketing trends? How often should this information be monitored?
3. Identify a negative trend in the hospitality industry and how you can exploit this opportunity for your selected hospitality organization.
4. How do you see new technology (for the next five years) impacting the segmentation process for your organization (or your selected hospitality organization)?

5. How will e-commerce impact your organization's use of the segmentation process and the selection of target segments?
6. How will the marketing trend toward greater personal security affect your organization's marketing strategy?
7. List four main methods of qualitative analysis used for forecasting in the hospitality industry. Critique the methods that your organization now uses.
8. How does your organization plan to adapt to the "green marketing" trends related to preserving the ecology and environment? Do you see this trend as attractive to your target segments?
9. How should your organization respond to the fact that the fastest growing demographic segment in the United States is currently mature travelers (fifty-five and older)? What changes need to be made to exploit this opportunity?
10. How do you see the trend toward globalization affecting your organization's marketing strategy?

Chapter 8

The Segmentation Process

Segmentation is a widely used marketing process which assumes that the firm can more effectively market to selected customer segments that have a good strategic fit with the resources, skills, goals, and external environment of the firm. This implies that the firm is in a position to best satisfy their particular needs and preferences. No single segmentation process is appropriate for all hospitality companies. This is because the segmentation process is dependent on the purpose of the segmentation process, the market structure, and the goals of the hospitality company. Many hospitality companies still use traditional a priori and product segmentation. Thus, airlines use the a priori product segments of coach, business, and first class, and hotels often use the a priori segments of corporate, transient, and groups.

The marketing literature traditionally describes the segmentation process as STP marketing as cited by Kotler (1999). This refers to the process of *segmenting* the market, *targeting* the most attractive segments, and *positioning* the firm with respect to these segments.

Selection of target segments is characterized by Wind (1978) as a complex "art" that should take into account such factors as the segment's expected response to marketing variables and management's resources and ability to implement strategy for the selected segments. This creative view of the segmentation process is supported by Sarabia (1996), who states that there is no specific model for segment selection that is in frequent use in segmentation studies.

Although an understanding of the segmentation process is fundamental to the understanding and implementation of segmentation, there is surprisingly little to date in the marketing and hospitality literature that goes beyond Kotler's STP approach.

Segmentation Strategies for Hospitality Managers
Published by The Haworth Press, Inc., 2007. All rights reserved.
doi:10.1300/5716_08

SEGMENTATION AS PART
OF A TARGET MARKETING STRATEGY

Segmentation has traditionally been viewed as the first step in the firm's target marketing strategy, which consists of the following sequential steps. It is important to note that segmentation strategy does not stand alone but is intimately related to both targeting and positioning. No segmentation strategy would be effective if it did not lead to a competitive positioning strategy. Nor would a segmentation strategy be effective if it did not result in a good "strategic fit" between the firm and its target segments. Strategic fit refers to the goal of a co-alignment of the firm's resources (funding, skills, core competencies, technology, corporate culture, values, location, etc.), goals, and external environment with the firm's target market and offerings (Morritt & Lawson, 2000).

Segmenting the Market

This is the process of dividing the market into distinct groups of buyers with common purchase behavior or needs who may require separate products or marketing mixes. Segmentation bases that are used for this purpose (such as benefit, usage, demographic, and psychographic segmentation) are reviewed in Chapter 2.

Targeting the Most Attractive Segments

This is the process of evaluating each segment's attractiveness and selecting the segments that are most attractive to the company relative to its resources, skills, goals, and external environment.

Segment attractiveness criteria include the following (Kotler, 2003, 2005).

Measurability

Marketers need to know segment metrics such as size, disposable income, purchase frequency, growth rate, and profitability of selected segments in order to select target segments.

Substantiality

The segment must be large enough in size and demand to be profitable. Thus, many hospitality companies target business travelers, an a priori segment which is large and profitable.

Accessibility

Members of the target segment must be reachable by your marketing communications (e.g., people who prefer to fly at the last minute may not be a reachable segment).

Differentiable

The segments are conceptually different and respond differently to different marketing mixes. Thus, U.S. female business executives may be an example of a hotel segment which may have special needs and preferences that are different from their male counterparts, such as having stronger preferences for security and clean comfortable bathrooms.

Actionable

This refers to the possibility of effective programs that can be developed to attract and serve the segment (Kotler, 2003).
Other criteria for effective segmentation include the following.

Defensibility

The marketer must be confident that the selected target segment(s) can successfully be defended from competitors. Although the segment of business travelers is an attractive segment, there is much competition for this segment. Thus, it may be a wise choice to select a smaller niche of this segment to which your firm can deliver superior relative value. For example, some hotels target convention travelers (Hilton Hotels) or independent business travelers who are not on expense accounts (Courtyard by Marriott).

Durability

The selected segment must be stable over time. Some segments are short-lived or have little growth potential. For example, attendees of nonrecurring events would probably not have sufficient durability.

Competitiveness

The segment must be evaluated to ensure that your company has some unique or differentiated product or service which will position you to best serve this segment. Thus, Singapore Airlines and Cathay Pacific Airlines target longhaul intercontinental passengers with the longest flights in the industry (fourteen to seventeen hours). Their reputation for delivering superior service is legendary. This is due to their world-class service by highly trained and friendly flight attendants, gourmet meals, interactive individual monitors, exercise sessions for passengers, as well as roomier accommodations for business and first-class passengers. My four visits to China on Cathy Pacific and one U.S. airline last year indicated a huge quality difference between this customer-oriented airline and U.S.-owned, cost-oriented airlines that have the same route.

Positioning the Firm's Offerings
for Your Selected Target Segments

This is the process of developing a competitive advantage for your firm's products/services by using the most effective marketing mix for your selected target segments. A product's positioning refers to the way the product is perceived in the mind of the company's target segments relative to the competition and also on attributes and benefits viewed as most important by these segments. Three steps in this process are (1) identifying a set of competitive advantages on which to build a position, (2) selecting the right competitive advantages, and (3) effectively communicating and delivering this position to the selected target market segments (Bowen & Shoemaker, 1998).

A good example of the STP process was the segmentation strategy of Negril Gardens Hotel in Jamaica. They segmented the market and selected the European travelers to Jamaica (English, German, & Dutch) for targeting. They offer a European vacation resort experience to these guests that includes no telephones, European-style high

service and cuisine, entertainment, a topless beach, and a European hotel manager. This hotel achieved a 90 percent occupancy rate in the 1990s under their excellent manager, Pat Morgan, when other hotels in their city had difficulty breaking even (Morritt, 1997).

The market segmentation process has been described as STP marketing (Segmenting, Targeting, and Positioning) (Kotler, Bowen, & Makens, 2003). The steps in the segmentation process include:

1. Segmenting your hospitality market using different segmentation bases (e.g., usage, geographical location, RFM analysis, demographics, best customer profiles, etc.). This may also include combinations of segmentation bases. This is an art rather than a science, and will take a combination of experience, creativity, and innovation.
2. Select the optimal target segments based on both the attractiveness of the segment (use measures such as segment profitability, growth, usage, accessability) and also segment fit with both company resources (i.e., skills goals, competencies, ability to achieve a sustainable competitive advantage with this segment) and external trends and opportunities.
3. Create a positioning strategy for each target segment and customize the marketing mix for that segment. Then you will need to communicate the superior value of your service to that target segment based on their researched needs and preferences. You will need to use the appropriate medium, message, value proposition, and language for each target segment.

Kotler (2003) later expanded on this approach to the segmentation process by endorsing a needs-based market segmentation approach by Roger Best, displayed as follows.

STAGE ONE: MARKET SEGMENTATION

Best (2000) has identified the following steps in the segmentation process:

1. *Needs-based segmentation:* Group customers into different groups or segments based on similar needs and benefits.

2. *Segment identification:* For each needs-based segment, determine which demographics, lifestyles, and usage behaviors make the segment distinct.
3. *Segment attractiveness:* Use predetermined segment attractiveness criteria (e.g., segment growth rate, disposable income, ability to provide superior service to the segment) to determine the overall attractiveness of each segment.
4. *Segment profitability:* Estimate segment profitability.
5. *Segment positioning:* Create a value proposition for each segment that contains benefits/features which provide superior perceived value for that segment.
6. *Segment "acid test":* Create segment storyboards to test the attractiveness of each segment's positioning strategy.
7. *Marketing mix strategy:* Customize the marketing mix (product, distribution, price, promotions) to include the benefits/features that promote superior value to each target segment and that are consistent with the value proposition for each segment.

Presumably there are two additional stages (omitted above) in which the most attractive segments are selected (stage 8) and the marketing mix is communicated and delivered to target customers (Stage 9).

Meyers (1996), in his classic text, *Segmentation and Positioning for Strategic Marketing Decisions,* elaborates on the Kotler model in his "General Procedure for Segmenting markets" (pp. 18-19):

1. Decide on optimal segmentation variables (e.g., benefits, lifestyles, demographics, usage)
2. Decide on data analysis methodology (e.g., clustering, data mining, etc.)
3. Apply methodology to identify the most attractive segments relative to a particular product/service
4. Profile/describe all segments using segmentation bases and other variables (This is a description usually involving geo-demographics, which describes a typical "best customer" for a particular product or service of your company.)
5. Select target segments to pursue (These target segments should be a good fit with the resources of your organization and also with external trends and opportunities.)

6. Develop a marketing mix for each target segment (Each target segment should be provided with a marketing mix that is customized to the needs and preferences of that segment and which together provide superior perceived value for members of that segment.)

Deciding on Segmentation Bases

Marketers are well advised to discuss with management the ultimate objectives of the segmentation study before deciding on segmentation variables. Common objectives cited by Myers (1996) are

1. identifying and characterizing (profiling) frequent users;
2. focusing advertising efforts for greater impact;
3. identifying likely targets for new technologies or new products;
4. improving existing product/service design;
5. looking for new product/service opportunities;
6. assessing the impact of a competitor's new offering; and
7. establishing a better brand or corporate image.

Different segmentation bases may be appropriate for different company objectives. For example, the identification of heavy/frequent users may be helpful for planning better advertising and promotions but may not be useful for improved product/service design unless the proper basis variables have been included in the study. Thus, clarification on the purpose of the segmentation study will ensure that the appropriate data are collected (Myers, 1996).

Bases used in targeting consumers usually include some combination of members of the following basis categories:

- Geographic (region, city, density, etc.)
- Demographic (age, family size, family life cycle, gender, income, etc.)
- Psychographic (lifestyle, personality)
- Behavioral (usage, loyalty, benefits, occasions, etc.)
- RFM analysis (recency, frequency, and monetary value of individual customer purchases)

It is well-known that hospitality companies also target business customers as well as individuals for their target marketing. Hotels target corporations for their conventions and business travelers by giving them special corporate rates. They also target tour operators and travel companies. Airlines target travel agents, Internet travel sites, and corporations for their frequent travelers. They also target food service companies that supply their planes. Restaurants may target food suppliers, uniform suppliers, and unions for their relationship marketing.

Business markets use some of the same segmentation bases as consumer markets but also use the following additional categories (Shapiro & Bonomo, 1984):

Demographics

- What kind of organizations should we target?
- What size organizations should we target?
- Which geographical areas should we target?

Operating Variables

- Technology: What customer technologies should we focus on?
- User or nonuser status: Should we serve heavy, medium, or light users, or nonusers?
- Customer capabilities: Should we serve customers needing many or few services?

Purchasing Approaches

- Purchasing-function organization: Should we serve companies with highly centralized or decentralized organizations?
- Power structure: Should we serve organizations that are engineering dominated, financially dominated, and so on?
- Nature of existing relationships: Should we serve companies with which we have good relationships or that have compatible values and goals?
- General purchase policies: Should we serve companies that prefer service contracts, heavy discounts, very large meeting facilities, catering services, etc.?
- Purchasing criteria: Should we service companies that are seeking quality, service, or low price?

Situational Factors

- Urgency: Should we serve companies that need quick and sudden services?
- Specific application: Should we focus on certain applications of our services rather than all applications (e.g., all-suite hotels)?
- Size of order: Should we focus on large or small purchases?

Personal Characteristics

- Buyer-seller similarity: Should we serve companies whose people and values are similar to our own?
- Attitudes toward risk: Should we serve risk-taking or risk-avoiding customers?
- Loyalty: Should we serve companies that show high loyalty to their suppliers?

Deciding on Data Analysis Methodology

Many data analysis techniques are used in segmentation studies. They range from simple tabulations and cross-tabulations to sophisticated statistical multivariate analysis tools that simultaneously consider many variables. The research analyst must choose the most appropriate techniques for the particular circumstances and objectives of the study. The principal multivariate techniques used for segmenting markets include partition clustering, hierarchical clustering and Q-type factor analysis, Automatic Interaction Detector (AID), Chi-squared Automatic Interaction Detector (CHAID), CART, and discriminant analysis for profiling segments (Myers, 1996).

Simpler techniques involve using a selected combination of basis variables (e.g., geographic, demographic, usage benefit) in order to target the most attractive segments for your hospitality company.

Applying Methodology to Identify Several Segments

The next step (after deciding on the purpose of the study and basis variables) is to identifying a reasonable number of potential target segments that fit the firm's criteria for attractive segments (see Stage Two, below) and also are a good "strategic fit" for the organization.

Thus, the hotel chain, Ritz-Carlton (given its resources, goals, and external environment) has a good strategic fit with its target segments of corporate upper management and the ultra wealthy but may not be a good fit for targeting truckers or students on vacation.

Profile/Describe All Segments Using Bases and Other Variables

After segments are formed, they are profiled or described using other information obtained from surveys, internal transaction records, customer databases, or merged data from other sources. This profiling can be done for selected target segments alone or for other attractive segments to assist in the selection of segments that are the best strategic fit for the company. Profiling can be accomplished by using (1) the basis variables used to form the segments, (2) demographics, (3) media usage, (4) product/service usage rates or patterns, (5) internal transaction records, (6) product features desired, and (7) other survey responses that would be useful in for market planning purposes (Myers, 1996).

For example, customer profile studies can contrast the demographics of light and heavy users for a particular restaurant, compare these profiles with competing restaurants, and monitor the changes in customer composition over time (Hsu & Powers, 2002).

Develop a Marketing Mix for Each Target Segment

After segments have been selected and profiled, a separate customized marketing mix (i.e., the four Ps of Product, Price, Promotion, and Place) needs to be developed for each target segment. Data obtained by market research into these target segments should provide the following information:

- What kind of product/services are desired/preferred by this segment.
- The price sensitivity of this segment and what price ranges are appropriate.
- What advertising and promotion approaches are likely to be most effective including the appropriate media, value proposi-

tion, language, and message to use for this segment. For example, the language, message, and media for ads targeting senior business executives are quite different from ads that may target the younger generation associated with the hip hop and rap music culture.

- And what are the optimal distribution channels for this segment? (For example, fast-food companies like McDonald's target college students by having mini kiosks located at various college dining halls.)

A FEEDBACK MODEL OF THE SEGMENTATION PROCESS

But is target marketing really a sequential process or is it rather an iterative process involving feedback among these three different steps?

It is clear that the segmentation process is mutually dependent on the positioning process. Just as target marketing requires that the firm positions its offerings to the needs and preferences of target segments, it is also true that segments must be chosen so that the firm can offer a superior value proposition to those same target segments.

An example will illustrate this point. Let us suppose that a restaurant initially chooses an attractive segment (upper-income families who dine out at least once a week) in their geographical area who are interested in casual dining in a moderately priced seafood restaurant. But they discover that they cannot position their restaurant so as to offer superior value to this target segment. They must then go back to the targeting stage to select another segment to whom they can offer superior value. This may result in the launching an Italian restaurant that offers homemade pasta which appeals to another customer segment.

Furthermore, if they cannot find such a segment, they may have to go back to the first stage of the process and identify different bases or customer profiles. Therefore, target marketing, as a segmentation strategy, is iterative with feedback loops as illustrated in Figure 8.1, which can lead to a successful positioning strategy.

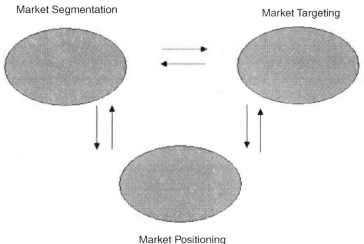

Market Segmentation   Market Targeting

Market Positioning

FIGURE 8.1. Feedback Model of Three-Stage Target Marketing Process

SEGMENTATION RESEARCH AND ANALYSIS TECHNIQUES

Hospitality marketers use research to obtain a range of data about their market, environmental forces, trends, target customers, and competitors. The data used for this purpose come from a mix of internal and external sources. For example, guest registration records are a common source of internal data for hotels. Additional information may require primary and/or secondary research. This research may be done in-house or by outside research specialists.

Hospitality companies have traditionally used a priori segmentation designs (product/price segments, leisure/corporate/transient hotel segments, coach/business class/first class airline segments). Research was dominated by both a priori and cluster-based segmentation designs. Survey results showed segments estimated size, demographic, psychographic, and other relevant characteristics. Cluster-based designs (such as benefit, need, and attitude segmentation) were used with similar survey goals (Wind, 1978).

More recently, service companies use a variety of computer-aided techniques such as structural equation modeling (SEM), neural net-

works, conjoint analysis, and multidimensional scaling. There is an increased interest in the use of data modeling techniques associated with database marketing. With the advent of new developments in technology, including centralized computer systems, database systems, and data mining, there is a blurring of the three STP functions. They are no longer distinct and separate steps in the segmentation process. For example, positioning techniques are often useful in choosing target markets. Thus, the segmentation process is nolonger a sequential process but is now an integrative process (Bowen, 1998).

The hospitality manager is especially concerned about the translation of segmentation studies into effective marketing strategy for their company. Unfortunately, this is recognized to be the most difficult aspect of any segmentation project (Wind, 1978). Suggestions offered by Wind are

- involving all the relevant users (e.g., hospitality managers, marketers, ad agencies, etc.) in the problem definition, research design, and interpretation stages.
- viewing segmentation data as one input to a total marketing information system and combining them with sales and other relevant data.
- using the segmentation data on a continuous basis. The reported study results should be viewed only as the *beginning* of a utilization program.

STAGE TWO: SEGMENT TARGETING

Targeting is the selection of the most attractive segments for the hospitality company. How do marketers decide which segment or combination of segments would be most profitable? This is accomplished after an analysis of segment attractiveness that includes the strategic fit of potential segments with the resources, goals, and external environment of the company. Criteria for effective targeting include the following requirements. The selected segment(s) should be

1. Measurable
2. Substantial
3. Accessible

4. Differentiable
5. Actionable
6. Compatible with other target segments of the company: One hotel company made the mistake of targeting both seniors for getaway weekends (who appreciate peace and quiet) and young singles who enjoyed loud parties and entertainment. This was clearly a recipe for disaster.
7. Stable: Segment stability is essential in order to minimize investment risk in this segment. Thus, the segment should have reasonable expectations of longevity.
8. Associated with an acceptable level of within-segment competition: This means that too much competition for a given segment is a contraindication for that segment. This should focus on both existing and potential competitors.
9. Have a good strategic fit with the hospitality company: The company needs to assess its business resources (capabilities, skills, culture, technology, brand equity, funding), corporate goals, and external environment to ensure a good strategic fit between the company and its selected target segments. As a rule, companies are more likely to succeed when they move into segments that are close to their existing brands or areas of expertise. A good strategic fit between the company and its target segment implies that the company has the appropriate resources, skills, and competencies, and a strategy that is aligned with the external environment to successfully position itself with respect to that target segment (Morritt, 1999).

It is common for hospitality companies like hotels, restaurants, and airlines to pursue traditional a priori segments such as corporate, transient, group, and leisure segments for hotels and coach, business class, and first class segments for airlines. Restaurants often pursue the traditional a priori segments of fast food, casual upscale, midscale, and fine dining.

But it is clear that these a priori segments do not have homogenous buyer behavior, and firms that rely on these a priori segments alone are vulnerable to competitors who use a post hoc research-based approach to market segmentation. These competitors will find strategic niches and micro segments which will erode the market share of companies who do not make use of current researched-based target mar-

ket segments which are found to be a good strategic fit with the hospitality company.

For example, discount airlines like Southwest and JetBlue have eroded the market share of major carriers like American Airlines and United Airlines by focusing on the price-sensitive leisure traveler niche. Another example is the very successful Marriott Courtyard product which took market share from other hotel companies who targeted the business segment and forced them to copy the marketing mix of this chain that targeted the subsegment of independent business travelers who were not on expense accounts and desired clean, comfortable, reasonably priced rooms with business amenities.

In simple segmentation studies (which often use only usage rates and demographics), companies often pursue the heavy users of their products/services (Myers, 1996). But the "heavy user" segment of business travelers, targeted by many hospitality firms, is an area of intense competition. And there are differences in this segment (e.g., needs, benefits sought, preferences) that leave niches of opportunity for perceptive marketers. Thus, Virgin Airlines targets business travelers who value having more seat room while traveling. And Ritz-Carlton of Jamaica targets business travelers who prefer the amenities of a destination resort in Montego Bay, including a world-class golf course, deep sea fishing, and beautiful beaches. A new catering concept is catching on in major cities (e.g., Boca Raton and San Francisco) that targets health-conscious upscale business people who enjoy having gourmet diet meals delivered to their door for about $50 per day.

It is generally agreed that the use of a single basis for market segmentation is inferior to studies which use multiple basis variables. For example, Wind states:

> In contrast to the theory of segmentation that implies that there is a single best way of segmenting a market, the range and variety of marketing decisions suggests that any attempt to use a single basis for segmentation (such as psychographics, brand preference, or product usage) for all marketing decisions may result in incorrect marketing decisions as well as a waste of resources. (Wind, 1978, p. 319)

Before the advent of computerized segmentation software, the process of segmentation and targeting were two distinct processes. But

with computerized methods, these two steps are often combined. These computerized models identify segments and segment values given the marketer input data used for targeting (Bowen, 1998)

Recently, hospitality marketers have been rightly concerned about the ethics and public relations risks of targeting specific segments for potentially harmful products (e.g., disadvantaged segments and products such as nonnutritious meals and addictive products such as gambling and alcohol (Bowen, 1998). Hospitality firms must be alert to the danger of having their target marketing perceived as exploitive or harmful to their selected segments.

PATTERNS OF TARGET MARKET SELECTION

Kotler cites five patterns of target market selection, which occurs after the firm has evaluated the attractiveness and strategic fit of potential target segments (adapted from Kotler, 2003, p. 299).

Single Segment Concentration

Volkswagen concentrates on the small car market and Motel 6 concentrates on the no-frills economy niche. Operating economies are possible through specialization, but this strategy is not without risk. Putting all your eggs in one basket can lead to disaster if your one segment becomes no longer profitable or if it is acquired by competitors.

Selective Specialization

Selective specialization is a multi-segment strategy that has the advantages of diversification. The firm chooses multiple segments that are each attractive and appropriate. One example is the Brinker's restaurant chain that includes Chili's, Corner Bakery, On the Border, Maggiano's, Big Bowl, and Rockfish. These restaurants all appeal to different target segments.

Product Specialization

This strategy involves selling a single product to several segments. One example would be McDonald's, which sells the same product to

adults and children, and (with minor modifications) to different countries.

Market Specialization

Here the firm focuses on serving the many needs of a particular customer segment. One example of this would be a hotel that specializes in serving the destination small meeting market.

Full Market Coverage

This is the strategy (usually adopted by only very large firms) of serving all customer segments with all the products they may need. IBM and GM are examples of this among product firms. An example of a hospitality firm that does this would be Marriott hotels, which offers a full spectrum of hospitality products to many markets. This includes a full range of price/quality hotels, resorts, extended-stay properties, time-shares, and assisted care facilities for seniors.

Market Partitioning

Another method of selecting new segments is to research the *hierarchy of attributes* consumers use in choosing a brand. For example, in the automobile industry, buyers may first choose they want to buy a Japanese car, then Toyota, and then the Camry model of Toyota. The hierarchy of attributes can reveal attractive consumer segments. For example, buyers who first decide on price are price dominant; those who first decide on the type of service (e.g., for hotels: all suites, luxury, midscale, etc.) are type dominant; those who first decide on brand (Marriott, Hilton, Best Western) are brand dominant. One can identify those who are type/price/brand dominant as making up a segment, and those that are quality/service/type dominant as making up another segment. Each segment can be profiled using demographics, psychographics, and media-graphics (Kotler, 2003).

Use of Multiple Segments

Usually hospitality firms target more than one segment, which has the advantages of diversifying risk and increasing revenue. Targeting

additional segments may also have the advantage of exploiting seasonal and other cyclical differences in demand and the type of customer being serviced. However, the targeting of multiple segments also has risks, and care must be taken that these different segments are compatible and seek similar benefits

For example, hotels in Jamaica often target local customers during the hot summer months when U.S. and European tourism is slower. Local guests are usually given rooms at half the price charged tourists by most hotels in Jamaica. Jamaican hotels, airlines, and government tourism agencies also target Japan and Germany in addition to the U.S. and European markets. One hotel in Jamaica, Negril Gardens, has thrived during past tourism recessions in Jamaica with an occupancy rate of more than 90 percent by specializing in European guests. They do this by offering these guest vacation accommodations closer to the European lifestyle, including no phones, a topless beach, and European-style cuisine and entertainment (Morritt, 1997).

But it is important that hospitality managers avoid conflicts in the multiple segments that are served. One hotel that suffered from slow weekend occupancy marketed to two additional segments with separate marketing plans for each segment. The segments were romantic couples who wanted to get away for a peaceful and quiet weekend and families with children who wanted a mini-vacation with lots of activities. It was inevitable that these two segments would collide head-on and lead to disastrous results for the hotel (Lewis & Chambers, 2000).

But the needs and behavior of different segments change over time. Since one of the advantages of segmentation is to stay close to the needs and preferences of target segments, it is important to maintain this advantage by continuously monitoring and researching target segments to anticipate changing, merging, and dividing segments.

SEGMENTATION SELECTION AND STRATEGIC FIT

There are several strategic factors (internal and external) that together determine optimal target segment selection and positioning. These strategic factors must be favorable to the company's choice of target segments in order for the company's segmentation and positioning strategy to be effective.

Market Structure

1. Is this an attractive hospitality service product segment (level of competition, present and future growth, profitability, power of buyers, suppliers, strength of entry and exit barriers)?
2. What are the market drivers of this service (e.g., price, service, value, quality, personalization, etc.)?
3. What is the current life-cycle stage of this product/service (e.g., is this hospitality product in a high growth stage or a declining mature market characterized by commoditization and price competition)?

Segment Attractiveness

1. Is this segment measurable, accessible, and durable?
2. Does this segment exhibit sufficient size, growth, and profitability?
3. What is the level of competition for this segment?
4. What about the threat of actual or potential substitute products?
5. Will members of this segment pay a premium for specialized services?
6. Is this segment vulnerable to microsegmentation or competition from broad-based companies?
7. How do current and projected marketing trends impact this segment?

Company Goals and Resources

1. Is the targeting of this segment(s) consistent with corporate strategic goals?
2. Are the company's financial resources adequate to serve this segment?
3. Are personnel and management skills adequate to serve this segment?
4. Is the company well positioned to satisfy the needs of this segment better than the competition?
5. What is the track record of this company in serving this or similar segments?

6. Does the company have adequate facilities to serve this segment?
7. What are the technological requirements for serving this segment?
8. Are employees sufficiently trained to serve this segment?

Impact of the Macroenvironment

What macroenvironmental forces are projected to significantly impact this segment over the next five years (legal, economic, political, ecological, sociological, cultural), including marketing and demographic trends?

Organizational Culture

1. Is the company culture sufficiently market oriented to effectively serve this segment?
2. Is the company culture sufficiently ethical to effectively serve this segment?
3. Is the company culture sufficiently service oriented to effectively serve this segment?

Competitive Advantage

The company should only enter segments where it can offer superior value and achieve a competitive advantage (Kotler, Bowen, & Makens, 1998).

Therefore, the following questions should be considered before deciding to target a specific segment:

1. How will this company achieve a sustainable competitive advantage with this segment (vis-à-vis its competitors)?
2. What is the company's value proposition for this segment and how do we know that this proposition is attractive to this segment?
3. Does the firm have adequate and current market research on this segment and segment competitors?
4. Can the company position itself effectively against competitors?

Branding Strategy

1. How is this brand perceived by target segments relative to competitors (awareness, quality, value, prestige, etc.)?
2. Should the hospitality product for this segment be under the main brand or should a new brand or brand extension be considered?
3. What needs to be done to increase brand awareness, brand positioning, and brand attractiveness?

USING GUEST DATABASES
FOR SEGMENT SELECTION

The customer database is well defined as "an organized collection of comprehensive information about individual customers or prospects that is current, accessible, and actionable for such marketing purposes as lead generation, lead qualification, sale of a product or service, or maintenance of a customer relationship" (Kotler & Keller, 2006, p. 163). Building a database usually involves investments in central and remote computer systems, data processing software, information enhancement programs, communication links, specialized personnel, use training, special program design, and so forth. The system should be user-friendly, cost-effective, and available to various departments (Kotler et al., 1998).

I remember visiting a new hotel in Montego Bay, Jamaica, that had invested heavily in a new computerized reservation system that would also allow the firm to keep track of guest preferences and financial information. I was told that the computer system had been installed two years ago at great cost but that it was not being used. On inquiry, I learned it was because the hotel staff had difficulty in using the system, which was complex and required extensive training. Clearly the needs of these employees had not been consulted before installing this system.

Benefits of Database Marketing

Some of the major benefits of database marketing are as follows:

1. It leverages the power of guest information (obtained from internal and external sources, to improve the quality and efficiency of customer communications.
2. It allows for customized treatment of guests using this personalized information.
3. It is necessary for building long-term relationships with the firm's best customers (used in the firm's relationship marketing programs).
4. It is useful for direct marketing of promotions and incentives for target customers.
5. It is valuable for segmentation research since a profile of the firm's current best customers can be used to target other people with a similar profile.

It has been acknowledged that the use of the guest database is a more powerful tool of segment selection than other means such as demographics and psychographics (Hsu & Powers, 2002). The actual preferences profiles and activities of your past and present guests are more reliable tools for segment selection than segment selection based on demographic and psychographic segmentation bases. This is because guests that fit the profile of your present guests are more likely to purchase your hospitality services.

It is common for hotel property management systems (PMS) to have extensive data on customers that have used the hotel in the past. This data is derived from the PMS, registrations cards, old invoices, credit card companies, and business cards. The list of past customers is put on a computer database like Paradox. Certain fields are then established to allow the segmentation of customers. Examples of such fields are location, date of check-out, market segment, amount spent per visit, guest preferences, etc. For example, hotels usually distinguish between leisure and corporate customers. Fields are established in the guest database that allow the hotel to select corporate or leisure customers to contact for marketing purposes.

For example, the marketing department may want to contact leisure customers who have the most frequent visits to the hotel in the past two years. Or they may wish to contact corporate customers who have the highest volume of purchases for the past three years. Since these two groups have very different needs and preferences, very different communications are designed for these two segments. The

next step would be to profile the firm's best customers under the assumption that similar customers purchase similar services. Profiling is done by the use of data from database fields such as age, location, income, type of car, purchase habits, media usage, number of visits to a restaurant monthly, and so forth.

For example, a hotel in Stamford, Connecticut, profiled its past customers and developed a profile for its leisure segment of a married couple with $75,000 per year household income, forty-nine years of age, two cars, stayed one night, ate breakfast and dinner in the hotel, and lived within forty-five miles of the hotel (Lewis & Chambers, 2000).

Measures Applied to Guest Databases

Hospitality firms use their customer database to identify the firm's most profitable customers for promotions, relationship marketing, and developing new customers through profiling. Measures used (see detailed explanation of these measures in Chapter 5) for identifying the firm's best customers include these:

1. recency,
2. frequency,
3. monetary value, and
4. lifetime value of the customer.

STAGE THREE: SEGMENT POSITIONING

I have stated earlier that a successful segmentation process results in the selection of a set of target segments to which the firm can successfully position its services. Positioning refers to the position of the company and its products in the mind of its target segments and relative to the competition. The goal of positioning is to achieve a competitive advantage by providing superior perceived value to these target segments. This is done by customizing the company's marketing mix (product, price, promotion, distribution, communications) to the needs and preferences of these target segments. In the next chapter we will discuss positioning strategies for hospitality companies.

SUMMARY

Many hospitality firms have not progressed far beyond the traditional a priori segmentation bases that were used decades ago. Thus, it is common for airlines to use the product segments of first class, business class, and coach. Restaurant companies use the product segments of fast food, fine dining, and casual dining. Hotels often use the a priori product segmentation of corporate, transients, and groups. But these traditional a priori product segmentation bases are vulnerable to innovative competitors who use post hoc market segmentation based on market research into the needs and preferences of their most attractive target segments.

One of the most important sources of customer segmentation data is the guest database, which should be maintained properly and mined for segmentation selection and maintenance purposes. It is also important to integrate hospitality firm databases so that customer information is available on an enterprise-wide basis to managers and customer contact personnel. This will also facilitate the use of effective relationship marketing programs to enhance customer loyalty and retention.

Segmentation processes differ for different firms and for different purposes. Segmentation selection is an iterative process that includes segment positioning. This is because firms should engage only segments for which they can achieve superior positioning leading to a competitive advantage. And this superior positioning should be communicated to both employees and target segments and integrated into the firm's marketing mix and branding strategy.

EXERCISES

1. Identify your main target segment and justify the selection of this segment for your hospitality organization.
2. List and evaluate the market research methods used by your organization in the selection of target segments.
3. Does your hospitality organization have (or can reasonably hope to achieve) a competitive advantage with its main target segments? Justify your answer.

4. What is the superior value offered by your hospitality organization to one of its main target segments? How do you know that this feature is highly valued by this target segment?
5. What segmentation bases are used by your organization to select target segments? Evaluate this selection.
6. Does your organizations marketing mix (4 P's) best satisfy the needs of its target segments? Explain why or why not.
7. List and justify the business segments targeted by your hospitality organization.
8. Evaluate the segment attractiveness of one target segment of your organization using the eight-point criteria cited above.
9. Which of Kotler's five patterns of target market selection (cited above) applies to your hospitality firm? Evaluate this choice.
10. How can your hospitality firm improve its present target market selection? Explain.

Chapter 9

Positioning Strategy

> Generally, in battle, use the common to engage the enemy
> and the uncommon to gain victory.
>
> Sun-Tsu, *The Art of War,* 500 BC

INTRODUCTION

Market positioning means creating an image in the mind of your target segments regarding your product (or service) and how it compares with the competition. Effective positioning requires that your target segments regard your product as unique, possessing valuable attributes, and having superior value. It means that your brand name will naturally come to the top of their mind as the product of choice when they are in need of your services.

In the twenty-first century, we, as consumers, are constantly being bombarded by advertising and promotion messages in an era of an oversupply of competing brands and products. For example, if I would like to stop at a fast-food restaurant for lunch, which of the dozens of fast-food brands do I patronize? And why? Most of this informational clutter associated with the mass media and the Internet is not remembered or not remembered in a way that is intended by the advertiser. Sometimes we remember an amusing or unusual ad but not what product or service it is intended to promote. Therefore, effective advertising and promotion implies that hospitality firms should differentiate their products and services in the minds of their target customers. But differentiation is not enough. You need to differentiate your product or service in a way that is perceived as valuable by your target segments.

Segmentation Strategies for Hospitality Managers
Published by The Haworth Press, Inc., 2007. All rights reserved.
doi:10.1300/5716_09

Thus, McDonald's is positioned as the leading family-oriented fast-food chain, offering high value, clean restaurants, fast service, and great-tasting food. Southwest Airlines is positioned as a discount direct route carrier that offers low prices, friendly and courteous service, and ticket-less reservations to selected destinations. And Motel Six is positioned as an economy motel that offers low prices for comfortable, clean accommodations with minimal services, in most cities.

The concept of positioning was first introduced in the widely read book, *Positioning: The Battle for Your Mind* (Ries & Trout, 2000). The five basic premises of positioning as set forth in this book are as follows:

1. Positioning takes place in the mind of your target customers.
2. Positioning is based on the concept that effective communication with target customers requires the right time, place, and circumstances.
3. Advertising is about being the first to get into the mind of your target customers.
4. You cannot position your company without first getting into the mind of your target customers.

It is important to determine why customers are purchasing a particular product or service before determining a positioning strategy. These "key drivers" of a particular market will help determine the positioning of the company. These drivers will also be used in the perceptual mapping of the firm's positioning (discussed later in this chapter).

COMPONENTS OF CONSUMER PRODUCT PERCEPTIONS

Three components of customer perceptions of a product or service are *salience, determinance,* and *importance.* Salience refers to what is on the "top of the mind" of customers when they select a product or service. Determinance attributes   actually determine the choice, while importance attributes are those that are important to the customer (Lewis & Chambers, 2000).

Salience

Salient attributes are those attributes that first come to mind for target customers when selecting a product or service but may or may not determine that choice. They are not determining if they are not the differentiating factor that is being sought by the consumer or are common throughout the service class. For example, location is a salient attribute for restaurants but may not be a determining attribute, especially if there are several restaurants in the same location. Price is a salient attribute for hotels but may not determine choice for a consumer who may have several choices within his or her price range (Lewis & Chambers, 2000).

Determinance

Determinance attributes are qualities used by target customers that actually determine choice, such as reputation, price/value, service, and product quality. These attributes are critical to the choice process. In positioning, we use determinance factors in positioning the firm. Thus, Ritz-Carlton hotels and Outback Steakhouse are positioned by high quality and reputation. Motel 6 and JetBlue Airlines are positioned by price/value. Reputation alone may be a determinance attribute for customers who have not yet experienced the product/service and therefore rely more on the image of the company or product (Morritt, 1999).

Importance

Importance attributes are those that are important to target customers in the choice process but may or may not be a determining attribute for those customers. For example, an important attribute may be bathroom amenities for a prospective hotel customer. It is important that these amenities exist, but they are usually not a determining attribute, except in their absence. This means that this feature would not be sufficient in itself to determine hotel choice, but its absence would be sufficient to reject the option of this hotel. Importance factors are often used to arouse interest in the product or service or create a benefit bundle that will lead to determinance.

ELEMENTS OF POSITIONING

Good positioning is said to depend on three criteria:

1. creating an image,
2. differentiating the product/service, and
3. making a promise (all based on salient and/or determining attribute and/or importance factors (Lewis & Chambers, 2000).

Positioning is a strategic marketing process that has as its objective the establishment of the perception of a highly valued, distinctive, and/or superior product in the minds of target customers. It is also about how the firm's products are perceived by these target customers relative to competitive products. The firm's positioning is determined, in part, by existing competing brands in the market. Customers first assign a position to the best-known or leading brand, which then becomes the standard against which other brands are compared.

Thus, McDonald's brings to mind the leading child-friendly, clean, fast-food, family restaurant specializing in tasty hamburgers and fries where customers can expect the same experience worldwide. Its well recognized global symbol is the "Golden Arches," found at all of its locations. Southwest Airlines has positioned itself as a friendly, no-frills airline with direct connections to its target cities at discount prices. Finally, Marriott Courtyard has positioned itself as a place where independent business travelers (who are not on an expense account) can find clean and comfortable accommodations with essential business facilities and good service at reasonable prices.

METHODS OF POSITIONING

Firms can position themselves for a competitive advantage by either differentiation, low-cost leadership, or focus (niche marketing using differentiation or low-cost leadership) (Porter, 1985). For example, a hotel brand like Embassy Suites can position itself by differentiation by offering all suites to its target customers. JetBlue Air-

lines has used the focus (niche) strategy to offer discount service with unexpected luxury (e.g., leather seats, programmable monitors) to its target niche of leisure travelers. McDonald's is a low-cost leader in fast food due to economies of scale associated with its size and automation methods.

Hospitality firms, according, to Kotler et al. (1998) can *differentiate* themselves in the following ways:

1. *Physical attributes:* Hotels may exhibit unusual architecture or furnishings.
2. *Service differentiation:* Some restaurants offer home delivery as a service differentiation.
3. *Personnel differentiation:* Cathay Pacific and Singapore Airlines are well differentiated by the courtesy and grace of their flight attendants.
4. *Location differentiation:* Mohonk Mountain House resort of New Paltz, New York, is uniquely situated near a large lake surrounded by mountains with hiking trails which provide unique recreational opportunities for guests.
5. *Image differentiation:* The Ritz-Carlton hotel attracts high-paying executives partly because of its image of being a premier upscale hotel with the highest service quality and patronized by the rich and powerful.
6. *Price differentiation:* Motel 6 differentiates itself by its low prices and no-frills, yet clean and comfortable, accommodations.
7. *Value differentiation:* By value we mean guest perceptions of what is received from the hospitality firm (benefits, convenience, service, etc.) versus what is paid for those services (money, time, stress, etc.). Thus, a hotel such as Marriott Courtyard, which is moderately priced and has business facilities but has minimal services, was perceived as a superior value by its target customers (independent business travelers) as indicated by the success of this brand and its many imitators. This hotel was successful because it filled a need that was discovered as a result of segmentation research by the Marriott hotel corporation into the needs and preferences of independent business travelers.

POSITIONING STRATEGIES

Hospitality companies can adopt one of the following positioning strategies (Kotler et al., 1998; Lewis & Chambers, 2000). Products can be positioned by:

1. *Specific attributes:* Disney advertises its theme and adventure parks.
2. *Needs or benefits:* Bennigans restaurants advertises itself as a fun place.
3. *Price/quality:* Motel 6 advertises a very low price for its accommodations.
4. *Use or application:* Enterprise car rentals has a service which picks up their clients at their present location.
5. *Users or class of users:* Some hotels advertise themselves as a women's hotel. The hotel Palomar in San Francisco caters to women travelers with a wide range of available women's amenities. This ranges from nylons delivered to the room to full-length mirrors, special makeup mirrors, and special bathroom lighting.
6. *Against the competition:* Burger King used the "flame broiled" method against McDonald's grill method of cooking burgers.
7. *Against a product class:* McDonald's "You deserve a break" campaign was positioned against home-cooked meals. Hotel conference centers have positioned themselves against hotels with conference facilities (Kotler et al., 1998).

EFFECTIVE POSITIONING

Effective positioning results in your product/service occupying a distinctive place in the minds of your target customers. Your company's position should be both highly valued and preferred relative to your competition. Thus, Wal-Mart has positioned itself as a superstore that has everyday low prices on quality brands and offers superior value to its global customers. JetBlue has positioned itself as a low-cost regional airline that offers great service and luxury accommodations to its singleclass passengers. Denny's is famous for its delicious value breakfasts, which include its famous "Grand Slam" breakfast for only $2.99. This includes two buttermilk pancakes, two

strips of bacon, two sausages, and two made-to-order eggs. This brings in customers for its other offerings for all three meals.

The Value Proposition

The end result of this positioning is held to be the successful creation of a customer-focused value proposition, a cogent reason why the target market should buy the product (Kotler, 2005). Webster (1994) defines the value proposition as "a verbal statement that matches up the firm's distinctive competencies with the needs and preferences of a carefully designed set of potential customers." What makes positioning effective is that this value proposition is both well-known by the target market and also motivates the customers in these target segments to purchase the firm's offerings over the competition.

Examples of hospitality industry value propositions are the following:

- Disneyland: The world's premier family theme park with rides and attractions for the whole family
- Southwest Airlines: A friendly airline with no frills and low prices
- Holiday Inn Express: A family-friendly, midpriced hotel with free meals and accommodations for children
- McDonald's: A kid-friendly, fast-food restaurant offering quality, cleanliness, and value

Three requirements for effective positioning cited by Myers (1996) are uniqueness, desirability, and believability.

Uniqueness

Uniqueness refers to positioning your product in a way that is perceived as important to your target customers. But it may also be more desirable to be different in features that are not as important than to be the same or slightly better than the competition in the most important attribute (Myers, 1996). Thus, it may not be the best positioning strategy to offer the best-tasting steak if all competing steakhouses are competing on the same attribute. Therefore, Outback Steakhouse is

positioned as a friendly steakhouse with an Australian theme and excellent service.

Importance/Desirability

Beverage firms such as Shenley and Coca-Cola have (in the past) differentiated their products by introducing colorless beverages. But these products did not succeed because this kind of uniqueness was not deemed important by target customers. Market research is employed by successful companies to find the "hotspot" that will turn prospects into buyers. These hotspots may refer to truly superior attributes, or, in their absence, imagery, celebrities, and humor are used to position products. This strategy is used to create positive feelings toward the brand (Myers, 1996).

Thus, United uses the "Fly the Friendly Skies" theme to advertise their airlines, which serves to create positive feelings toward their brand but does not capitalize on a truly superior attribute. On the other hand, Burger King's advertising focusing on its flame-broiled burgers and having your burger "your way" are examples of citing ways in which the company was superior to its competition in the minds of its target customers.

Believability

General claims of superiority or being the best in some attribute or product without evidence are not usually believable to consumers. That is one reason for the success of the J.D. Powers customer satisfaction ratings for various products and services. It is one thing to say that Dell has the best service and quite another to announce that Dell has won the J.D. Powers Best Service award for computer firms three years in a row! Dell also received (along with Apple) the highest customer satisfaction ratings of the American Consumer Satisfaction Index (ACSI) for 2004 as cited in *Computer World* magazine (www .computerworld.com/managementtopics/management/helpdesk/story/ 0,10801,95454,00.html).

This objective third-party evaluation renders the claim of superior service more believable in the minds of prospective customers. In general, the more reasonable the claim and the more objective the support for that claim, the more believable it will be.

Another error is to confuse the customer by making too many positioning claims for a product or service. Your hospitality company should ask the following questions (cited by Myers) prior to creating their positioning:

1. Is my (current or proposed) positioning already taken by another company?
2. Does my positioning depend on features or benefits perceived as important to my target segments?
3. If so, what is the valid market research results that supports this claim?
4. How credible is our positioning to our target segments?
5. What evidence can we present to make our positioning claims more credible?
6. What is the cost and scope of the promotions that will succeed in planting our positioning in the minds of our target segments (Myers, 1996)?

Effective positioning must promise the benefits that the customer will receive, create realistic expectations, and offer a solution to the customer's problem. And that solution, if possible, should appear to be different and better than the competition's. It is also important to deliver on customer promises if you want your positioning to be successful over the long term (Lewis & Chambers, 2000). The following are examples of hospitality positioning advertising slogans that promise a distinctive benefit:

- McDonald's: "We do it all for you"
- Burger King: "Have it your way"
- Holiday Inn: "Start with someone you know"
- Embassy Suites: "Twice the hotel"
- Marriott: "Service: The ultimate luxury"
- United Airlines: "Fly the friendly skies of United"

But positioning goes beyond advertising and slogans. It involves the integration of the firm's positioning strategy into the marketing brochures, facilities, policies, training, customer relations, and complaint management. Consistency in service delivery strengthens the desired image of the brand, and integration of the firm's positioning

statement into the marketing mix, and functional departments ensures this consistency.

BRANDING AND POSITIONING

Branding is a major factor in the positioning of hospitality companies. Hotels and restaurants often go through a process of brand switching due to changes in ownership and repositioning.

Increasing consolidation of the fragmented hospitality industry during the past two decades resulted in the increasing dominance of major global hospitality brands who have the advantages of economies of scale, the best managers, global diversity, and the funding to provide the latest technology and mass marketing. Names like Marriott, McDonald's, and American Airlines dominate their respective industries. Currently the airline industry is witnessing the increasing power of niche discount airlines like Southwest and JetBlue, which are thriving due to being unencumbered by the expensive infrastructure and labor costs of the older carriers. These discount airlines have successfully positioned themselves as friendly, low-cost, regional carriers.

Multiple Branding Strategy

Major hospitality companies often develop multiple brands for the purposes of expanding market coverage, developing new market niches, and reducing business risks by diversification. Thus, Marriott developed the Courtyard (midprice segment), the Fairfield Inn (budget segment), Residence Inn (extended-stay segment), Marriott Marquis (convention hotel), Marriott Suites (luxury all-suite hotels), and the J. W. Marriott and Ritz-Carlton as upscale luxury hotels.

Brinker International, a premiere casual dining company, has been called the "mutual fund of casual dining." This stems from a powerful portfolio of outstanding concepts which target different customer segments, including Chili's Grill & Bar, Romano's Macaroni Grill, On The Border Mexican Grill & Cantina, Maggiano's Little Italy, Corner Bakery Cafe, Big Bowl Asian Kitchen, and Rockfish Seafood Grill.

Two major problems associated with a multiple branding strategy are cannibalization and image dilution. A multiple branding strat-

egy sometimes protects against the erosion of the brand by niche competitors by self-cannibalization. That is, the company develops niches that may cannibalize its own brands to defend against competitors doing the same thing. But it is even better if the multiple brands in the company's portfolio are differentiated (like the Brinker portfolio) and thus involve minimal cannibalization. In addition, using different names for its different products avoids the image confusion associated with using the same name for very different hospitality products (Lewis & Chambers, 2000).

One advantage of using the same name (used by some international hospitality companies like Hilton Hotels) is the integrated advertising message and the halo effect of a strong brand such as Sheraton Hotels supporting weaker properties. An advantage of using different names is the ability to clearly position each brand to focus on different target segments.

Niche Marketing As an Effective Positioning Strategy

In recent years, traditional hotels with a ballroom, conference rooms, full-service restaurants, a bar, and exercise room have had their market share eroded by niche hotels with specialized products such as conference centers, specialized food and beverage (F&B) outlets, fitness centers, bed and breakfast places (B&Bs), all-suite lodging, condo hotels, and meeting rooms within convention centers. When two or more firms pursue the same position, further differentiation is required. Each company must develop a unique bundle of valued attributes that appeals to a substantial subgroup within the segment. This sub-positioning is an example of niche marketing (Kotler et al., 1998).

PERCEPTUAL MAPPING: A POSITIONING TOOL

Perceptual maps, sometimes known as positioning maps, are a way of mapping your own and competitors' positioning in your particular market based on research of consumer perceptions. Techniques used in the construction of perceptual maps range from simple plotting on an arbitrary scale to statistical methods such as multidimensional scaling and discriminant analysis.

Perceptual maps can be used to find gaps or niches in the market and also to correct poor positioning due to overcrowded markets. They often are focused on a small number of industry drivers: features that are most important to members of target segments. The perceptual map displays in two-dimensional space the physical positioning of several competing companies with respect to these critical industry drivers. Thus, the first perceptual map (Figure 9.1) focuses on the hotel industry drivers of price and service, while the second

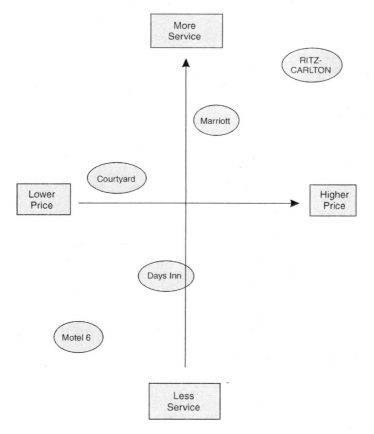

FIGURE 9.1. Positioning Map of Service versus Price of some U.S. Hotels (*Source:* Author)
Note: It can be seen that there is a market gap (assuming that these are the only hotels in the market) for lower price, higher service hotels.

(Figure 9.2) focuses on the fast-food restaurant drivers of convenience and cleanliness.

Perceptual maps are tools used to measure and evaluate the positioning of a brand. They are used to identify opportunities for creating the desired image that differentiates the firm from the competition in a way that is valued by target segments. Perceptual maps vary from a simple two-dimensional map (using two attributes) to more sophisticated multidimensional maps using several attributes.

From the two-dimensional perceptual map of competing hotels seen in Figure 9.1 we can see that the Grand and Regency are both positioned as expensive high-service hotels, so they must do something to differentiate themselves. The Airport Plaza, on the other hand, is

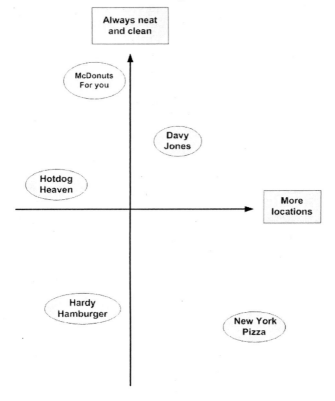

FIGURE 9.2. Perceptual Map of Hypothetical Fast-Food Restaurants (*Source:* Author)

well positioned as the only inexpensive low-service hotel in this group. The Italia hotel is well positioned as a value hotel that is relatively inexpensive but offers moderately high services. On the other hand, the Palace hotel may have to think about repositioning, since it is a fairly expensive hotel that does not rate high on services.

Looking at the horizontal axis of the perceptual map in Figure 9.2, McDonald's is perceived as having the most convenient locations and Wendy's is perceived as having the fewest. Looking at the vertical axis, Wendy's has a slight advantage over most of its competition as far as being neat and clean. Jack in the Box was rated the lowest in this category (Myers, 1996). But there seems to be an opening here for a fast-food restaurants that have higher "neat and clean" ratings.

Figure 9.3 displays a hypothetical positioning map that shows how four airlines might be positioned relating the variables of on-time service and customer service. (It is assumed that consumer research has determined that these two variables are important to travelers.) This map indicates that JetBlack is performing best on these two variables, with Fly-by-Night doing the worst on these two variables. Possible strategies for NorthAmerican and Fly-by-Night are to improve customer service or on-time service, or to excel on another variable that is important to their target market (e.g., price, airline miles, etc.).

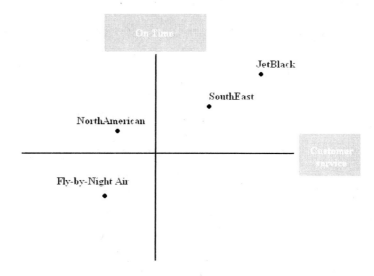

FIGURE 9.3. Hypothetical Airline Positioning Map (based on fictitious data)

THE POSITIONING PROCESS

Lovelock (1991) diagrams the overall segmentation-positioning process, which begins with market analysis and ends with the positioning of the firm's products/services and the implementation of that positioning in an action plan.

Steps in this process include the following:

1. *Market analysis:* Market analysis is required to identify attractive segments, identify significant environmental forces and marketing trends, assess demand, and understand the purchase behavior of target segments.
2. *Internal corporate analysis:* This requires the identification of corporate resources (financial, management skills, core competencies, corporate goals, and strengths and weaknesses of the company).
3. *Competitive analysis:* This refers the identification and analysis of current and potential competitors for the company's target segments, including their positioning, strengths, and weaknesses.
4. *Development of a positioning statement:* The outcome of this process is a positioning statement. This statement should indicate a position that is distinctive, highly valued by target segments, and, preferably, associated with a competitive advantage for your company. The positioning of your company needs to be continuously monitored by market research in the light of changing environmental conditions. At times, these changing conditions will require the repositioning of the company.
5. *Marketing action plan:* This refers to the integration of the company's positioning statement into its marketing mix to ensure integration of this positioning into its strategy and global communications.

SUMMARY

In Chapter 8 we reviewed the STP (Segment, Target, and Position) segmentation process, which has, as its major objective, the successful positioning of the firm's product/services to its target segments. In this chapter we have defined positioning, discussed its strategic im-

plications, and reviewed different positioning strategies of hospitality companies.

The hospitality industry of the twenty-first century is a global marketplace of many companies and brands competing for different target segments. The competitive hospitality company must achieve brand awareness and preference in a fiercely competitive dynamic environment amid the hypercluttered advertising barrage of modern mass media. To achieve this goal, the hospitality firm must employ effective market research (including segmentation studies, market trend analysis, competitive intelligence, environmental scanning, and an internal assessment [SWOT analysis]) to determine the most attractive target segments. These are the most attractive segments which have the best strategic fit relative to the resources and competencies of the firm, significant environmental forces, and sustainable marketing trends

Research into the needs, preferences, and purchase behavior of target segments and the external environment, together with a competitive analysis, will form the basis for a distinctive and customer-oriented positioning statement that will best position the firm relative to the needs of its target segments. Perceptual maps are a useful analytical tool in this research which can reveal not only current positioning relative to competitors but also indicate gaps in the market that can be effectively filled by your hospitality firm.

An effective positioning for your firm is one that is well differentiated, is highly valued by and credible to your target segments, delivers superior perceived value, and is supported by what is actually delivered by your product and services. This is also the kind of positioning that will build the brand of your hospitality company.

EXERCISES

1. Draw a perceptual map of the positioning of your company with respect to two variables that are most important to your customers. Include circles that represent at least three competitors. Do you think that your company is well positioned or not? Explain.
2. Evaluate the positioning of your hospitality firm with regard to uniqueness and what is important to your target customers.
3. It has been said that effective positioning means that the company is perceived by its target customers as providing *superior*

value relative to competitive services and products. Evaluate your hospitality company with respect to the value of your services as perceived by a major target segment. What is your justification for this claim?

4. How often do you think that a hospitality company should reposition its products/services and why? What factors would indicate the need for repositioning?

5. What is the relationship between brand equity and effective positioning? Can you have effective positioning without a strong brand? Why or why not?

6. Why should hospitality companies be interested in niche marketing strategies, and how is this related to positioning?

7. Explain and evaluate the positioning of two leading hospitality companies (e.g., McDonald's, Southwest Airlines, Marriott hotels). How could this positioning be improved?

8. You have been hired as a marketing consultant to your hospitality company to develop a new positioning statement for your company. State this positioning statement and why you believe this to be an effective one.

9. Choose a hospitality company in your sector (hotels, airlines, restaurants, etc.) that is not doing well now and suggest how you would reposition this company to be competitive again.

10. You are marketing director for a hospitality company that has been directed to explore niche opportunities in China for a hotel or restaurant. Identify what you believe to be a lucrative niche for your company and position your company for this niche market.

Chapter 10

The Impact of New Technology on the Hospitality Industry

> Information technology is the single greatest force affecting change in the hospitality industry. Everywhere one looks, one can find evidence of how technology is reshaping the industry and changing the very dynamics of competition and customer/ employee interaction. *Going forward, technology will be the most important competitive weapon for any hospitality company.* If hospitality organizations want to compete successfully, they must do so by using technology to drive value to both the customer and to the firm. These are among the conclusions of two Technology Think Tank sessions sponsored by the International Hotel and Restaurant Association (IH&RA). (Olsen & Connolly, 1999, p. 29, emphasis added)

This citation represents the conclusions of two "think tank" conference of top worldwide hospitality managers hosted by the International Hotel and Restaurant Association (H&RA) in 1997 and 1998 (Olsen & Connolly, 1999). Eight major drivers of IT change (identified by these managers) affecting the hospitality industry can be found in Exhibit 10.1.

In this chapter we will be discussing elements of new technology, both technology currently used by hospitality firms and exciting new technology that is anticipated for the future. Examples range from the plans for a hotel in space by the Hilton hotel corporation to touch screen, self-service, point-of-sale systems for fast-food restaurants, facial recognition database security systems for air travel, and data-mining technology using artificial intelligence to discover new target segments for hospitality companies. We will also be discussing the impact of new technology on hospitality segmentation strategy.

Segmentation Strategies for Hospitality Managers
Published by The Haworth Press, Inc., 2007. All rights reserved.
doi:10.1300/5716_10

EXHIBIT 10.1.
Eight Drivers of IT Change
Impacting the Hospitality Industry

1. The Internet	This refers to the emergence of the Internet giving customers instant access to information about companies, products, and prices.
2. Real-time communications	Customers demand faster response times due to real-time processing using faster servers.
3. Regulating cyberspace	Future government regulations will change the current freewheeling nature of the Internet.
4. Data warehousing and data mining	Data warehousing and data mining technologies are gaining in popularity as tools for managing customer information, including lifetime value of the customer.
5. Segments of one	One-to-one relationship marketing is enabled by computer technology. This will facilitate customized marketing to the firm's best customers.
6. Archaic technology	Hospitality firms are hampered by old legacy systems that are single purpose (e.g., property management systems). Integrated systems and a multimedia approach to customer transactions will become the rule.
7. IT for executives	Because IT lies at the core of hospitality strategy, all managers will need to be familiar with the potential advantages and limitations of IT applications. And employees will need to be better trained in their use.
8. Cost of technology	Resistance to IT hospitality investments comes mainly from independent owners and franchisees. Part of the problem is that expensive IT systems must be designed to be more cost-benefit effective in the future.

Source: Adapted from Olsen & Connolly, 2000.

TECHNOLOGY AS A MARKETING TOOL

Despite widespread enthusiasm for the opportunities related to new technology, it is clear that technology should not be viewed as a quick fix for marketing problems. Hospitality managers and owners are rightly concerned about the cost-effectiveness of expensive new technology for their particular operations, as well as the value of and perceived satisfaction with this new technology for their target segments. Thus, new technology can be viewed as a two edged sword. This marketing weapon has the potential to provide hospitality firms with improved positioning and customer satisfaction. But it may also be a weapon that cuts against your firm by decreasing customer and employee satisfaction with expensive new technology that is often expensive, difficult to use, and does not meet customer or employee needs or preferences.

Information technology (IT) has become a primary marketing tool of hospitality companies. It is a tool that can improve the segmentation and positioning activities of your company, improve brand image, and may also lead to competitive advantage.

Hospitality researchers have held that one premise of the new information age is that hospitality firms in the future will build their success on how much they know about their customers and how they use technology to provide customers with both useful information and effective delivery of services. (Olsen & Connolly, 2000)

Liabilities of New Technology

But there are also liabilities associated with the use of technology. Results of a 1998 hospitality marketing think tank involving thirty marketing experts indicated that technology was one of the major challenges facing the hospitality industry in the future. This challenge is related to the great expense of keeping up with technology, the overload of data, as opposed to useful information, and the complexities of its use. IT systems were found to be opaque to many senior managers who need to be familiar with how it works, how it benefits the company, and what value it adds to the customer experience (Dev & Olsen, 2000). It was recommended that marketers be given additional technical training, that IT employees should report to the marketing department, that companies should clearly define knowl-

edge needs and use technology to transfer data into useable knowledge, and, finally, that hospitality companies should broaden the technical understanding of all employees. These marketing experts saw IT as absorbing immense amounts of money for which hospitality companies were often receiving too little value (Dev & Olsen, 2000).

I remember visiting a brand new beachfront hotel at Montego Bay, Jamaica, some years ago which was deserted with the exception of a single uniformed clerk working behind a counter. I asked her why this new hotel was not open for business for the past few months. She told me that the owner had purchased an expensive new computerized reservation system but the employees had no training for this system and they did not know how to use it. Lack of adequate training and preparation postponed the opening of this hotel for six months during the busiest tourist season.

DRIVERS OF IT CHANGE

E-Commerce on the Internet

The first driver of IT change cited above is the Internet. The Internet is a major channel of distribution for the hospitality industry and produces the highest online revenues of any industry type, including personal computers and books. The American Hotel and Motel Association estimated that online hotel bookings were $2.9 billion in 2002 (Lewis & Chambers, 2000).

One positioning advantage is that advertising, reservations, and sales pages on the Net can be customized to different target segments and integrated with the company's enterprise-wide reservations systems and yield management systems to obtain optimal pricing structures.

Another advantage is the use of the Internet by hospitality companies for the purposes of collecting data for segmentation studies, advertising services, e-commerce, and other market research. Data is often automatically collected from the keystrokes of customers who enter the Web sites and also the "cookies" that have been deposited on the customer's computer by different Internet companies.

The increased use of the World Wide Web, where more than 60 percent of U.S. households have access to the Internet, has enhanced the use of the Internet for the following purposes:

- Collecting detailed information on customers through the use of "cookies" which record the activity of users, including where they go on the company Web site, how long they stay, and their interests, activities, and purchases. Information is also collected from customers registering for the Web site or various promotions.
- Using this information, the company can create customized ads and promotions to target segments. Furthermore, the Web site can be made to have a different appearance that is customized to the needs of different target segments.
- Use of the Internet for hospitality companies for advertising, reservations, sales, and market research.

However, Internet reservation booking using "booking engines" like Travelocity.com and Priceline.com are eroding the former dominance of the GDS (Global Distribution System) bookings and compelling travel companies like American Airlines to develop their own booking engines such as Orbitz.com.

Internet Shopbots

Shopbots are software-based virtual shopping robots that allow the consumer to use a specialized and customized search engine that is programmed with the individual's shopping criteria. These shopbots simultaneously search multiple search engines and Web sites, and present the consumer with the results of this search in a compact and concise manner that allows comparisons with different offerings of the desired product or service. An example would be Anderson Consulting's Bargainfinder (Jasco, 1998).

Farechaser is another vendor of shopbot software and targets corporations. The shopbot software of this company retrieves travel inventory from over 150 Web sites and presents the results in a standardized format that is sorted by preset parameters such as price, supplier, number of connections, and so forth. This enables client companies to obtain significant savings by easy and fast access to the

lowest prices. The program guides the user through a fast and easy booking process by taking the user directly to the transaction page on the travel provider Web site and prepopulates all the relevant information. The program also tracks and reports all user activity for its corporate licensees for accounting and analysis purposes (www .farechaser.com).

Although wholesale use of travel shopbots is still in its early stages, the implications of the use of such shopping robots are critical to the segmentation and positioning strategies of hospitality managers. This kind of technology clearly reduces the pricing power of hospitality companies and makes it imperative for companies to continuously monitor the competition to ensure that their company is not offering an inferior value to its target customers. Some companies (who presumably do not wish to compete on price) will decide not to allow their Web sites to be used by these shopbots. For example, one-third of music Web sites decided not to allow Bargainfinder access to their stores (Jasco, 1998).

Global Distribution Systems (GDS)

GDS systems are the primary technological platform of airlines and hotel reservations systems and are used by hospitality companies, travel agents, and online customers for booking and reservations. The major providers of GDS systems are Amadeus, Sabre, Galileo, and Worldspan, which together represent 89 percent of the market.

Special advertising functions are available on GDS for hotels and airlines to market special promotions. GDS originally represented a closed dedicated connection of terminals in travel agencies displaying information about airlines, hotels, car rentals, and other travel products. This system has since expanded to include additional products, advertising, and functions (Lewis & Chambers, 2000).

Central Reservation Systems (CRS)

Central reservation systems (CRS) allow booking from an 800 number and are integrated with the firm's GDS and PMS systems. This system was created so that individual hotel chains could provide global access to their global inventories through toll-free phone calls.

If the system is working correctly, the customer will get the same rates available on the GDS and at the hotel property.

Property Management Systems (PMS)

This is the primary technology at a hotel. This ranges from a simple check in and check out system to an integrated system that is linked to all the revenue areas of a hotel and produces instant reports on all these areas. The marketing advantage of a PMS relates to guest history, as most PMS build guest history files based on previous stays. Hotels like Ritz-Carlton have an enterprise-wide guest database, which allows a guest leaving a hotel in New York to obtain the same amenities in a hotel in Montego Bay. Future applications of PMS systems are projected to include menu selections, gift shop purchases, movie rentals, and other purchases made during the guest's stay at the hotel.

Point-of-Sale Systems

Point-of-sale systems are the food and beverage functions of the PMS, but they are also used by stand-alone restaurants. Valuable information relating to menu preferences, average waiting time per entree, and availability of entrees are available to waiters and managers. For hotel restaurants, all transactions for cash, house accounts, credit cards, and room service are instantly available to the guest for balance inquiries. POS systems are also being used to manage frequent diner programs. Customer preferences can also be stored in a way similar to the PMS system. This allows the guest to be greeted by name, seated at a favorite table, and offered the appropriate menu, which is managed by the new software available on POS. QSR (fast-food restaurant) customers can place their orders on a touch screen computer terminal (Hsu & Powers, 2002).

In-Room Technology

Many hotels now offer in-room business services that were once offered only by the hotel's business center. These include the ability to e-mail, print, and fax documents, two-line telephones, modems, and broadband access.

Also part of new technology offerings are in-room electronic safes, in-safe recharging for portable computers, bedside electronic lighting controls, and interactive TVs with pay-per-view movies. Interactive guest concierge systems allow the guest to retrieve information on local dining and shopping, attractions, recreations, entertainment, and hotel services. Future offerings will include personal computers, which will allow guests to travel with only a compact memory stick or a portable hard drive which weighs less than two pounds (available at leading online computer stores like Buy.com).

Database Marketing

Database marketing is defined by the National Center for Database marketing as

> managing a computerized relationship database system, in real time, of comprehensive, up-to-date, relevant information on customers, inquiries, prospects and suspects, to identify our most responsive customers for the development of high quality, long-standing relationships of repeat business by developing predictive models which enable us to send and receive desired messages at the right time, to and from the right persons giving us the result of pleasing our customers, increasing response rates . . . lowering our cost per order . . . and building our profitable business. (www.dbmktg.com/basic_principles.html)

Customer information derived from company records and market research is usually stored in central databases or *database warehouses,* which are often linked to other hospitality systems such as decision support systems, reservation systems, yield management systems, expert systems, and property management systems. The use of customer databases enables hospitality companies to track and retrieve customer preferences and purchase behavior that can be used for the purpose of

- segmenting and profiling current customers,
- finding new customers that fit the same profile,
- positioning the firm to these target segments, and
- facilitating promotions to target customers.

Your company's customer database is arguably the most reliable tool for segmenting, and positioning to your firm's best customers. This is because your current frequent customers, and those who fit the same profile as these customers, have the highest probability of purchasing your services. Hair et al. (2003) cite the most important objectives of the customer database to be

1. improving the efficiency of market segment construction,
2. increasing the probability of repeat purchase behavior, and
3. enhancing sales and media effectiveness.

The main benefits to the firm are facilitating the exchange of information with customers, determining heavy users (frequent guests), determining lifetime customer value, and building segmentation profiles. Using a "lifetime value of the customer" algorithm, the customer database allows the firm to project what business it can expect from, and calculate the optimal amount to spend on marketing to a particular customer to keep her satisfied and loyal (Hair et al., 2003).

Data-Mining

Data mining is the process of finding hidden patterns and relationships among variables/characteristics stored in the data warehouse. Data mining is a data analysis procedure known primarily for the recognition of significant patterns of data as they pertain to particular customers or customer groups (Exhibit 10.2). For example, management in a casino business may ask the question: what attributes characterize the gaming customer who has the largest gambling budget in our casino last year. Data mining techniques would be employed to search the data warehouse, capture the relevant data, categorize the significant attributes, and form a profile of the high-budget gambler.

Hospitality firms are faced with the buildup of vast amounts of information in their customer databases in their efforts to segment and effectively position themselves to their target segments. Using these stores of information, data-mining technology is an automated process that extracts meaningful patterns and builds predictive customer behavior models that aid in decision making. Data-mining as opposed to statistical modeling builds new models rather than verifying the theory-driven hypotheses of researchers. For example, Marriott

EXHIBIT 10.2.
Uses of Data Mining Techniques in Hotel Marketing

1. Creation of direct mail campaigns; including recipients, timing, and placement of ads
2. Seasonal promotions
3. Construction of new segmentation bases using new models
4. Calculation of reserved rooms for organizational customers
5. Calculation of most profitable market segments using new models
6. Construction of personalized ads
7. Yield management models

Source: Adapted from Magnini et al., 2003.

Vacation Club International, using data-mining technology, reduced the volume of direct mail needed to reach target sales levels by correlating response rates to specific vacation offerings and specific consumer characteristics (Magnini, Honeycutt, & Hodge, 2003).

Data-mining has the advantages of high gains in performance, speed of use, user friendliness, and the ability to handle large and complex databases. Highly specialized knowledge of statistical principles is not required by managers who use this technology, although a basic knowledge of statistical principles is necessary. Data-mining also helps managers spot trends more quickly (Magnini et al., 2003).

Some Limitations of Data-Mining

1. Data-mining analyzes only data from current customers derived from reservation, POS, and guest loyalty systems. Therefore, data-mining provides no information about market segments not found in the company's database systems. For similar reasons, no information is obtainable from market segments contained in competitors' databases.
2. Databases used in the mining process are usually brand specific. Therefore, companies employing a multibrand strategy are obliged to create a data warehouse and conduct data-mining for each brand.

3. Data-mining does not segment customers by psychographic attributes or provide information about customer thought processes. This is one reason why this type of research needs to be supplemented with other traditional types of market research (Magnini et al., 2003)

Concierge Expert Systems

Concierge expert systems are computer systems that can communicate with guests, providing them with personalized information. One example would be responding to a guest that would like to visit an Italian restaurant that has a romantic atmosphere and singing waiters.

Limitations of Concierge Expert Systems

Limitations of existing concierge expert systems to date (2006) include the following:

1. The concierge system is unable to *make* the reservations.
2. The concierge system is unable to learn through experience, although the integration of artificial intelligence systems may enable this feature in the future.
3. The concierge system is unable to give reasons for its repsonses and recommendations.
4. The concierge system is unable to make conversation or reply to questions.
5. The concierge system is unable to give directions.
6. The concierge system is not customized to the needs and preferences of target customers.
7. The concierge system is unable to make reservations based on experience (Cho, Sumiclirast & Olsen, 1996).

However, the potential of such expert systems is important for obtaining guest information for guest databases, customizing services for guest preferences, improving positioning to the firm's best customers, and increasing guest satisfaction while freeing personnel for other duties. It is also a source of income from advertising fees for products and services contained in the concierge system. Finally, the

use of the latest technology can improve the brand image and perceived quality of your brand for its target segments.

Security Issues

At no time has personal security been a more important concern to global hospitality customers than after the 2001 attack on the World Trade Center in New York City. This act of terrorism had a devastating, if short term, effect on travel and tourism in the United States. Security systems are critical in the post-9/11 era, where today's hospitality customers are concerned about terrorism, disease epidemics (e.g., AIDS, SARS, bird flu), kidnapping of executives, and crime.

Security systems used range from advanced x-ray systems and facial recognition systems to protect travelers at airports to keyless smart card entry systems and hidden cameras at hotels, which also monitor the guest rooms and hallways for safety and conservation issues.

Wireless Internet Access (WI-FI)

The number of hotels in the world with WI-FI (Wireless Internet access) has grown from 2,500 in 2003 to an estimated 6,000 in 2004, and is expected to reach 35,000 by 2008. In addition, the number of WI-FI users is growing from 12 million globally to an estimated 707 million by 2008. Intel announced that 90 percent of all laptops shipped in 2004 will have Centrino mobile wireless technology. Major chains such as Marriott, Hyatt, Starwood, and Omini hotels are using this technology to improve their positioning to the lucrative business traveler segment (Valhouli, 2004).

The advantage of using WI-FI is that it provides more flexibility and mobility. Guests can leave their hotel room and can work at another location in the hotel. This could be a meeting room, restaurant, or even outside by the pool or the ocean. The only requirement is that the guest be in the hot spot or reception zone for WI-FI access.

Virtual Reality (VR)

Virtual reality is a computer simulation of a real or imaginary environment which provides the illusion of actually being there, including a three-dimensional visual field which the user can navigate. Fu-

ture VR programs may add other sense perceptions (such as sound, smell, and taste) to this illusion to afford greater verisimilitude.

Virtual reality technology is now being used by hospitality marketers on their Web sites to provide prospective guests with a 360 degree interactive view of hotel facilities, resorts, cruise ships, and adventure tours. One example is the use of Ipex-enabled 360 degree VR tours of the Grand Canyon, Washington, DC, and Cuba by the Travel Channel's *Virtual Journeys*.

But this is only the beginning. Virtual reality technology promises a future where guests can take vacations to distant lands, tour exotic destinations, and engage in exciting adventures from the comfort of their living room. This technology, which is already being used to allow prospective guests to tour hotels, cruise ships, and vacation sites, has wide-ranging implications for the travel and tourism business.

Examples of the exciting potential of this technology cited in the following text range from getting the view of a football game from the perspective of the traveling football on the way to the receiver to receiving all the sensations of walking along a beach in Tahiti.

VR Applications in the Hospitality Industry

Digital Tech Frontier, a pioneer in the design, development, and delivery of virtual reality systems, has created immersive reality systems for national hospitality customers such as Disney, and Hyatt Regency ("Launch of virtual vacations," 2003). DTF has created virtual vacations that provide the viewer with 360 degree spherical environment as well as intuitive navigation allowing the viewer to navigate in this virtual environment. The first three titles in this virtual vacation series are tours of the ancient Mesa Verde Anasazi culture in a remote region of Colorado, Machu Picchu of the ancient Inca culture in Peru, and the engineering marvel the Hoover Dam, located on the border between Colorado and Arizona. VR "guests" report the advantages of the low cost, diminished risks, avoidance of sore muscles, and going at their own pace.

Although VR technology provides distinctive benefits to the hospitality industry over the short term, it also poses the clear threat of substitute services over the long term. There is no near-term threat, but the quality of VR experiences will tend to approach the quality of

real experiences while having the following advantages over reality-based travel experiences:

1. Significantly lower costs in money, effort, and time.
2. Avoids the stress and inconvenience of traveling long distances
3. Avoids physical stress for older and handicapped travelers
4. Greater convenience, since this represents instant gratification
5. Avoids physical risks such as SARS, AIDS, hepatitis, injury, crime, sexual assault, and kidnapping experienced by real-world travelers
6. Will be programmable to the traveler's tastes and preferences (Some of you may have seen the memorable VR Mars vacation experienced by Arnold Schwarzenegger in the 1990 film *Total Recall*).
7. The foods "consumed" will not be fattening or result in indigestion
8. The experiences can be censored or restricted for children

The challenge to hospitality marketers in the foreseeable future will be to exploit this new technology in a way that enhances their products and services for target customers while defending against the future threats that this technology represents. For example, this medium offers exciting potential in the area of targeted advertising and promotions.

Future VR Hospitality Applications

It is probable that hospitality companies in the future will diversify by adding virtual tours to their portfolio of travel and tourism services. This would be especially true for high-risk and/or high-cost products, such as vacationing on other planets, mountain climbing expeditions, and undersea vacations. Hotel and airline guests of the future may enjoy VR entertainment which would allow them to "participate" in VR movies that would engage the five senses. Future hospitality guests may purchase VR-accredited educational instruction allowing them to "participate" in the classrooms of the world's finest professors and technical experts while a guest at your resort, hotel, or airline. And of course, the same technology would allow the illusion of being present at concerts and a variety of conventions, plays, and shows.

PITFALLS AND LIABILITIES

It is no secret that there are many problems and pitfalls associated with the adoption of new technology. There are many stories both anecdotal and in the literature of expensive failures in the attempt to install new technology by hospitality companies. Sometimes this means that a new system is not used by employees who have inadequate training or that the system is overly complicated. Sometimes the system is already obsolete or it cannot be integrated with present legacy systems. Sometimes parts or service are not available for this system, so that it is often off-line. And sometimes the system cannot be expanded or ported to fit other platforms.

Hospitality managers and owners need to ask themselves the following questions before committing to a new and expensive technology system:

1. What specific problems does this new system solve for us?
2. Is this system expandable and portable to other platforms?
3. Is this system going to be cost-effective for this particular operation?
4. What is the break-even point for this investment?
5. Is this system going to be accepted by our employees because it is easy to use, practical, and solves present problems?
6. Do we have an implementation plan that allows for employee input, customization, and buy-in on this system?
7. Is this system going to produce more satisfied target customers who value what this system provides to them?
8. Can this system be integrated with present information and communication (legacy) systems and our current marketing strategy?
9. How will this system affect our positioning with respect to our target segments?
10. What have been the results obtained by other hospitality companies who have used this system? (Do we really want to be the first to use this system?)
11. Has this system undergone enough tests and trials to ensure that all of the bugs and problems have been removed?
12. What is the reputation and track record of the company installing this system?

13. What will be ancillary costs involved (of maintaining this system, training costs, and integrating this system with existing systems)?
14. How will this system affect the brand image of this company?
15. Are there any obstacles to the implementation of this system posed by the present infrastructure (important for third world locations)?
16. What is the expected life of this system (time till obsolescence) and is it upgradeable?
17. Who will be doing the servicing and tech support on this system, and what is their reputation for service quality and dependability?
18. Can attractive financing be negotiated for this purchase?
19. Are there adequate warranties including on-site dependable tech support for this system?
20. Do we have a backup system in case this new system fails?
21. How long will it take to make this system operational?
22. What are the recommendations of at least two IT managers or consultants on this system? Do they have a personal interest in adopting this system (to avoid a biased report)?

SUMMARY

It is clear that new technology and the market implications of this technology are, and will continue to be in the future, a critical factor in the brand positioning and competitiveness of your hospitality company. This includes the increasing power of the consumer (individual and corporate) who can use the Internet to instantly compare the services and prices of your competitors in a world of instant information and communications. The advent and diffusion of shopping robots (shopbots) will leverage that power geometrically and force sea changes in the way that hospitality companies price and position their brands.

Effective use of new technology will result in more effective management of target customer information, including storage, enterprise-wide access and dissemination, and use of this information to customize services and facilities to the needs and preferences of these customers. But the gathering and application of this information

should be seamless and noninvasive, respecting your guests' need for privacy and the peaceful enjoyment of your facilities and services.

Thus, the use of advanced security systems will make your customers feel safer and increase satisfaction if the same system does not also produce increased waiting time, unnecessary searches, and invasive procedures. The ideal security system is one that the customer knows is working without being aware that it is working. For example, future security systems may use noninvasive facial recognition software using hidden cameras so that the screening process does not disturb customers.

Other uses of new technology will add convenience, knowledge, or entertainment to enhance the guest experience. This includes WI-FI access for the PDAs and notebook computers of business travelers, interactive TV, VR systems, and customer touch screen ordering systems, in-room business facilities, and concierge systems. These systems should be easy to use, cost-effective, and of practical importance to your target customers and their guests. (Marketers have been slow to learn that spouses, children, and significant others play an important part in purchase decisions.)

In order to avoid being left behind in the race for competitive advantage, hospitality firms need to monitor current and future technological trends (as part of an annual environmental study) and anticipate how these trends will impact target segments and competitors over both the short and long term.

In particular, hospitality managers need to understand how to effectively select and implement new and cost-effective information technology that will further corporate goals. This technology should be capable of being integrated with corporate strategy, external environmental factors and trends, current information systems, and the needs and preferences of target segments. Hospitality managers need to ask the right questions and have access to the right research and IT consultants in order to effectively manage new technology.

EXERCISES

1. Using your hospitality firm, or a selected firm, give one example of new technology that was recently implemented by this firm. Evaluate this new technology for cost-effectiveness, em-

ployee satisfaction, customer satisfaction, and perceived value.

2. What changes would you propose to improve the effects of this new technology on target customers and the firm?

3. Is it ever necessary to abandon new technology after it has been implemented? Explain.

4. Does the implementation of a specific new technology by your competition necessitate the adoption of this same technology by your hospitality firm? Explain why or why not.

5. What current or new technology do you believe is most important in achieving a competitive advantage for your hospitality firm? Justify your position.

6. How important is employee buy-in and training on the adoption of new technology? Explain.

7. You are the general manager of an independent hotel or restaurant and you have a disagreement with the owner about acquiring a specific new technology that you believe is necessary for your operation. The owner does not believe the expense is justified. What do you do and why?

8. What is a data warehouse and how would this help your firm in selecting and communicating with target customers?

9. What security systems are in place to protect the guests of your hospitality firm? Do these security systems accomplish security goals without a negative impact of guest satisfaction? Explain.

10. What useful information is now being provided by your firm's information systems? What information would you like to receive and but not now obtaining from this system? How could this be corrected?

Chapter 11

Your Segmentation Battle Plan: Strategy and Tactics

Those skilled in strategy achieve cooperation in a group so that directing the group is like directing a single individual . . . Employing the entire force is like employing a single individual. Strategy is a problem of coordination, not of masses.

Sun Tsu, *The Art of War,* 500 BC

YOUR STRATEGIC SEGMENTATION PLAN

Strategy implies having both a goal and a plan on how to achieve this goal. The goal of your strategic segmentation plan is a sustained competitive advantage (SCA) for the target segments of your hospitality company. Your strategic plan is a segmentation-based plan for the effective employment and coordination of the resources at your disposal to achieve this goal.

This means that executive management team of your company, with the information and guidance provided by market research and marketing segmentation consultants, will develop a plan to employ company resources to select the optimal combination of market segments for your company and position the company to best achieve a competitive advantage with these target segments.

Having a competitive advantage implies that your target segments prefer your services to those of the competition because of a perceived superiority in benefits received, superior perceived value, and/or service quality. A *sustained* competitive advantage is one that is not easily copied by the competition (Porter, 1985).

Segmentation Strategies for Hospitality Managers
Published by The Haworth Press, Inc., 2007. All rights reserved.
doi:10.1300/5716_11

Examples of hospitality companies that have achieved a competitive advantage with their target segments are McDonald's (families with children), Southwest Airlines (leisure travelers on a budget), and Marriott Courtyard hotels (independent business travelers who are not on expense accounts).

Segmentation Plan Options

Hospitality managers who are developing a new segmentation plan have the following options, which have different levels of effectiveness in achieving the goal of competitive advantage (see Table 11.1).

TABLE 11.1. Segmentation Plan Options: Levels of Effectiveness

Level	Power	Effectiveness for competitive advantage
1. Copying a segmentation plan from a competitor in the same market	Low	Negligible
2. Using (only) your guest database to target the most profitable customer segments	Low to medium	Low
3. Using your guest database, together with a database of customers with similar profiles	Low to medium	Low
4. Using market research and analysis to target the most attractive current segments first in addition item #3	Medium to strong	Low to medium
5. Adding the co-alignment of company skills, resources, core competencies, and external environmental factors to the segment selection process to #4	Medium to strong	Medium
6. Using statistical analysis, perceptual mapping, and/or data mining technology to uncover neglected or underserved niche markets in addition to item #5	Medium to strong	Medium to strong
7. Adding to item #6, assistance of a segmentation marketing consultant, development of a customized segmentation plan, and the integration of this plan into the company culture	Strong	Strong

Strategic Fit

The concept of *strategic fit*, or *alignment* of the company's strategy and resources with the external environment, has long been a traditional but controversial part of the corporate strategy literature. The strength of this concept lies in the insights of business strategists who understand that the *resources* of the company (e.g., strengths, skills, core competencies, financial resources, and goals) must be aligned (have the right "fit") with its target markets, customer segments, opportunities, and external environment (Olsen, West, & Ching-Yick Tese, 1998).

The weakness of this concept is related to the lack of clarity and consensus of an operational definition of *strategic fit* (or its synonym, co-alignment) as well as the related scarcity of empirical studies which corroborate this theory.

Contributions to this area include market research results of a positive correlation between business performance and co-alignment of environment-strategy using the PIMS database* (Venkatraman & Prescott, 1990). Venkatraman and Prescott have also argued for the integration of IT strategy and infrastructure into the conceptual framework of strategic fit, highlighting the importance of new technology and information systems as driving forces in the global business environment of the twenty-first century (Henderson & Venkatraman, 1999; Olsen et al., 1998; Venkatraman & Prescott, 1990).

A review of the strategy literature reveals four major types of strategic fit for corporations (Morritt & Lawson, 2001).

External Alignment

This is the alignment between the internal factors of the firm (including its mission, resources, skills, goals, core competencies, and competitive portfolio) and external factors such as driving forces, op-

---

*The PIMS (Profit Impact of Market Strategy) of the Strategic Planning Institute is a large scale study designed to measure the relationship between business actions and business results. The project was initiated and developed at the General Electric Co. from the mid-1960s and expanded upon at the Management Science Institute at Harvard in the early 1970s; since 1975, the Strategic Planning Institute has continued the development and application of the PIMS research.

portunities, and target markets (Olsen et al., 1998; Venkatraman & Prescott, 1990) (see Figure 11.1). For example, the recent selection and delivery of services attractive to target segments located in China have proven a good fit for the resources and market strengths of the fast-food chains of McDonald's and KFC.

Functional Alignment

This is the alignment between the firm's strategy and the traditional functions of the firm (e.g., operations, finance, IT, marketing, HR, and R&D) (Arthur Young International, 1989; Porter, 1996) (see Figure 11.2). For example, the successful development and maintenance of an effective yield management system (discussed in Chapter 6) requires the alignment of the strategy and resources of the functional departments of marketing, IT, operations, and finance.

Strategic Fit with Other Businesses

This is the strategic fit that looks for synergies and cost-reduction opportunities between the firm and other firms being considered for acquisition, merger, strategic alliances, and divesture (Thompson & Strickland, 1992). An example of this would be the growing trend of

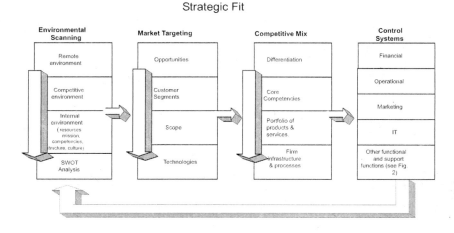

FIGURE 11.1. Co-Alignment of the Firm's Strategy with Its External Environment

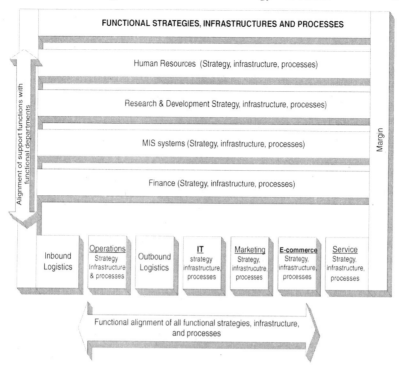

FIGURE 11.2. Functional Alignment: Alignment of Functional Elements and Support Functions for Consistency, Reinforcement, and Optimization (includes a modified version of Porter's value chain)

placing mini-versions (kiosks) of brand-name fast-food restaurants and coffee houses, like Burger King, Pizza Hut, and Starbucks, within the facilities of hotels, universities, bookstores, and Wal-Marts. Recently, I patronized a popular Starbucks coffee house kiosk located in the renowned Cleveland Clinic (hospital and outpatient clinic) located in Weston, Florida.

Business and IT Functional Alignment

This is the functional alignment between business and IT strategy and their infrastructures and processes. A major contribution to the area was the Landmark MIT Study (Arthur Young International,

1989). Another important contribution was the strategic alignment model provided by Henderson and Venkatraman (1991, 1999). This model was defined in terms of the four fundamental domains of strategic choice: business strategy, IT strategy, organizational infrastructure and processes, and IT.

An example of the alignment of segmentation strategy (marketing) and IT strategy (information systems) is the use of enterprise-wide data warehouses and central database systems combined with datamining techniques to assist in target segment selection and direct marketing to these segments (see Chapter 10).

Strategic Tool Box for Hospitality Managers

In the area of military strategy, no one battle plan can be used for all military units under all circumstances and fields of battle. An effective battle plan requires accurate, relevant, and timely information, including current knowledge of battle conditions, strengths and weaknesses of your own forces and those of the competition, along with creativity, insight, innovation, and the ability to quickly adapt and respond to changing conditions.

Clearly, there is no single formula or segmentation strategy that will be appropriate for all hospitality companies or even all hospitality companies of a certain type. Creativity, innovation, managerial experience, and market research need to be combined with the assistance of trained segmentation consultants to achieve the optimal segmentation and positioning plan for your particular company or brand.

However, just as there are well-known strategic principles that are helpful in preparing a battle plan (such as the value of surprise, well-trained troops, coordination of your forces, and making best use of the existing battle terrain), there are strategic principles that are helpful in formulating your segmentation battle plan. Some of these principles are enumerated below.

Strategic Principles

Gap Analysis

Look for gaps in your market for segments that are not being served or are underserved by your competitors. These may be profit-

able niche markets for you. The use of perceptual mapping techniques is helpful here. Some hospitality firms now target the niche segments of women business travelers, gay cruises, and spa guests.

Synergistic Segment Combinations

Compatible combinations of segment bases are preferable to a single segmentation base. Look for segmentation bases that are appropriate and synergistic for your hospitality company. For example, geographical segmentation is usually part of the segmentation base combination for restaurants, since most restaurants serve a defined geographical area. This can be combined with usage segmentation base to attract heavy users in your geographical radius. Hotels often target local corporate segments for meetings and conventions.

Value Chain Analysis

Look for better ways to deliver your services by an analysis of your company's value chain for your target segments (Porter, 1985). Better, faster, more convenient, and cheaper are some of the ways to improve delivery. Thus, some restaurants use touch screen, point-of-sale systems to allow guests to see and order their own meals. Virgin Atlantic Airlines offers greater legroom to passengers who must occupy cramped quarters for extended flights.

Market Drivers

Learn what the current *drivers* are in your different product and brand markets. What are the most important current reasons why your target customer segments purchase your brands and why? For example, fast-food customers may value convenience, speed, cleanliness, and healthy, tasty food.

SWOT Analysis

Build on the strengths of your company and minimize its weaknesses in the serving of target segments. Forecast and exploit future opportunities and minimize or exploit future threats. Thus, hotels

with advanced security systems have minimized the threat caused by increased post-9/11 terrorism.

Environmental Scanning

Develop an annual environmental scanning report so that you can plan for forecasted opportunities and threats that will significantly impact the success of your company. These include regulatory, technological, economic, legal, political, ecological, social and cultural factors as well as changing market trends. It also includes competitive intelligence. How will these factors impact on the selection of, and service of target segments? For example, restaurant chains like Taco Bell, Chilis, and Polle Tropical have exploited demographic trends in the United States that include the fast-growing Hispanic segment.

Employee Focus

Never underestimate the importance and impact of your hospitality employees, who are the face of your company and arguably your *most important customers*. Regularly monitor employee satisfaction, retention, and compensation plans. Find out what your employees value most and least about your company. Review the effectiveness of your employee selection process. It is more cost-effective to select the right (people oriented) employees in the first place than to try to improve them later with expensive training, conferences, and retraining, due to high turnover rates. It is difficult to have happy loyal customers if your customers are being served by disgruntled, unmotivated, and low morale employees.

Virgin Airlines, for example, has high employee satisfaction ratings despite relatively low employee compensation. This is attributed to the culture of this company, which empowers its employees to solve problems and encourages innovation and risk taking.

Service Recovery Plan

A service recovery plan is critical in retaining your best customers. It has been shown that the successful and timely resolution of customer problems can result not only in customer retention but also in creating more loyal customers (Morgan & Hunt, 1994; Zeithaml & Bitner, 2003). Speedy correction of complaints by empowered,

friendly, and customer-oriented employees will go a long way to keeping your target customers on board.

Delivering Superior Value

It is critical to competitive advantage that your target customer segments perceive your hospitality services as having superior value relative to competitive services. This may be differently defined by different segments. This does not necessarily mean that your service must be less expensive but rather that what is received in valued benefits is perceived as more than can be obtained by competitors for the same price. Perceived quality is an important factor in perceived value. Customers are willing to pay more for what they perceive to be higher quality or more convenience or services that are more customized to their preferences (Weinstein & Johnson, 1999). For example, customers pay more for FedEx services because of the perceived dependability and speed of this service, which is translated into high perceived value by target customers. The same can be said of Disney theme resorts located inside of Disney World in Orlando. The combination of unique themes and convenient location delivers superior value to target customers and premium prices to the Disney Corporation.

Segment Diversification

Serving multiple segments has been found to be more profitable if they are compatible segments. It also has the advantage of diversifying your customer segment portfolio (Enz, Potter, & Siguaw, 1999).

TQM

Quality assurance programs like TQM have been found to be useful in the continuous improvement of service quality in a way that integrates TQM into the culture of the organization An example would be Baldridge Award winner Ritz-Carlton hotels, which implements a TQM program with the opening of every new hotel. This includes active support of upper management, quality teams, a quality deployment program, and continuous improvement. The TQM program is infused as part of the culture of these new locations and is best done at

the inception of a property. This is because there is less employee resistance related to changing old habits and old cultures.

Brand Equity

It is important that your brand be perceived as unique by your target segments in a way that is highly valued by these segments. Your brand should be on of the top three brands that come to mind when your services are needed by your target segments. Examples are McDonald's in the fast-food sector, Cathay Pacific Airlines to U.S. travelers to Hong Kong and points East, and Hertz rental cars.

Develop a Market-Oriented, Segment-Based Culture

This includes customer orientation, competitor orientation, and interfunctional coordination in a company culture that is focused on the research-based selection and development of optimal target segments and the delivery of superior value to these segments (Narver & Slater, 1990; Weinstein, 2004). This is difficult to accomplish with already existing properties and requires the full moral and financial support of upper management for at least three years. It also requires delegation of the necessary authority to accomplish these goals.

Cost-Effective Use of Technology

Cost-effective use of technology can add value and convenience to target segments. An example is in-room interactive TV delivery of hotel bills and automated check-in/check-out services to reduce guest waiting time. Another example is fast and convenient booking on on-line Web sites that offer competitive prices and which may be customized with rewards for frequent guests.

Focus on Your Best Customers

Develop a relationship marketing program which focuses on your most profitable customers. Use measures such as RFM analysis and lifetime value analysis to accurately identify your most profitable customers, become experts on their expectations and needs, and develop mutually beneficial relationships with them which will result in

greater revenues per customer and also valuable word-of-mouth advertising (see Chapter 5 on RM).

Checklist for Market Research Needed for Your Segmentation Plan

The following checklist is provided to help ensure that your company has the necessary and timely market research reports to support your segmentation plan:

Annual Internal Analysis of Your Company

This includes resources, goals, and core competencies. This includes a SWOT analysis, financial condition analysis, corporate vision and mission, and market research on how your brand and services are perceived by your target segments relative to the competition.

Annual Environmental Scan

This market research report includes external factors (macro and micro) that have or are projected to have a significant effect on the success of your company in the next five to ten years. This includes macro environmental factors such as legal, regulatory, economic, political, sociological, technological, environmental, and marketing trends. It also includes micro factors such as market structure, relative strength, and positioning of competitors.

Baseline Market Segmentation Study

This market research report is a study which outlines the market structure (including current and prospective market segments and products) and market drivers, including segment attractiveness, growth, and forecasted demand. This type of study is intensive, usually takes several months, is relatively expensive, and uses a larger sample size (>1,000). A large amount of time is spent on the analysis of this study, exploring alternative segmentation approaches. This study should be done every two to three years.

Annual Market Segmentation Report

This is a market research report on the current status, expectations, perceptions, and preferences of target and potential target segments. It highlights changes in the attractiveness, forecasted demand, profitability, price elasticity, and expectations of these target segments. It is usually preceded by a baseline segmentation study.

Annual Strategic Analysis Report

This is a market research executive report on how well the company is achieving its segmentation and positioning goals, including the following sections:

- Financial report on revenues, profitability, market share by product segment and location
- Employee satisfaction and retention report by location and product segment
- Target segment customer retention, satisfaction, and perceived value report by target segment and location
- Brand equity report which includes brand awareness, perceived quality, reputation, and service quality of the brand by target segments and location
- Positioning effectiveness report, a report on the effectiveness of current positioning, including perceived differentiation, value, and differentiation importance to target segments
- Technology effectiveness report, a report on the cost-effectiveness of current technology and how well this technology satisfies employee and target customer needs (from the viewpoint of these customers).

Semi-Annual Competitive Intelligence Report

This is a semi-annual market research executive report on the firm's major competitors, their strategies, offerings, target segments, brands, and market share. This report should indicate the relative performance of the firm compared to these competitors.

A Ten-Point Strategy for Building an Effective Segmentation Plan

Although it is clear that no magic formula is applicable to all hospitality companies in a given service sector (hotels, restaurants, airlines, etc), there are certain strategic guidelines that my be helpful in developing an effective segmentation plan for your hospitality company:

1. Who are your most profitable customer segments? Acquire the goal of becoming an expert on these target segments: their needs, preferences, and within-segment trends.

Examine and research your guest database to discover and help select your most profitable customers. Use measures such as RFM analysis and lifetime value of the customer. Construct profiles of these customers and look for meaningful patterns and similarities that link them into identifiable segments. Research the best combination of segmentation bases that can be used to best screen for this type of customer. Find out why these segments choose your hospitality firm and what makes them stay or switch. What are the industry *drivers* for these customers (features which are most important in their purchase decisions)? What are the *satisfiers* for these customers (features which would deter customers by their absence but which are not sufficient for a purchase decision). For example, more spacious seating may be a *driver* for airline business customers but on-time departures of flights may be a *satisfier.* Research what other competitors they use, how often, and how you are rated in comparison to these competitors.

2. What are the strengths and weaknesses of your organization?

What are those resources that enable your organization to compete effectively in your markets in relation to your target segments? Consider your company's core competencies, managerial expertise, marketing skills, financial resources, ability to innovate, brand equity, location, service quality, partnerships, market-oriented company culture, and employee performance and satisfaction. Consider also

the weaknesses of your organization and what need to be improved or changed to compete effectively in your markets.

3. Examine the fit of your organization's profile (SWOT analysis) with the needs and preferences of your target segments.

If the fit is good, plan how to build on these strengths to more effectively serve these segments and increase target segment satisfaction and retention. If the fit is poor, examine closely to discover if this problem can be fixed or whether you need to select different or modified target segments that are a better fit with your firm's profile. Remember that your optimal target segments are those that are most profitable, compatible with one another, and also afford your company the best opportunity to build a competitive advantage by best satisfying their actual and anticipated needs and preferences.

4. Once you verified a good fit with your target segments, continue to develop the optimal combination of segmentation bases and profiles that can be used to screen future prospects and customers that resemble these target segments.

This is a continuous process due to changing markets and trends. You will probably be able to purchase additional information on these customers from outside research firms and credit card companies, which will help fill out the profiles of your target segments. Primary and secondary market research, and an effective guest database system can be used to more clearly define these target segments. The goal is the continuous gathering of relevant information on these target segments which is then stored in your guest database and then applied to better serve these customers by your critical customer contact personnel, managers, and marketing team.

5. Research and analyze your current market and main competitors for their strengths, weaknesses, and their target segments.

This information can be obtained in house or purchased by an outside research firm. The use of perceptual maps can be helpful in seeing the relative positioning of your firm relative to the main drivers of your target market and also gaps in your market coverage. Look for weaknesses in your competitors' strategy, new marketing trends,

forecasted demand for target and potential target segments, and positioning gaps that can be exploited by your company.

6. Position your company with reference to its target segments and main competitors. Take the best advantage of, and exploit, your strengths, competitor weaknesses, target customers' actual and anticipated needs and preferences, and gaps in your market.

Construct your company's value proposition which will differentiate it from the competition in a way that is highly valued by target segments. Find ways to offer *superior value* to your target segments. And look for profitable niche markets or sub-segments which may be underserved by the competition. For example, regular service business hotels like Hilton and Marriott that have offered a wide range of services to their guests have had their market share reduced by niche brands that specialize in a subset of these services such as convention centers, hotels that cater to women business travelers, extended-stay residences, and hotels that cater to top executives and the wealthy. Defenses against this type of market encroachment include the developing of brand extensions and also new brands that specialize in these profitable niches. Continuous market research including trend analysis and demand forecasting is essential here to anticipate changes in segment buyer behavior and competitive strategies and new offerings.

"All-inclusive" hotel chains in the Caribbean (like the ubiquitous Sandals chain) have increased the loyalty of their "all-inclusive" market segments (like honeymooners, couples, sports-minded yuppies) by offering guests the attractive opportunity to spread their stay over their member hotels at no extra cost. And Cathay Pacific now offers U.S. travelers to Hong Kong direct flights from New York as well as superior service and cuisine.

7. Build a market-oriented culture that is able to maintain long-term relationships with your firm's most profitable customers.

A market-oriented culture (correlated with superior business performance) is one that is characterized by customer focus, competitor orientation, and interfunctional coordination (Narver & Slater, 1990). Customer focus means being experts on the current and anticipated

needs and preferences of employees *(perhaps your best customers)*, suppliers, and intermediaries as well as your target guest segments and using this information to better satisfy these needs at a profit.

Competitor orientation means that your company systematically and regularly obtains relevant, timely, and actionable market research on the marketing practices, target segments, and strategies of your major competitors. Interfunctional coordination refers to the ability of the organization to effectively share and communicate this information to all functional departments in a manner that allows consistent and timely application of this research to the organization's marketing mix in the effort to achieve a maintain a sustainable competitive advantage. A market-oriented culture is also important because it is one of the few features of a company that is difficult to copy by competitors.

Find out how you can add value to these important customers so that you can exceed their expectations and discourage switching behavior. Your relationship marketing program will require the full support and example of upper management, effective use of capital-intensive company database systems, and the continuous measurement, monitoring, and customizing of your services to the needs of these target segments. The use of benchmarking, TQM, and research on the "best practices" of your industry will also help. Relentless and continuous improvement of your services (as perceived by target customers) should be the mantra of your organization.

8. Develop an effective, segment based, branding strategy.

This should be designed to support and leverage your present market strengths and your ability to develop chains, franchises, IPOs, and management companies.

This will include not only the development and enhancement of present segment-focused brands but also the development of new brands or brand extensions to exploit new or developing market trends. But it also includes an effective brand management program intended to differentiate and effectively position your brand in the minds of your target customers. Who will be the optimal target segments for your brand, what will be its value proposition, and who are its main competitors? How will you consistently promote this brand

in your advertising, promotions, and facilities that form the tangible servicescape of your organization?

9. Use tested, flexible, and cost-effective technology that will leverage your ability to serve the needs of your target segments and employees.

Advantages and liabilities of new technology were discussed in Chapters 6 and 10. The adoption of new technology and information systems should be screened for the ability of new systems to leverage the organization's ability to more effectively and efficiently serve its target segments at a reasonable rate of return. They should also be compatible with existing and legacy systems, including the company's database, reservation systems, and integrated information and property systems. Finally, they should be customized to the needs of the employees, managers, suppliers, and guests who will be using these systems.

10. Market research, monitoring, and controls.

Continuously research, monitor, and control your market structure, target segments, market trends, and environmental factors to ensure the integrity and increased performance of your segmentation-based marketing strategy.

Construct five- and ten-year strategic segmentation plans to exploit and anticipate future projected changes in your current and projected markets. Construct measurable and quantifiable controls to measure the critical variables of your segmentation and positioning plan.

MARKET SEGMENTATION AUDIT

The following is a market segmentation audit tool which can be used to check on the extent to which your company has incorporated and implemented an effective strategic segmentation plan into its culture (adapted from Goetsch, 1983; Kotler, 1991; Weinstein, 2004).

Sales History

1. How do sales break down for your different products and brands?
2. Do you know where your sales are coming from—segments and customer classifications?
3. Which products and segments are not meeting potential and why?

Marketing Commitment

1. Do all marketing managers use formal marketing planning?
2. Do you implement a segment-based marketing plan, set objectives, measure performance, and adjust for deviation?
3. Is the marketing plan largely based on segmentation research findings?

Marketing Environment

1. What major developments and trends pose opportunities or threats to the company in the next five years? What actions have been taken, or are planned, to respond to these opportunities and threats?
2. What major technological changes are anticipated over the next five to ten years? What plans are in place to respond to these changes?
3. What are you major competitor's positions in the market (SWOT analysis. competitive intelligence, strategies)?
4. What is happening to market size, growth rate, profitability, and customer satisfaction and retention for your target product and customer segments?

Market Segments

1. Who are your major customer segments for each product and brand?
2. How do the different customer segments make their buying decisions? What are the major drivers (most influential factors for target customers) of your different product and brand markets?
3. Who are the major (attractive) potential customer segments for your product?

4. Are target segments identified, measured, and monitored on a regular basis?
5. Are small but profitable segments being overlooked?
6. How do you presently segment the market for each of your brands?
7. Is the present segmentation approach effective? By what measures of effectiveness?
8. Have you developed detailed customer profiles of major market segments?
9. Should your company contract from or withdraw from any target segment? What would be the short- and long-term consequences of this decision?
10. Are market segmentation definitions based on market research (preferable)?

Product and Brand Lines

1. What are the product and brand line objectives?
2. Is the company well organized to gather, generate, and screen new product/brand ideas on a regular basis?
3. Does the company carry out adequate product and brand market research before launching new product/brands?
4. Is there a well-defined program to weed out unprofitable products and add new ones?
5. Is there a systematic sharing of information with R&D and other key departments in this company (interfunctional communications) so that managers are kept current on current plans, research, and activities?

Marketing Management

1. Are marketing activities optimally structured along functional, product, end user, and territorial lines?
2. Are marketing resources optimally allocated to market segments, products, and locations?
3. Is an adequate, accurate, and timely marketing intelligence/marketing information system in place?
4. Is the marketing research being effectively used by company decision makers?

5. Does management regularly analyze the profitability of products, brands, markets, territories, and related issues?

SUMMARY

The goal of a strategic segmentation plan is to develop a sustainable competitive advantage for your hospitality company. There are five major steps in this process, which are not necessarily sequential. This is accomplished first by the selection of an optimal combination of research-based, target segments which have a good strategic fit with the profile of your company (resources, goals, strengths, and weaknesses) and whose selection is also supported by an environmental scan, marketing trends, and forecasted demand from these segments.

Positioning the firm's service to these target segments is a critical second step, which involves the use of competitive intelligence, and market research into the needs, preferences, and buying behavior of these segments. Perceptual maps are useful tools of positioning analysis, positioning your brands in the minds of your target segments as both unique and having superior perceived value to these segments.

A third step in the process is to deliver those features and attributes to your target segments (via your marketing mix) in a way that is most valued by these segments, will deliver on your positioning and service promises, and will eventually result in a sustained competitive advantage with these segments. This includes the communication of this positioning, including your value proposition, to your target and prospective customers and your employees.

The next (fourth) step in the process is that your market-oriented, research-based segmentation plan must be integrated into your company culture to be effective. This includes the alignment of your segmentation strategy with the functional departments of your company. Segment-based programs such as yield management, relationship marketing, and market orientation all require the coordination and alignment of all functional departments as well as coordination with suppliers and intermediaries.

Finally, the fifth and last step in the process is the use of measures, monitoring, and controls to maintain and continuously improve your strategic plan. This includes the used of periodic market research into your changing market structure, demographics, market trends, opti-

mal target segments, competitive strategies and offerings, forecasted demand, and the perceived value, quality, and customer satisfaction ratings of your target segments. It also includes an annual environmental scan, which includes the projection of future threats and opportunities for your company. Remember that your customers also include your employees, suppliers, intermediaries, and partners.

EXERCISES

1. How can perceptual mapping be used to evaluate your segmentation plan?
2. Identify an important target segment of your hospitality company. Is this segment a good strategic fit with your company? Explain.
3. Why does using an a priori product segmentation plan (e.g., segmenting guests into transient, corporate, and business) leave your company vulnerable to the competition?
4. Explain why your company's segmentation strategy should be aligned with the strategy with its functional departments (IT, marketing, operations, finance, HR, etc.).
5. What is the value of gap analysis in the selection of target segments?
6. What are the market drivers of one of the major target segments of your company with regard to one service or brand?
7. What is the relationship between environmental scanning and the choice of target segments?
8. Who are the most profitable segments of your firm and how is this measured?
9. How does your firm plan to deliver superior value to these target segments (cited in #8) and how is this measured?
10. Why is developing strong brands an important part of your segmentation plan?
11. Who should be involved in the construction of your firm's segmentation plan and why?
12. What is a baseline segmentation study and why is it important?
13. What is the role of competitive intelligence in developing your segmentation plan?

14. Why does an effective segmentation strategy involve a change in the culture of your organization?
15. Why is the use of measures important in maintaining the effectiveness of your segmentation plan?

References

Ante, S. (2004, January 12). Shifting work offshore? Outsourcer beware. *Business Week*, pp. 36-37.

Arthur Young International. (1989). *The landmark MIT study: Management in the 1990s*. Boston, MA: Arthur Young International.

Auty, S. (1992). Consumer choice and segmentation in the restaurant industry. *The Services Industry Journal, 12*(3), 324-339.

Bergeron, B. (2002). *Essentials of CRM*. New York: John Wiley & Sons.

Best, R. (2000). *Market based management*. Upper Saddle River, NJ: Prentice Hall.

Bowen, J. (1998). Market segmentation in hospitality research: No longer a sequential process. *International Journal of Contemporary Hospitality Management, 10*(7), 289-296.

Bowen, J., & Shoemaker, S. (1997). Relationship in the luxury hotel segment: A strategic perspective. Research paper. Center for Hospitality Research, Cornell University, Ithaca, NY.

Bowen, J., & Shoemaker, S. (1998). Loyalty: A strategic commitment. *Cornell Hotel and Restaurant Administration Quarterly, 39*(1), 12-25.

Brown, P., & Stange, K. (2002). Investment in information technology: The multi-billion dollar game of chance. *Hospitality Business Review, 4*(1), 28-38.

Cahill, D. (1997). *How consumers pick a hotel: Strategic segmentation and target marketing*. Binghamton, NY: The Haworth Press.

Camacho, F. (1988). Meeting the needs of senior citizens through life care communities: Marriott's approach to the development of a new service business. *Journal of Service Marketing, 2*(1), 49-53.

Cho, W., Sumiclirast, R., & Olsen, M. (1996). Expert-system technology for hotels: Concierge application. *Cornell Hotel and Restaurant Administration Quarterly, 37*(1), 54-60.

Clemons, E., & Weber, E. (1994). Segmentation, differentiation, and flexible pricing: Experiences with information technology and segment tailored strategies. *Journal of Management Information Systems, 11*(2), 9-36.

Club Med. (1995). Annual report. Retrieved from www.clubmed.com

Conlon, G. (1996). True romance. *Sales and Marketing Management, 148*(5), 85-90.

Connolly, D., & Olsen, M. (2001). An environmental assessment of how technology is reshaping the hospitality industry. *Tourism and Hospitality Research, 3*(1), 73-93.

Segmentation Strategies for Hospitality Managers
Published by The Haworth Press, Inc., 2007. All rights reserved.
doi:10.1300/5716_12

Copulsky, J., & Wolf, M. (1991, July 8). Relationship marketing: Positioning for the future. *Journal of Business Strategy, 11,* 16-26.

Coviello, N., Brodie, R., & Munro, H. (1997). Understanding contemporary marketing: Development of a classification scheme. *Journal of Marketing Management, 13*(6), 501-522.

Cunningham, M., & Dev, S. (1992, August). Strategic marketing: A lodging "end run." *Cornell Hotel and Restaurant Administration Quarterly,* pp. 36-43.

Dalgic, T., & Leeuw, M. (1994). Niche marketing revisited: Concept, applications, and some European cases. *European Journal of Marketing, 28*(4), 39-56.

Davis, S., and Davidson, W. (1991). *2020 vision.* New York: Simon and Schuster.

Deneen, S. (1993). Marketing mother nature. *Hotel and Motel Management, 208*(5), 25.

Dev, C., & Hubbard, J. (1989). A strategic analysis of the lodging industry. *Cornell Hotel and Restaurant Administration Quarterly, 30*(1), 19-23.

Dev, C., and Klein, S. (1993). Strategic alliances in the hotel industry. *Cornell Hotel and Restaurant Administration Quarterly, 34*(4), 42-45.

Dev, C. & Olsen, M. (2000). Marketing challenges for the next decade. *Cornell Hotel and Restaurant Administration Quarterly, 41*(1), 41-47.

Dowling, W. (1980, September). Creating the right identity for your hotel. *Lodging,* p. 58.

Durocher, J., & Niman, N. (1991). Automated guest relations that generate hotel reservations. *Information Strategy: The Executives Journal, 7*(3), 27-30.

Durr, J. (1989). The value of the guest register. *Direct Marketing, 52*(5), 48-55.

Eisman, R. (1993). Corporate spotlight: The Ritz-Carlton. *Incentive, 167*(1), 24-27.

Enz, C., Potter, G., & Siguaw, J. (1999). Serving more segments and offering more products. *Cornell Hotel and Restaurant Administration Quarterly, 40*(6), 54-62.

Escalera, K. (1994). Special-interest marketing builds business for hotels. *Hotel and Motel Management, 209*(19), 50.

Francese, P., & Renaghan, L. (1990). Database marketing: Building consumer profiles. *Cornell Hotel and Restaurant Administration Quarterly, 31*(1), 60-63.

Global Marketing Services. (1993). Raw survey of Jamaican smaller resort hotels.

Goetsch, H. (1983, March 18). Conduct a comprehensive marketing audit to improve market planning. *Marketing News,* Section 2, p. 14.

Goodrich, J. (1994). Health tourism: A new positioning strategy for tourist destinations. *Journal of International Consumer Marketing, 6*(3/4), 227-238.

Green. P. (1977). A new approach to market segmentation. *Business Horizons, 20,* 61-73.

Hair, J., Bush, R., & Ortinau, D. (2003). *Marketing research within a changing information environment.* New York: McGraw-Hill Irwin.

Hanks, R., Cross, R., & Noland, R. (1992). Discounting in the hotel industry: A new approach. *Cornell Hotel and Restaurant Administration Quarterly, 33*(1), 15-23.

Henderson, J., and Venkatraman, N. (1991). Understanding strategic alignment. *Business Quarterly, 55*(3), 72-78.

Henderson, J., and Venkatraman, N. (1999). Strategic alignment: Leveraging information technology for transforming organizations. *IBM Systems Journal, 38*(2-3), 472-483.

Hsu, C., & Powers, T. (2002). *Marketing hospitality* (3rd ed.). New York: John Wiley and Sons.

Hughes, A. (1994). *Strategic database marketing.* Chicago: Probus Publishing.

Jasco, P. (1998). Shopbots: Shopping robots for electronic commerce. *Online, 22*(4), 14-19.

Jones, C. (1998). Applications of database marketing in the tourism industry. Economics Research Associates. Retrieved June 30, 2004, from www.hotel-on-line.com/Trends/ERA/ERADataBaseTourism.html. ·

Jones, P., & Hamilton, D. (1992). Yield management: Putting people in the big picture. *Cornell Hotel and Restaurant Administration Quarterly, 33*(1), 89-94.

Kimes, S.E. (1989). The basics of yield management. *Cornell Hotel and Restaurant Administration Quarterly, 30*(3), 14-19.

Kimes, S.E. (2003, October-December). Revenue management: A retrospective. *Cornell Hotel & Restaurant Administration Quarterly,* pp. 131–138.

Kotler. P. (1991). *Marketing management* (7th ed.). Englewood Cliffs, NJ: Prentice Hall.

Kotler, P. (1996). *Marketing management* (8th ed.). Upper Saddle River, NJ: Prentice Hall.

Kotler, P. (1999). *Marketing management* (9th ed.). Upper Saddle River, NJ: Prentice Hall.

Kotler, P. (2003). *Marketing management* (10th ed.). Upper Saddle River, NJ: Prentice Hall.

Kotler, P. (2005). *Marketing management* (11th ed.). Upper Saddle River, NJ: Prentice Hall.

Kotler, P., Bowen, J., & Makens, J. (1998). *Marketing for hospitality and tourism* (2nd ed.). Upper Saddle River, NJ: Prentice Hall.

Kotler, P., Bowen, J., & Makens, J. (2003). *Marketing for hospitality and tourism* (3rd ed.). Upper Saddle River, NJ: Prentice Hall.

Kotler, P., Haider, D.H., & Rein I. (1993). *Marketing places.* New York: Free Press.

Kotler, P., & Keller, K. (2006). *Marketing management* (12th ed.). Upper Saddle River, NJ: Pearson Prentice Hall.

Launch of virtual vacations transports armchair adventurers to exotic destinations around the world. (2003, August 5). *PR Newswire,* p. 1.

Levey, R. (2004, June 1). Database marketing: Diamond customers are forever. *Direct Marketing Business Intelligence.* Retrieved Jun 25, 2004, from directmag .com/microsites/magazinearticle.asp?mode=print&magazinearticleid=198712 & releaseid=&srid=11317&magazineid=151&siteid=2.

Levitt, T. (1983, September/October). After the sale is over. *Harvard Business Review, 61,* 87-93.

Lewis, R. (1989). Hospitality marketing: The internal approach. *Cornell Hotel and Restaurant Administration Quarterly, 30,* 41-45.

Lewis, R., & Chambers, R. (2000). *Marketing leadership in hospitality* (3rd ed.). New York: John Wiley and Sons, Inc.

Lovelock, C. (1991). *Services marketing.* Upper Saddle River, NJ: Prentice Hall.

Lovelock, C. (1996). *Services marketing.* Upper Saddle River, NJ: Prentice Hall.

Magnini, V., Honeycutt, E., & Hodge, S. (2003). Data mining for hotel firms: Use and limitations. *Cornell Hotel and Restaurant Administration Quarterly, 44*(2), 94-105.

Mann, I. (1993). Marketing to the affluent: A look at their expectations and service standards. *Cornell Hotel and Restaurant Administration Quarterly, 34*(5), 54-58.

McKinsey & Company. (2002). Marketing practice (August). Retrieved January 20, 2005, from marketing.mckinsey.com.

Mene, P. (1994). The winning practices of Baldridge award winner Ritz-Carlton Hotel Co. *Tapping the Network Journal, 5*(1), 10-14.

Mondy, R., & Hollingsworth, W. (1984). Getting the most from your Club Survey. *Cornell Hotel and Restaurant Administration Quarterly, 24*(4), 77-80.

Morgan, R., & Hunt, S. (1994). The commitment trust theory of relationship marketing. *Journal of Marketing, 58*(3), 20-38.

Morrison, A. (1996). *Hospitality and travel marketing* (2nd ed.). Boston: Delmar Publishers.

Morritt, R. (1995). Jamaica's resort hotels: Competing in the 90's. In Lawson S. & Little, C. (eds.), *Caribbean and Latin American economic development: Progress and challenges* (pp. 261-270). New York: St. Johns University Press.

Morritt, R. (1997). Niche marketing for hotel managers. *Journal of Segmentation in Marketing, 1*(2), 103-119.

Morritt, R. (1999). Perceived price effects on service repurchase intention: Toward a disconfirmation model of price, quality, satisfaction, value, and brand name. *Dissertation Abstracts International,* A 60/07, AAT 9938275, p. 2589, January 2000.

Morritt, R., & Lawson, S. (1999). Using TQM to build competitive advantage in the hotel industry. In C. Little, N. Delener, & S. Lawson (eds.), *Emerging global issues in the next millennium* (pp. 205-216). Global Business and Technology Association. New York: St. Johns University Press.

Morritt, R., & Lawson, S. (2000). What is strategic fit? Innovation in business practices. "The Role of Globalization, Regional Integration, and Technology," Global Business and Technology Association Conference, June 15-18, Rio De Janeiro, Brazil.

Morritt, R., & Lawson, S. (2001). An update on strategic fit for the new millennium. "Economies and Business in Transition: Facilitating Competitiveness and Change in the Global Environment," Global Business and Technology International Conference Proceedings (pp. 618-627), July 11-15, Istanbul, Turkey.

Myers, J. (1996). *Segmentation and positioning for strategic marketing decisions.* Chicago: American Marketing Association.

Narver, J., & Slater, S. (1990). The effect of a market orientation on business profitability. *Journal of Marketing, 54*(4), 20-36.

Nowakowski, J. (1991). The jewels of technology. *Lodging Hospitality, 4*(2), 66-68.

Olsen, M., & Connolly, D. (1999). Antecedents of technological change in the hospitality industry. *Tourism Analysis, 4*(1), 29-46.

Olsen, M., & Connolly, D. (2000). Experience-based travel. *Cornell Hotel and Restaurant Administration Quarterly, 41*(1), 30-40.

Olsen, M., West, J., & Ching-Yick Tese, E. (1998). *Strategic management in the hospitality industry* (2nd ed.). New York: John Wiley & Sons.

Omni Hotels plays up to aficionados. (1995). *Hotel and Motel Management, 210*(6), 53.

O'Neill, J.W. (2006, January). Defining segments. Lodging Hospitality: Ideas for Hotel Developers and Operators. Available online at www.lhonline.com/article/10889.

O'Neill, J., & Lloyd-Jones, A. (2002). One year after 9/11—hotel values and strategic implications. *Cornell Hotel and Restaurant Administration Quarterly, 43*(5), 53-64.

Parasuraman, A., Berry, L., & Zeithaml, V. (1988). SERVQUAL: A multiple item scale for measuring customer perceptions of service quality. *Journal of Retailing, 64*, 12-40.

Partlow, C. (1993). How Ritz-Carlton applies TQM. *Cornell Hotel and Restaurant Administration Quarterly, 34*(4), 16-24.

Payne, A., Christopher, M., Clark, M., & Peck, H. (2000). *Relationship marketing for competitive advantage.* Boston: Butterworth Heinemann.

Porter, M. (1980). *Competitive strategy.* New York: Free Press.

Porter, M. (1985). *Competitive advantage: Creating and sustaining superior performance.* New York: Free Press.

Porter, M. (1996, November/December). What is strategy? *Harvard Business Review,* pp.61-77.

Reich, A.Z. (1997). *Marketing management in the hospitality industry: A strategic approach.* New York: John Wiley.

Reichheld, K. (2002). Letters to the editor. *Harvard Business Review, 80*(11), 126.

Reichheld, K., and Sasser, W. (1990). Zero defects: Quality comes to services. *Harvard Business Review, 68*(5), 105-111.

Reinhartz, W., & Kumar, V. (2002). The mismanagement of customer loyalty. *Harvard Business Review, 80*(7), 86-98.

Rich rule online roost. (2002). *Road and Travel.* Available online at www.roadandtravel.com/newsworthy/newsworthy2002/richonline.htm.

Ries, A., & Trout J. (2000). *Positioning: The battle for your mind.* New York: McGraw Hill.

Ritz-Carlton employees go for the gold. (2002). *Human Resource Management International Digest, 10*(7), 23-25.

Rodeway seniors get choice room. (1993, November). *Lodging Hospitality*, p. 11.

Rushmore, S. (1993). Beyond recycling: The ecotel. *Lodging Hospitality, 49*(9), 20.

Rushmore, S. (1997, July). Hotel chain class survey. *Lodging Hospitality*, p. 18.

Sarabia, F. (1996). Model for market segments evaluation and selection. *European Journal of Marketing, 30*(4), 58-74.

Schultz, R. (1994). A decade of segmentation. *Lodging Hospitality, 50*(10), 20.

Selwitz, R. (1992a). Boston Four Seasons cultural connections help boost business. *Hotel and Motel Management, 207*(3), 2, 16.

Selwitz, R. (1992b). New York boutique hotel fills extended-stay market niche. *Hotel and Motel Management, 207*(21), 3, 39.

Shani, D., & Chalasani, S. (1992). Exploiting niches using relationship marketing. *Journal of Service Marketing, 6,* 43-52.

Shapiro, B., & Bonomo, T. (1984). How to segment industrial markets. *Harvard Business Review, 62*(3), 104-110.

Sparks, B. (1993). Guest history: Is it being utilized? *International Journal of Contemporary Hospitality Management, 5*(1), 22-27.

Struhl, S. (1992). *Market segmentation: An introduction and review.* Chicago: American Marketing Association.

Thompson, A., & Strickland, A. (1992). *Strategic management.* Boston: Irwin Press.

Valhouli, C. (2004, April 29). The best Wi-Fi hotels. Retrieved from Forbes.com.

Venkatramen, N., & Prescott, J.E. (1990). Environment-strategy coalignment: An empirical test of its performance implications. *Strategic Management Journal, 11,* 1-23.

Webster, F. (1994). Defining the new marketing concept (part 1). *Marketing Management, 2*(4), 22-31.

Weinstein, A. (1994). *Market segmentation: Using demographics, psychographics, and other niche marketing techniques to predict customer behavior.* Chicago: Probus Publishing Company.

Weinstein, A. (1995). *Market segmentation: Using demographics, psychographics, and other niche marketing techniques to predict and model customer behavior.* Chicago: Probus Publishing.

Weinstein, A. (2004). *Handbook of market segmentation: Strategic Targeting for Business and Technology Firms* (3rd ed.). Binghamton, NY: The Haworth Press, Inc.

Weinstein, A., & Johnson W. (1999). *Designing and delivering superior customer value: Concepts, cases, and applications.* Boca Raton, FL: CRC Press.

Whelihan, W. (1991). Resort marketing trends of the 90s. *Cornell Hotel and Restaurant Administration Quarterly, 32*(2), 56-59.

Whitford, M. (1998). Cornell poll predicts trends 24 years out. *Hotel and Motel Management, 213,* 14- 15.

Wind, Y. (1978). Issues and advances in segmentation research. *Journal of Marketing Research, 15,* 317-337.

Winer, R. (2001). A framework for customer relationship management. *California Management Review, 43*(4), 89-106.

Worcester, B. (1998). Online locks may set trend toward real-time security. *Hotel and Motel Management, 213*(3), 53.

Yesawich, P. (1988). Planning: The second step in market development. *Cornell Hotel and Restaurant Administration Quarterly, 28*(4), 71-81.

Yesawich, P. (1991). Who are the guests of the 90's? *Lodging Hospitality, 47* (10), 50.

Yesawich, P. (2000). Ten trends shaping the future of leisure travel. *Hotel and Motel Management, 215,* 26-27.

Zeithaml, V., & Bitner, M. (2003). *Services marketing.* New York: McGraw-Hill Irwin.

Index

Page numbers followed by the letter "f" indicate figures; those followed by "t" indicate tables.

Segmentation Strategies for Hospitality Managers
Published by The Haworth Press, Inc., 2007. All rights reserved.
doi:10.1300/5716_13

Starwood hotels, 76, 174
Statistical analysis, 182t
Statistical modeling, 171
Statistical principles, knowledge of, 172
Statistical techniques, segmentation based on, 87
Steakhouses, 79, 116
STP marketing, 119, 122, 123, 131, 159
Strategic alliances, 45, 184
Strategic analysis reports, 192
Strategic choice, 186
Strategic control, 114
Strategic fit, 120, 127-128, 132, 133, 134, 130-139, 160, 183-186, 200
Strategic goals, 137
Strategic market, 36
Strategic market alliances, 59
Strategic plan, monitoring, 200-201
Strategic principles, 186-191
Strategic tool box, 186
Strategy (defined), 181
Strengths
    assessment of, 200
    building on, 187, 195
    identifying, 193-194
Strivers (consumer lifestyle segment), 25
Structural attractiveness, 56
Structural equation modeling (SEM), 130
Sub-positioning, 155
Sub-segments, 33, 195
Submarkets, 40-41, 42
Substitute products, 137
Substitute services, 175-176
Superior customer value (SCV), 61. *See also* Value: superior
Suppliers
    as customers, 201
    food and security, 66
    power of, 137
    preferences of, 196
    relationships with, 81
    retaining, 32
    segmentation plan coordination with, 200

Surveys
    customer, 69, 77, 115, 117
    of experts, 115
    guest, 47, 52, 76
Survivors (consumer lifestyle segment), 25
Sushi bars, 7-8, 12, 31, 110
Sustained competitive advantage (SCA), 181, 200. *See also* Competitive advantage
Sutton (firm), 49
SUVs, gas-guzzling, 116
Switchers, 23
Switching behavior, 92, 196
Switching costs, 63, 66, 68, 75, 81
SWOT analysis, 160, 187-188, 191, 194, 198
Synergistic segment combinations, 187

Taco Bell restaurants, 14, 15, 108, 188
Tall people, hotels catering to, 87
Target customers
    mind of, 146, 150, 152, 153, 196, 200
    preferences of, 7-8, 14, 15, 34
Target markets
    business customers as, 126
    channel segments matching profiles of, 25
    competitiveness in, 6, 122
    nature of, 129
    potential, identifying, 125
    selecting, 36, 40, 47, 134-136
    strategic fit with, 183, 184
Target segments and segmentation
    adapting, 58
    brand positioning to focus on different, 155
    changes in, 57, 58
    of competitors, 194
    customer profile-based, 69, 140
    drivers of, 93
    examining, 194
    expertise on, 193

Patricia Ticineto Clough is professor of sociology, women's studies, and intercultural studies at the Graduate Center of the City University of New York and Queen's College. Her previous books include *The End(s) of Ethnography: From Realism to Social Criticism* and *Feminist Thought: Desire, Power, and Academic Discourse.*